STUDENT SOLUTIONS MANUAL

TO ACCOMPANY

Elementary Linear Algebra with Applications

NINTH EDITION

Howard Anton
Chris Rorres
Drexel University

Prepared by
Christine Black
Seattle University

Blaise DeSesa
Kutztown University

Molly Gregas
Duke University

Elizabeth M. Grobe
Charles A. Grobe, Jr.
Bowdoin College

JOHN WILEY & SONS, INC.

Cover Photo: ©John Marshall/Stone/Getty Images

To order books or for customer service call 1-800-CALL-WILEY (225-5945).

ISBN-13 978- 0-471-43329-3
ISBN-10 0-471-43329-2

Printed in the United States of America

10 9 8 7 6 5 4 3 2 1

Printed and bound by Bind-Rite Graphics, Inc.

TABLE OF CONTENTS

Chapter 1

Exercise Set 1.1 1
Exercise Set 1.2 3
Exercise Set 1.3 13
Exercise Set 1.4 21
Exercise Set 1.5 31
Exercise Set 1.6 39
Exercise Set 1.7 47
Supplementary Exercises 1 51

Chapter 2

Exercise Set 2.1 61
Exercise Set 2.2 65
Exercise Set 2.3 73
Exercise Set 2.4 77
Supplementary Exercises 2 81
Technology Exercises 2 87

Chapter 3

Exercise Set 3.1 89
Exercise Set 3.2 95
Exercise Set 3.3 97
Exercise Set 3.4 101
Exercise Set 3.5 107

Chapter 4

Exercise Set 4.1 111
Exercise Set 4.2 115
Exercise Set 4.3 119
Exercise Set 4.4 125

Chapter 5

Exercise Set 5.1 131
Exercise Set 5.2 135
Exercise Set 5.3 141
Exercise Set 5.4 145
Exercise Set 5.5 151
Exercise Set 5.6 155
Supplementary Exercises 5 157

Chapter 6

Exercise Set 6.1 159
Exercise Set 6.2 165
Exercise Set 6.3 171
Exercise Set 6.4 179
Exercise Set 6.5 185
Exercise Set 6.6 189
Supplementary Exercises 6 191

Chapter 7

Exercise Set 7.1 195
Exercise Set 7.2 203
Exercise Set 7.3 207
Supplementary Exercises 7 209

Chapter 8

Exercise Set 8.1 215
Exercise Set 8.2 219
Exercise Set 8.3 223
Exercise Set 8.4 227
Exercise Set 8.5 231
Exercise Set 8.6 239
Supplementary Exercises 8 243

Chapter 9

Exercise Set 9.1 251
Exercise Set 9.2 261
Exercise Set 9.3 265
Exercise Set 9.4 267
Exercise Set 9.5 271
Exercise Set 9.6 275
Exercise Set 9.7 281
Exercise Set 9.8 287
Exercise Set 9.9 291

Chapter 10

Exercise Set 10.1. 299
Exercise Set 10.2. 303
Exercise Set 10.3. 309
Exercise Set 10.4. 315
Exercise Set 10.5. 321
Exercise Set 10.6. 329
Supplementary Exercises 10 339

Chapter 11

Exercise Set 11.1 . . . 343
Exercise Set 11.2 . . . 347
Exercise Set 11.3 . . . 351
Exercise Set 11.4 . . . 355
Exercise Set 11.5 . . . 357
Exercise Set 11.6 . . . 365
Exercise Set 11.7 . . . 369
Exercise Set 11.8 . . . 373
Exercise Set 11.9 . . . 375
Exercise Set 11.10 . . . 379
Exercise Set 11.11 . . . 381
Exercise Set 11.12 . . . 387
Exercise Set 11.13 . . . 391
Exercise Set 11.14 . . . 401
Exercise Set 11.15 . . . 405
Exercise Set 11.16 . . . 417
Exercise Set 11.17 . . . 425
Exercise Set 11.18 . . . 429
Exercise Set 11.19 . . . 431
Exercise Set 11.20 . . . 433
Exercise Set 11.21 . . . 435

EXERCISE SET 1.1

1. **(b)** Not linear because of the term x_1x_3.

(d) Not linear because of the term x_1^{-2}.

(e) Not linear because of the term $x_1^{3/5}$.

7. Since each of the three given points must satisfy the equation of the curve, we have the system of equations

$$ax_1^2 + bx_1 + c = y_1$$

$$ax_2^2 + bx_2 + c = y_2$$

$$ax_3^2 + bx_3 + c = y_3$$

If we consider this to be a system of equations in the three unknowns a, b, and c, the augmented matrix is clearly the one given in the exercise.

9. The solutions of $x_1 + kx_2 = c$ are $x_1 = c - kt$, $x_2 = t$ where t is any real number. If these satisfy $x_1 + \ell x_2 = d$, then $c - kt + \ell t = d$, or $(\ell - k)t = d - c$ for all real numbers t. In particular, if $t = 0$, then $d = c$, and if $t = 1$, then $\ell = k$.

11. If $x - y = 3$, then $2x - 2y = 6$. Therefore, the equations are consistent if and only if $k = 6$; that is, there are no solutions if $k \neq 6$. If $k = 6$, then the equations represent the same line, in which case, there are infinitely many solutions. Since this covers all of the possibilities, there is never a unique solution.

EXERCISE SET 1.2

1. **(e)** Not in reduced row-echelon form because Property 2 is not satisfied.

(f) Not in reduced row-echelon form because Property 3 is not satisfied.

(g) Not in reduced row-echelon form because Property 4 is not satisfied.

5. **(a)** The solution is

$$x_3 = 5$$

$$x_2 = 2 - 2\,x_3 = -8$$

$$x_1 = 7 - 4\,x_3 + 3x_2 = -37$$

(b) Let $x_4 = t$. Then $x_3 = 2 - \mathrm{t}$. Therefore

$$x_2 = 3 + 9t - 4x_3 = 3 + 9t - 4(2 - t) = -5 + 13t$$

$$x_1 = 6 + 5t - 8x_3 = 6 + 5t - 8(2 - t) = -10 + 13t$$

7. **(a)** In Problem 6(a), we reduced the augmented matrix to the following row-echelon matrix:

$$\begin{bmatrix} 1 & 1 & 2 & 8 \\ 0 & 1 & -5 & -9 \\ 0 & 0 & 1 & 2 \end{bmatrix}$$

By Row 3, $x_3 = 2$. Thus by Row 2, $x_2 = 5x_3 - 9 = 1$. Finally, Row 1 implies that $x_1 = -x_2 - 2\,x_3 + 8 = 3$. Hence the solution is

$$x_1 = 3$$

$$x_2 = 1$$

$$x_3 = 2$$

(c) According to the solution to Problem 6(c), one row-echelon form of the augmented matrix is

$$\begin{bmatrix} 1 & -1 & 2 & -1 & -1 \\ 0 & 1 & -2 & 0 & 0 \\ 0 & 0 & 0 & 0 & 0 \\ 0 & 0 & 0 & 0 & 0 \end{bmatrix}$$

Row 2 implies that $y = 2z$. Thus if we let $z = s$, we have $y = 2s$. Row 1 implies that $x = -1 + y - 2z + w$. Thus if we let $w = t$, then $x = -1 + 2s - 2s + t$ or $x = -1 + t$. Hence the solution is

$$x = -1 + t$$

$$y = 2s$$

$$z = s$$

$$w = t$$

9. (a) In Problem 8(a), we reduced the augmented matrix of this system to row-echelon form, obtaining the matrix

$$\begin{bmatrix} 1 & -3/2 & -1 \\ 0 & 1 & 3/4 \\ 0 & 0 & 1 \end{bmatrix}$$

Row 3 again yields the equation $0 = 1$ and hence the system is inconsistent.

(c) In Problem 8(c), we found that one row-echelon form of the augmented matrix is

$$\begin{bmatrix} 1 & -2 & 3 \\ 0 & 0 & 0 \\ 0 & 0 & 0 \end{bmatrix}$$

Again if we let $x_2 = t$, then $x_1 = 3 + 2x_2 = 3 + 2t$.

11. **(a)** From Problem 10(a), a row-echelon form of the augmented matrix is

$$\begin{bmatrix} 1 & -2/5 & 6/5 & 0 \\ 0 & 1 & 27 & 5 \end{bmatrix}$$

If we let $x_3 = t$, then Row 2 implies that $x_2 = 5 - 27t$. Row 1 then implies that $x_1 = (-6/5)x_3 + (2/5)x_2 = 2 - 12t$. Hence the solution is

$$x_1 = 2 - 12t$$

$$x_2 = 5 - 27t$$

$$x_3 = t$$

(c) From Problem 10(c), a row-echelon form of the augmented matrix is

$$\begin{bmatrix} 1 & 2 & 1/2 & 7/2 & 0 & 7/2 \\ 0 & 0 & 1 & 2 & -1 & 4 \\ 0 & 0 & 0 & 1 & -1 & 3 \\ 0 & 0 & 0 & 0 & 0 & 0 \end{bmatrix}$$

If we let $y = t$, then Row 3 implies that $x = 3 + t$. Row 2 then implies that

$$w = 4 - 2x + t = -2 - t.$$

Now let $v = s$. By Row 1, $u = 7/2 - 2s - (1/2)w - (7/2)x = -6 - 2s - 3t$. Thus we have the same solution which we obtained in Problem 10(c).

13. **(b)** The augmented matrix of the homogeneous system is

$$\begin{bmatrix} 3 & 1 & 1 & 1 & 0 \\ 5 & -1 & 1 & -1 & 0 \end{bmatrix}$$

This matrix may be reduced to

$$\begin{bmatrix} 3 & 1 & 1 & 1 & 0 \\ 0 & 4 & 1 & 4 & 0 \end{bmatrix}$$

If we let $x_3 = 4s$ and $x_4 = t$, then Row 2 implies that

$$4x_2 = -4t - 4s \quad \text{or} \quad x_2 = -t - s$$

Now Row 1 implies that

$$3x_1 = -x_2 - 4s - t = t + s - 4s - t = -3s \quad \text{or} \quad x_1 = -s$$

Therefore the solution is

$$x_1 = -s$$

$$x_2 = -(t + s)$$

$$x_3 = 4s$$

$$x_4 = t$$

15. **(a)** The augmented matrix of this system is

$$\begin{bmatrix} 2 & -1 & 3 & 4 & 9 \\ 1 & 0 & -2 & 7 & 11 \\ 3 & -3 & 1 & 5 & 8 \\ 2 & 1 & 4 & 4 & 10 \end{bmatrix}$$

Its reduced row-echelon form is

$$\begin{bmatrix} 1 & 0 & 0 & 0 & -1 \\ 0 & 1 & 0 & 0 & 0 \\ 0 & 0 & 1 & 0 & 1 \\ 0 & 0 & 0 & 1 & 2 \end{bmatrix}$$

Hence the solution is

$$I_1 = -1$$

$$I_2 = 0$$

$$I_3 = 1$$

$$I_4 = 2$$

(b) The reduced row-echelon form of the augmented matrix is

$$\begin{bmatrix} 1 & 1 & 0 & 0 & 1 & 0 \\ 0 & 0 & 1 & 0 & 1 & 0 \\ 0 & 0 & 0 & 1 & 0 & 0 \\ 0 & 0 & 0 & 0 & 0 & 0 \end{bmatrix}$$

If we let $Z_2 = s$ and $Z_5 = t$, then we obtain the solution

$$Z_1 = -s - t$$

$$Z_2 = s$$

$$Z_3 = -t$$

$$Z_4 = 0$$

$$Z_5 = t$$

17. The Gauss-Jordan process will reduce this system to the equations

$$x + 2y - 3z = 4$$

$$y - 2z = 10/7$$

$$(a^2 - 16)z = a - 4$$

If $a = 4$, then the last equation becomes $0 = 0$, and hence there will be infinitely many solutions—for instance, $z = t$, $y = 2\,t + \frac{10}{7}$, $x = -2\,(2t + \frac{10}{7}) + 3t + 4$. If $a = -4$, then the last equation becomes $0 = -8$, and so the system will have no solutions. Any other value of a will yield a unique solution for z and hence also for y and x.

19. One possibility is

$$\begin{bmatrix} 1 & 3 \\ 2 & 7 \end{bmatrix} \rightarrow \begin{bmatrix} 1 & 3 \\ 0 & 1 \end{bmatrix}$$

Another possibility is

$$\begin{bmatrix} 1 & 3 \\ 2 & 7 \end{bmatrix} \to \begin{bmatrix} 2 & 7 \\ 1 & 3 \end{bmatrix} \to \begin{bmatrix} 1 & 7/2 \\ 1 & 3 \end{bmatrix} \to \begin{bmatrix} 1 & 7/2 \\ 0 & 1 \end{bmatrix}$$

21. If we treat the given system as linear in the variables $\sin \alpha$, $\cos \beta$, and $\tan \gamma$, then the augmented matrix is

$$\begin{bmatrix} 1 & 2 & 3 & 0 \\ 2 & 5 & 3 & 0 \\ -1 & -5 & 5 & 0 \end{bmatrix}$$

This reduces to

$$\begin{bmatrix} 1 & 0 & 0 & 0 \\ 0 & 1 & 0 & 0 \\ 0 & 0 & 1 & 0 \end{bmatrix}$$

so that the solution (for α, β, γ between 0 and 2π) is

$$\sin \alpha = 0 \Rightarrow \alpha = 0, \pi, 2\pi$$

$$\cos \beta = 0 \Rightarrow \beta = \pi/2, 3\pi/2$$

$$\tan \gamma = 0 \Rightarrow \gamma = 0, \pi, 2\pi$$

That is, there are $3 \cdot 2 \cdot 3 = 18$ possible triples α, β, γ which satisfy the system of equations.

23. If $\lambda = 2$, the system becomes

$$\begin{aligned} -x_2 &= 0 \\ 2x_1 - 3x_2 + x_3 &= 0 \\ -2x_1 + 2x_2 - x_3 &= 0 \end{aligned}$$

Thus $x_2 = 0$ and the third equation becomes -1 times the second. If we let $x_1 = t$, then $x_3 = -2t$.

25. Using the given points, we obtain the equations

$$d = 10$$

$$a + b + c + d = 7$$

$$27a + 9b + 3c + d = -11$$

$$64a + 16b + 4c + d = -14$$

If we solve this system, we find that $a = 1$, $b = -6$, $c = 2$, and $d = 10$.

27. **(a)** If $a = 0$, then the reduction can be accomplished as follows:

$$\begin{bmatrix} a & b \\ c & d \end{bmatrix} \to \begin{bmatrix} 1 & \frac{b}{a} \\ c & d \end{bmatrix} \to \begin{bmatrix} 1 & \frac{b}{a} \\ 0 & \frac{ad-bc}{a} \end{bmatrix} \to \begin{bmatrix} 1 & \frac{b}{a} \\ 0 & 1 \end{bmatrix} \to \begin{bmatrix} 1 & 0 \\ 0 & 1 \end{bmatrix}$$

If $a = 0$, then $b \neq 0$ and $c \neq 0$, so the reduction can be carried out as follows:

$$\begin{bmatrix} 0 & b \\ c & d \end{bmatrix} \to \begin{bmatrix} c & d \\ 0 & b \end{bmatrix} \to \begin{bmatrix} 1 & \frac{d}{c} \\ 0 & b \end{bmatrix} \to \begin{bmatrix} 1 & \frac{d}{c} \\ 0 & 1 \end{bmatrix} \to \begin{bmatrix} 1 & 0 \\ 0 & 1 \end{bmatrix}$$

Where did you use the fact that $ad - bc \neq 0$? (This proof uses it twice.)

29. There are eight possibilities. They are

(a)

$$\begin{pmatrix} 1 & 0 & 0 \\ 0 & 1 & 0 \\ 0 & 0 & 1 \end{pmatrix}, \begin{pmatrix} 1 & 0 & p \\ 0 & 1 & q \\ 0 & 0 & 0 \end{pmatrix}, \begin{pmatrix} 1 & p & 0 \\ 0 & 0 & 1 \\ 0 & 0 & 0 \end{pmatrix},$$

$$\begin{pmatrix} 0 & 1 & 0 \\ 0 & 0 & 1 \\ 0 & 0 & 0 \end{pmatrix}, \begin{pmatrix} 1 & p & q \\ 0 & 0 & 0 \\ 0 & 0 & 0 \end{pmatrix}, \begin{pmatrix} 0 & 1 & p \\ 0 & 0 & 0 \\ 0 & 0 & 0 \end{pmatrix},$$

$$\begin{pmatrix} 0 & 0 & 1 \\ 0 & 0 & 0 \\ 0 & 0 & 0 \end{pmatrix},$$ where $p, q,$ are any real numbers,

and

$$\begin{pmatrix} 0 & 0 & 0 \\ 0 & 0 & 0 \\ 0 & 0 & 0 \end{pmatrix}$$

(b)

$$\begin{pmatrix} 1 & 0 & 0 & 0 \\ 0 & 1 & 0 & 0 \\ 0 & 0 & 1 & 0 \\ 0 & 0 & 0 & 1 \end{pmatrix}, \begin{pmatrix} 1 & 0 & 0 & p \\ 0 & 1 & 0 & q \\ 0 & 0 & 1 & r \\ 0 & 0 & 0 & 0 \end{pmatrix}, \begin{pmatrix} 1 & 0 & p & 0 \\ 0 & 1 & q & 0 \\ 0 & 0 & 0 & 1 \\ 0 & 0 & 0 & 0 \end{pmatrix},$$

$$\begin{pmatrix} 1 & p & 0 & 0 \\ 0 & 0 & 1 & 0 \\ 0 & 0 & 0 & 1 \\ 0 & 0 & 0 & 0 \end{pmatrix}, \begin{pmatrix} 0 & 1 & 0 & 0 \\ 0 & 0 & 1 & 0 \\ 0 & 0 & 0 & 1 \\ 0 & 0 & 0 & 0 \end{pmatrix}, \begin{pmatrix} 1 & 0 & p & q \\ 0 & 1 & r & s \\ 0 & 0 & 0 & 0 \\ 0 & 0 & 0 & 0 \end{pmatrix},$$

$$\begin{pmatrix} 1 & p & 0 & q \\ 0 & 0 & 1 & r \\ 0 & 0 & 0 & 0 \\ 0 & 0 & 0 & 0 \end{pmatrix}, \begin{pmatrix} 1 & p & q & 0 \\ 0 & 0 & 0 & 1 \\ 0 & 0 & 0 & 0 \\ 0 & 0 & 0 & 0 \end{pmatrix}, \begin{pmatrix} 0 & 1 & 0 & p \\ 0 & 0 & 1 & q \\ 0 & 0 & 0 & 0 \\ 0 & 0 & 0 & 0 \end{pmatrix},$$

$$\begin{pmatrix} 0 & 1 & p & 0 \\ 0 & 0 & 0 & 1 \\ 0 & 0 & 0 & 0 \\ 0 & 0 & 0 & 0 \end{pmatrix}, \begin{pmatrix} 0 & 0 & 1 & 0 \\ 0 & 0 & 0 & 1 \\ 0 & 0 & 0 & 0 \\ 0 & 0 & 0 & 0 \end{pmatrix}, \begin{pmatrix} 1 & p & q & r \\ 0 & 0 & 0 & 0 \\ 0 & 0 & 0 & 0 \\ 0 & 0 & 0 & 0 \end{pmatrix},$$

$$\begin{pmatrix} 0 & 1 & p & q \\ 0 & 0 & 0 & 0 \\ 0 & 0 & 0 & 0 \\ 0 & 0 & 0 & 0 \end{pmatrix}, \begin{pmatrix} 0 & 0 & 1 & p \\ 0 & 0 & 0 & 0 \\ 0 & 0 & 0 & 0 \\ 0 & 0 & 0 & 0 \end{pmatrix}, \begin{pmatrix} 0 & 0 & 0 & 1 \\ 0 & 0 & 0 & 0 \\ 0 & 0 & 0 & 0 \\ 0 & 0 & 0 & 0 \end{pmatrix}, \text{ and } \begin{pmatrix} 0 & 0 & 0 & 0 \\ 0 & 0 & 0 & 0 \\ 0 & 0 & 0 & 0 \\ 0 & 0 & 0 & 0 \end{pmatrix}$$

31. **(a)** False. The reduced row-echelon form of a matrix is unique, as stated in the remark in this section.

(b) True. The row-echelon form of a matrix is not unique, as shown in the following example:

$$\begin{bmatrix} 1 & 2 \\ 1 & 3 \end{bmatrix} \to \begin{bmatrix} 1 & 2 \\ 0 & 1 \end{bmatrix}$$

but

$$\begin{bmatrix} 1 & 2 \\ 1 & 3 \end{bmatrix} \to \begin{bmatrix} 1 & 3 \\ 1 & 2 \end{bmatrix} \to \begin{bmatrix} 1 & 3 \\ 0 & -1 \end{bmatrix} \to \begin{bmatrix} 1 & 3 \\ 0 & 1 \end{bmatrix}$$

(c) False. If the reduced row-echelon form of the augmented matrix for a system of 3 equations in 2 unknowns is

$$\begin{bmatrix} 1 & 0 & a \\ 0 & 1 & b \\ 0 & 0 & 0 \end{bmatrix}$$

then the system has a unique solution. If the augmented matrix of a system of 3 equations in 3 unknowns reduces to

$$\begin{bmatrix} 1 & 1 & 1 & 0 \\ 0 & 0 & 0 & 1 \\ 0 & 0 & 0 & 0 \end{bmatrix}$$

then the system has no solutions.

(d) False. The system can have a solution only if the 3 lines meet in at least one point which is common to all 3.

EXERCISE SET 1.3

1. **(c)** The matrix AE is 4×4. Since B is 4×5, $AE + B$ is not defined.

(e) The matrix $A + B$ is 4×5. Since E is 5×4, $E(A + B)$ is 5×5.

(h) Since A^T is 5×4 and E is 5×4, their sum is also 5×4. Thus $(A^T + E)D$ is 5×2.

3. **(e)** Since $2B$ is a 2×2 matrix and C is a 2×3 matrix, $2B - C$ is not defined.

(g) We have

$$-3(D + 2E) = -3\left(\begin{bmatrix} 1 & 5 & 2 \\ -1 & 0 & 1 \\ 3 & 2 & 4 \end{bmatrix} + \begin{bmatrix} 12 & 2 & 6 \\ -2 & 2 & 4 \\ 8 & 2 & 6 \end{bmatrix}\right)$$

$$= -3\begin{bmatrix} 13 & 7 & 8 \\ -3 & 2 & 5 \\ 11 & 4 & 10 \end{bmatrix} = \begin{bmatrix} -39 & -21 & -24 \\ 9 & -6 & -15 \\ -33 & -12 & -30 \end{bmatrix}$$

(j) We have $tr(D - 3E) = (1 - 3(6)) + (0 - 3(1)) + (4 - 3(3)) = -25$.

5. **(b)** Since B is a 2×2 matrix and A is a 3×2 matrix, BA is not defined (although AB is).

(d) We have

$$AB = \begin{bmatrix} 12 & -3 \\ -4 & 5 \\ 4 & 1 \end{bmatrix}$$

Hence

$$(AB)C = \begin{bmatrix} 3 & 45 & 9 \\ 11 & -11 & 17 \\ 7 & 17 & 13 \end{bmatrix}$$

(e) We have

$$A(BC) = \begin{bmatrix} 3 & 0 \\ -1 & 2 \\ 1 & 1 \end{bmatrix} \begin{bmatrix} 1 & 15 & 3 \\ 6 & 2 & 10 \end{bmatrix} = \begin{bmatrix} 3 & 45 & 9 \\ 11 & -11 & 17 \\ 7 & 17 & 13 \end{bmatrix}$$

(f) We have

$$CC^T = \begin{bmatrix} 1 & 4 & 2 \\ 3 & 1 & 5 \end{bmatrix} \begin{bmatrix} 1 & 3 \\ 4 & 1 \\ 2 & 5 \end{bmatrix} = \begin{bmatrix} 21 & 17 \\ 17 & 35 \end{bmatrix}$$

(j) We have $tr(4E^T - D) = tr(4E - D) = (4(6) - 1) + (4(1) - 0) + (4(3) - 4) = 35$.

7. **(a)** The first row of A is

$$A_1 = [3 \quad -2 \quad 7]$$

Thus, the first row of AB is

$$A_1B = [\,3 \; -2 \; 7] \begin{bmatrix} 6 & -2 & 4 \\ 0 & 1 & 3 \\ 7 & 7 & 5 \end{bmatrix}$$

$$= [67 \;\; 41 \;\; 41]$$

(c) The second column of B is

$$B_2 = \begin{bmatrix} -2 \\ 1 \\ 7 \end{bmatrix}$$

Thus, the second column of AB is

$$AB_2 = \begin{bmatrix} 3 & -2 & 7 \\ 6 & 5 & 4 \\ 0 & 4 & 9 \end{bmatrix} \begin{bmatrix} -2 \\ 1 \\ 7 \end{bmatrix} = \begin{bmatrix} 41 \\ 21 \\ 67 \end{bmatrix}$$

(e) The third row of A is

$$A_3 = [0 \quad 4 \quad 9]$$

Thus, the third row of AA is

$$A_3A = [0 \quad 4 \quad 9] \begin{bmatrix} 3 & -2 & 7 \\ 6 & 5 & 4 \\ 0 & 4 & 9 \end{bmatrix}$$

$$= [24 \quad 56 \quad 97]$$

9. **(a)** The product $\mathbf{y}A$ is the matrix

$$[y_1a_{11} + y_2a_{21} + \cdots + y_ma_{m1} \qquad y_1a_{12} + y_2a_{22} + \cdots + y_ma_{m2} \cdots$$
$$y_1a_{1n} + y_2a_{2n} + \cdots + y_ma_{mn}]$$

We can rewrite this matrix in the form

$$y_1\,[a_{11}\; a_{12} \cdots a_{1n}] + y_2\,[a_{21}\; a_{22} \cdots a_{2n}] + \cdots + y_m\,[a_{m1}\;\; a_{m2} \cdots a_{mn}]$$

which is, indeed, a linear combination of the row matrices of A with the scalar coefficients of $\mathbf{y}$.

(b) Let $y = [y_1, y_2, \cdots, y_m]$

and $A = \begin{bmatrix} A_1 \\ A_2 \\ \vdots \\ A_m \end{bmatrix}$ be the m rows of A.

by 9a, $yA = \begin{bmatrix} y_1 & A_1 \\ y_2 & A_2 \\ & \vdots \\ y_m & A_m \end{bmatrix}$

Taking transposes of both sides, we have

$$(yA)^T = A^T y^T = (A_1 \mid A_2 \mid \cdots \mid A_m) \begin{bmatrix} y_1 \\ \vdots \\ y_m \end{bmatrix}$$

$$= \begin{bmatrix} y_1 & A_1 \\ y_2 & A_2 \\ \vdots & \\ y_m & A_m \end{bmatrix}^T = (y_1A_1 \mid y_2A_2 \mid \ldots \mid y_mA_m$$

11. Let f_{ij} denote the entry in the ith row and j^{th} column of $C(DE)$. We are asked to find f_{23}. In order to compute f_{23}, we must calculate the elements in the second row of C and the third column of DE. According to Equation (3), we can find the elements in the third column of DE by computing DE_3 where E_3 is the third column of E. That is,

$$f_{23} = [3 \;\; 1 \;\; 5] \left(\begin{bmatrix} 1 & 5 & 2 \\ -1 & 0 & 1 \\ 3 & 2 & 4 \end{bmatrix} \begin{bmatrix} 3 \\ 2 \\ 3 \end{bmatrix} \right)$$

$$= [3 \;\; 1 \;\; 5] \begin{bmatrix} 19 \\ 0 \\ 25 \end{bmatrix} = 182$$

15. **(a)** By block multiplication,

$$AB = \left[\begin{array}{c|c} \begin{bmatrix} -1 & 2 \\ 0 & -3 \end{bmatrix}\begin{bmatrix} 2 & 1 \\ -3 & 5 \end{bmatrix} + \begin{bmatrix} 1 & 5 \\ 4 & 2 \end{bmatrix}\begin{bmatrix} 7 & -1 \\ 0 & 3 \end{bmatrix} & \begin{bmatrix} -1 & 2 \\ 0 & -3 \end{bmatrix}\begin{bmatrix} 4 \\ 2 \end{bmatrix} + \begin{bmatrix} 1 & 5 \\ 4 & 2 \end{bmatrix}\begin{bmatrix} 5 \\ -3 \end{bmatrix} \\ \hline \begin{bmatrix} 1 & 5 \end{bmatrix}\begin{bmatrix} 2 & 1 \\ -3 & 5 \end{bmatrix} + \begin{bmatrix} 6 & 1 \end{bmatrix}\begin{bmatrix} 7 & -1 \\ 0 & 3 \end{bmatrix} & \begin{bmatrix} 1 & 5 \end{bmatrix}\begin{bmatrix} 4 \\ 2 \end{bmatrix} + \begin{bmatrix} 6 & 1 \end{bmatrix}\begin{bmatrix} 5 \\ -3 \end{bmatrix} \end{array}\right]$$

$$= \left[\begin{array}{c|c} \begin{bmatrix} -8 & 9 \\ 9 & -15 \end{bmatrix} + \begin{bmatrix} 7 & 14 \\ 28 & 2 \end{bmatrix} & \begin{bmatrix} 0 \\ -6 \end{bmatrix} + \begin{bmatrix} -10 \\ 14 \end{bmatrix} \\ \hline \begin{bmatrix} -13 & 26 \end{bmatrix} + \begin{bmatrix} 42 & -3 \end{bmatrix} & \begin{bmatrix} 14 \end{bmatrix} + \begin{bmatrix} 27 \end{bmatrix} \end{array}\right]$$

$$= \left[\begin{array}{cc|c} -1 & 23 & -10 \\ 37 & -13 & 8 \\ \hline 29 & 23 & 41 \end{array}\right]$$

17. **(a)** The partitioning of A and B makes them each effectively 2×2 matrices, so block multiplication might be possible. However, if

$$A = \begin{bmatrix} A_{11} & A_{12} \\ A_{21} & A_{22} \end{bmatrix} \quad \text{and} \quad B = \begin{bmatrix} B_{11} & B_{12} \\ B_{21} & B_{22} \end{bmatrix}$$

then the products $A_{11}B_{11}$, $A_{12}B_{21}$, $A_{11}B_{12}$, $A_{12}B_{22}$, $A_{21}B_{11}$, $A_{22}B_{21}$, $A_{21}B_{12}$, and $A_{22}B_{22}$ are *all* undefined. If even *one* of these is undefined, block multiplication is impossible.

21. **(b)** If $i > j$, then the entry a_{ij} has row number larger than column number; that is, it lies below the matrix diagonal. Thus $[a_{ij}]$ has all zero elements below the diagonal.

(d) If $|i - j| > 1$, then either $i - j > 1$ or $i - j < -1$; that is, either $i > j + 1$ or $j > i + 1$. The first of these inequalities says that the entry a_{ij} lies below the diagonal and also below the "subdiagonal" consisting of all entries immediately below the diagonal ones. The second inequality says that the entry a_{ij} lies above the diagonal and also above the entries immediately above the diagonal ones. Thus we have

$$[a_{ij}] = \begin{bmatrix} a_{11} & a_{12} & 0 & 0 & 0 & 0 \\ a_{21} & a_{22} & a_{23} & 0 & 0 & 0 \\ 0 & a_{32} & a_{33} & a_{34} & 0 & 0 \\ 0 & 0 & a_{43} & a_{44} & a_{45} & 0 \\ 0 & 0 & 0 & a_{54} & a_{55} & a_{56} \\ 0 & 0 & 0 & 0 & a_{65} & a_{66} \end{bmatrix}$$

23.

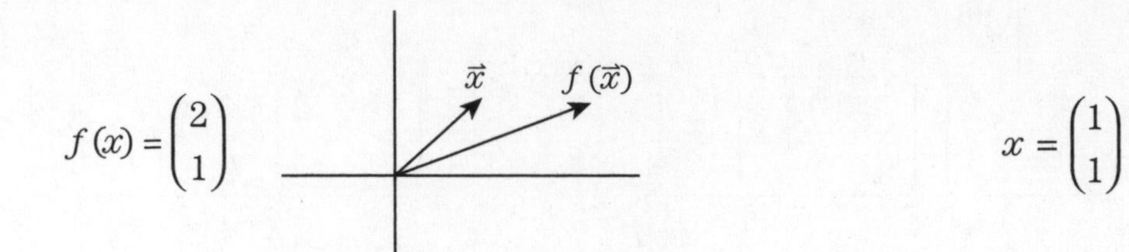

$f(x) = \begin{pmatrix} 2 \\ 0 \end{pmatrix}$ $\quad f(\vec{x}) = \vec{x}$ $\quad x = \begin{pmatrix} 2 \\ 0 \end{pmatrix}$

x $\quad f(x)$

$f(x) = \begin{pmatrix} 7 \\ 4 \end{pmatrix}$ $\quad x = \begin{pmatrix} 4 \\ 3 \end{pmatrix}$

(a)

$f(x) = \begin{pmatrix} 0 \\ -2 \end{pmatrix}$ $\quad x = \begin{pmatrix} 2 \\ -2 \end{pmatrix}$

(b)

$f(x)$ $\quad x$

$$f\begin{pmatrix} x_1 \\ x_2 \end{pmatrix} = \begin{pmatrix} x_1 + x_2 \\ x_2 \end{pmatrix}$$

(c)

27. The only solution to this system of equations is, by inspection,

(d)

$$A = \begin{bmatrix} 1 & 1 & 0 \\ 1 & -1 & 0 \\ 0 & 0 & 0 \end{bmatrix}$$

29. **(a)** Let $B = \begin{bmatrix} a & b \\ c & d \end{bmatrix}$. Then $B^2 = A$ implies that

$$(*) \qquad \begin{array}{ll} a^2 + bc = 2 & ab + bd = 2 \\ ac + cd = 2 & bc + d^2 = 2 \end{array}$$

One might note that $a = b = c = d = 1$ and $a = b = c = d = -1$ satisfy $(*)$. Solving the first and last of the above equations simultaneously yields $a^2 = d^2$. Thus $a = \pm d$. Solving the remaining 2 equations yields $c(a + d) = b(a + d) = 2$. Therefore $a \neq -d$ and a and d cannot both be zero. Hence we have $a = d \neq 0$, so that $ac = ab = 1$, or $b = c = 1/a$. The first equation in $(*)$ then becomes $a^2 + 1/a^2 = 2$ or $a^4 - 2a^2 + 1 = 0$. Thus $a = \pm 1$. That is,

$$\begin{bmatrix} 1 & 1 \\ 1 & 1 \end{bmatrix} \quad \text{and} \quad \begin{bmatrix} -1 & -1 \\ -1 & -1 \end{bmatrix}$$

are the only square roots of A.

(b) Using the reasoning and the notation of Part (a), show that either $a = -d$ or $b = c = 0$. If $a = -d$, then $a^2 + bc = 5$ and $bc + a^2 = 9$. This is impossible, so we have $b = c = 0$. This implies that $a^2 = 5$ and $d^2 = 9$. Thus

$$\begin{bmatrix} \sqrt{5} & 0 \\ 0 & 3 \end{bmatrix} \begin{bmatrix} -\sqrt{5} & 0 \\ 0 & 3 \end{bmatrix} \begin{bmatrix} \sqrt{5} & 0 \\ 0 & -3 \end{bmatrix} \begin{bmatrix} -\sqrt{5} & 0 \\ 0 & -3 \end{bmatrix}$$

are the 4 square roots of A.

Note that if A were $\begin{bmatrix} 5 & 0 \\ 0 & 5 \end{bmatrix}$, say, then $B = \begin{bmatrix} 1 & r \\ 4/r & -1 \end{bmatrix}$ would be a square root of A for every nonzero real number r and there would be infinitely many other square roots as well.

(c) By an argument similar to the above, show that if, for instance,

$$A = \begin{bmatrix} -1 & 0 \\ 0 & 1 \end{bmatrix} \text{ and } B = \begin{bmatrix} a & b \\ c & d \end{bmatrix}$$

where $BB = A$, then either $a = -d$ or $b = c = 0$. Each of these alternatives leads to a contradiction. Why?

31. **(a)** True. If A is an $m \times n$ matrix, then A^T is $n \times m$. Thus AA^T is $m \times m$ and A^TA is $n \times n$. Since the trace is defined for every square matrix, the result follows.

(b) True. Partition A into its row matrices, so that

$$A = \begin{bmatrix} r_1 \\ r_2 \\ \vdots \\ r_m \end{bmatrix} \text{ and } A^T = \begin{bmatrix} r_1^T & r_2^T & \cdots & r_m^T \end{bmatrix}$$

Then

$$AA^T = \begin{bmatrix} r_1r_1^T & r_1r_2^T & \cdots & r_1r_m^T \\ r_2r_1^T & r_2r_2^T & \cdots & r_2r_m^T \\ \vdots & \vdots & & \vdots \\ r_mr_1^T & r_mr_2^T & \cdots & r_mr_m^T \end{bmatrix}$$

Since each of the rows r_i is a $1 \times n$ matrix, each r_i^T is an $n \times 1$ matrix, and therefore each matrix $r_i\, r_j^T$ is a 1×1 matrix. Hence

$$\operatorname{tr}(AA^T) = r_1\, r_1^T + r_2\, r_2^T + \cdots + r_m\, r_m^T$$

Note that since $r_i\, r_i^T$ is just the sum of the squares of the entries in the ith row of A, $r_1\, r_1^T + r_2\, r_2^T + \cdots + r_m\, r_m^T$ is the sum of the squares of *all* of the entries of A.

A similar argument works for A^TA, and since the sum of the squares of the entries of A^T is the same as the sum of the squares of the entries of A, the result follows.

31. **(c)** False. For instance, let $A = \begin{bmatrix} 0 & 1 \\ 0 & 1 \end{bmatrix}$ and $B = \begin{bmatrix} 1 & 1 \\ 1 & 1 \end{bmatrix}$.

(d) True. Every entry in the first row of AB is the matrix product of the first row of A with a column of B. If the first row of A has all zeros, then this product is zero.

EXERCISE SET 1.4

1. **(a)** We have

$$A+B=\begin{bmatrix} 10 & -4 & -2 \\ 0 & 5 & 7 \\ 2 & -6 & 10 \end{bmatrix}$$

Hence,

$$(A+B)+C=\begin{bmatrix} 10 & -4 & -2 \\ 0 & 5 & 7 \\ 2 & -6 & 10 \end{bmatrix}+\begin{bmatrix} 0 & -2 & 3 \\ 1 & 7 & 4 \\ 3 & 5 & 9 \end{bmatrix}$$

$$=\begin{bmatrix} 10 & -6 & 1 \\ 1 & 12 & 11 \\ 5 & -1 & 19 \end{bmatrix}$$

On the other hand,

$$B+C=\begin{bmatrix} 8 & -5 & -2 \\ 1 & 8 & 6 \\ 7 & -2 & 15 \end{bmatrix}$$

Hence,

$$A+(B+C)=\begin{bmatrix} 2 & -1 & 3 \\ 0 & 4 & 5 \\ -2 & 1 & 4 \end{bmatrix}+\begin{bmatrix} 8 & -5 & -2 \\ 1 & 8 & 6 \\ 7 & -2 & 15 \end{bmatrix}=\begin{bmatrix} 10 & -6 & 1 \\ 1 & 12 & 11 \\ 5 & -1 & 19 \end{bmatrix}$$

1. **(c)** Since $a + b = -3$, we have

$$(a+b)C = (-3)\begin{bmatrix} 0 & -2 & 3 \\ 1 & 7 & 4 \\ 3 & 5 & 9 \end{bmatrix} = \begin{bmatrix} 0 & 6 & -9 \\ -3 & -21 & -12 \\ -9 & -15 & -27 \end{bmatrix}$$

Also

$$aC + bC = \begin{bmatrix} 0 & -8 & 12 \\ 4 & 28 & 16 \\ 12 & 20 & 36 \end{bmatrix} + \begin{bmatrix} 0 & 14 & -21 \\ -7 & -49 & -28 \\ -21 & -35 & -63 \end{bmatrix} = \begin{bmatrix} 0 & 6 & -9 \\ -3 & -21 & -12 \\ -9 & -15 & -27 \end{bmatrix}$$

3. **(b)** Since

$$(A+B)^T = \begin{bmatrix} 10 & -4 & -2 \\ 0 & 5 & 7 \\ 2 & -6 & 10 \end{bmatrix}^T = \begin{bmatrix} 10 & 0 & 2 \\ -4 & 5 & -6 \\ -2 & 7 & 10 \end{bmatrix}$$

and

$$A^T + B^T = \begin{bmatrix} 2 & 0 & -2 \\ -1 & 4 & 1 \\ 3 & 5 & -4 \end{bmatrix} + \begin{bmatrix} 8 & 0 & 4 \\ -3 & 1 & -7 \\ -5 & 2 & 6 \end{bmatrix} = \begin{bmatrix} 10 & 0 & 2 \\ -4 & 5 & -6 \\ -2 & 7 & 10 \end{bmatrix}$$

the two matrices are equal.

3. **(d)** Since

$$(AB)^T = \begin{bmatrix} 28 & -28 & 6 \\ 20 & -31 & 38 \\ 0 & -21 & 36 \end{bmatrix}^T = \begin{bmatrix} 28 & 20 & 0 \\ -28 & -31 & -21 \\ 6 & 38 & 36 \end{bmatrix}$$

and

$$B^T A^T = \begin{bmatrix} 8 & 0 & 4 \\ -3 & 1 & -7 \\ -5 & 2 & 6 \end{bmatrix} \begin{bmatrix} 2 & 0 & -2 \\ -1 & 4 & 1 \\ 3 & 5 & 4 \end{bmatrix} = \begin{bmatrix} 28 & 20 & 0 \\ -28 & -31 & -21 \\ 6 & 38 & 36 \end{bmatrix}$$

the two matrices are equal.

5. **(b)**

$$(B^T)^{-1} = \begin{bmatrix} 2 & 4 \\ -3 & 4 \end{bmatrix}^{-1} = \frac{1}{20}\begin{bmatrix} 4 & -4 \\ 3 & 2 \end{bmatrix}$$

$$(B^{-1})^T = \left[\frac{1}{20}\begin{bmatrix} 4 & 3 \\ -4 & 2 \end{bmatrix}\right]^T = \frac{1}{20}\begin{bmatrix} 4 & 3 \\ -4 & 2 \end{bmatrix}^T = \frac{1}{20}\begin{bmatrix} 4 & -4 \\ 3 & 2 \end{bmatrix}$$

7. **(b)** We are given that $(7A)^{-1} = \begin{bmatrix} -3 & 7 \\ 1 & -2 \end{bmatrix}$. Therefore

$$7A = ((7A)^{-1})^{-1} = \begin{bmatrix} -3 & 7 \\ 1 & -2 \end{bmatrix}^{-1} = \begin{bmatrix} 2 & 7 \\ 1 & 3 \end{bmatrix}$$

Thus,

$$A = \begin{bmatrix} 2/7 & 1 \\ 1/7 & 3/7 \end{bmatrix}$$

7. **(d)** If $(I + 2A)^{-1} = \begin{bmatrix} -1 & 2 \\ 4 & 5 \end{bmatrix}$, then $I + 2A = \begin{bmatrix} -1 & 2 \\ 4 & 5 \end{bmatrix}^{-1} = \begin{bmatrix} -\frac{5}{13} & \frac{2}{13} \\ \frac{4}{13} & \frac{1}{13} \end{bmatrix}$. Hence

$$2A = \begin{bmatrix} -\frac{5}{13} & \frac{2}{13} \\ \frac{4}{13} & \frac{1}{13} \end{bmatrix} - \begin{bmatrix} 1 & 0 \\ 0 & 1 \end{bmatrix} = \begin{bmatrix} -\frac{18}{13} & \frac{2}{13} \\ \frac{4}{13} & -\frac{12}{13} \end{bmatrix}, \text{ so that } A = \begin{bmatrix} -\frac{9}{13} & \frac{1}{13} \\ \frac{2}{13} & -\frac{6}{13} \end{bmatrix}.$$

9. **(b)** We have

$$\begin{aligned} p(A) &= 2\begin{bmatrix} 3 & 1 \\ 2 & 1 \end{bmatrix}^2 - \begin{bmatrix} 3 & 1 \\ 2 & 1 \end{bmatrix} + 1\begin{bmatrix} 1 & 0 \\ 0 & 1 \end{bmatrix} \\ &= 2\begin{bmatrix} 11 & 4 \\ 8 & 3 \end{bmatrix} - \begin{bmatrix} 3 & 1 \\ 2 & 1 \end{bmatrix} + \begin{bmatrix} 1 & 0 \\ 0 & 1 \end{bmatrix} \\ &= \begin{bmatrix} 22 & 8 \\ 16 & 6 \end{bmatrix} - \begin{bmatrix} 3 & 1 \\ 2 & 1 \end{bmatrix} + \begin{bmatrix} 1 & 0 \\ 0 & 1 \end{bmatrix} \\ &= \begin{bmatrix} 20 & 7 \\ 14 & 6 \end{bmatrix} \end{aligned}$$

11. Call the matrix A. By Theorem 1.4.5,

$$\begin{aligned} A^{-1} &= \frac{1}{\cos^2\theta + \sin^2\theta} \begin{bmatrix} \cos\theta & -\sin\theta \\ \sin\theta & \cos\theta \end{bmatrix} \\ &= \begin{bmatrix} \cos\theta & -\sin\theta \\ \sin\theta & \cos\theta \end{bmatrix} \end{aligned}$$

since $\cos^2\theta + \sin^2\theta = 1$.

13. If $a_{11}a_{22} \cdots a_{nn} \neq 0$, then $a_{ii} \neq 0$, and hence $1/a_{ii}$ is defined for $i = 1,2, \ldots, n$. It is now easy to verify that

$$A^{-1} = \begin{bmatrix} 1/a_{11} & 0 & \cdots & 0 \\ 0 & 1/a_{22} & \cdots & 0 \\ \vdots & \vdots & & \vdots \\ 0 & 0 & \cdots & 1/a_{nn} \end{bmatrix}$$

15. Let A denote a matrix which has an entire row or an entire column of zeros. Then if B is any matrix, either AB has an entire row of zeros or BA has an entire column of zeros, respectively. (See Exercise 18, Section 1. 3.) Hence, neither AB nor BA can be the identity matrix; therefore, A cannot have an inverse.

17. Suppose that $AB = 0$ and A is invertible. Then $A^{-1}(AB) = A^{-1}0$ or $IB = 0$. Hence, $B = 0$.

19. **(a)** Using the notation of Exercise 18, let

$$A = \begin{bmatrix} 1 & 1 \\ -1 & 1 \end{bmatrix} \text{ and } B = \begin{bmatrix} 1 & 1 \\ 1 & 1 \end{bmatrix}$$

Then

$$A^{-1} = \frac{1}{2}\begin{bmatrix} 1 & -1 \\ 1 & 1 \end{bmatrix}$$

so that

$$\begin{aligned} C &= -\frac{1}{4}\begin{bmatrix} 1 & -1 \\ 1 & 1 \end{bmatrix}\begin{bmatrix} 1 & 1 \\ 1 & 1 \end{bmatrix}\begin{bmatrix} 1 & -1 \\ 1 & 1 \end{bmatrix} \\ &= -\frac{1}{4}\begin{bmatrix} 0 & 0 \\ 4 & 0 \end{bmatrix} \\ &= \begin{bmatrix} 0 & 0 \\ -1 & 0 \end{bmatrix} \end{aligned}$$

Thus the inverse of the given matrix is

$$\begin{bmatrix} \frac{1}{2} & -\frac{1}{2} & 0 & 0 \\ \frac{1}{2} & \frac{1}{2} & 0 & 0 \\ 0 & 0 & \frac{1}{2} & -\frac{1}{2} \\ -1 & 0 & \frac{1}{2} & \frac{1}{2} \end{bmatrix}$$

21. We use Theorem 1.4.9.

(a) If $A = BB^T$, then

$$A^T = (BB^T)^T = (B^T)^T B^T = BB^T = A$$

Thus A is symmetric. On the other hand, if $A = B + B^T$, then

$$A^T = (B + B^T)^T = B^T + (B^T)^T = B^T + B = B + B^T = A$$

Thus A is symmetric.

(b) If $A = B - B^T$, then

$$\begin{aligned} A^T &= (B - B^T)^T = [B + (-1)B^T]^T = B^T + [(-1)B^T]^T \\ &= B^T + (-1)(B^T)^T = B^T + (-1)B = B^T - B = -A \end{aligned}$$

Thus A is skew-symmetric.

23. Let

$$A^{-1} = \begin{bmatrix} x_{11} & x_{12} & x_{13} \\ x_{21} & x_{22} & x_{23} \\ x_{31} & x_{32} & x_{33} \end{bmatrix}$$

Then

$$AA^{-1} = \begin{bmatrix} 1 & 0 & 1 \\ 1 & 1 & 0 \\ 0 & 1 & 1 \end{bmatrix} \begin{bmatrix} x_{11} & x_{12} & x_{13} \\ x_{21} & x_{22} & x_{23} \\ x_{31} & x_{32} & x_{33} \end{bmatrix}$$

$$= \begin{bmatrix} x_{11}+x_{31} & x_{12}+x_{32} & x_{13}+x_{33} \\ x_{11}+x_{21} & x_{12}+x_{22} & x_{13}+x_{23} \\ x_{21}+x_{31} & x_{22}+x_{32} & x_{23}+x_{33} \end{bmatrix}$$

Since $AA^{-1} = I$, we equate corresponding entries to obtain the system of equations

$$\begin{array}{lllllllllll}
x_{11} & & & & & & +x_{31} & & & =1 \\
& x_{12} & & & & & & +x_{32} & & =0 \\
& & x_{13} & & & & & & +x_{33} & =0 \\
x_{11} & & & +x_{21} & & & & & & =0 \\
& x_{12} & & & +x_{22} & & & & & =1 \\
& & x_{13} & & & +x_{23} & & & & =0 \\
& & & x_{21} & & & +x_{31} & & & =0 \\
& & & & x_{22} & & & +x_{32} & & =0 \\
& & & & & x_{23} & & & +x_{33} & =1
\end{array}$$

The solution to this system of equations gives

$$A^{-1} = \begin{bmatrix} 1/2 & 1/2 & -1/2 \\ -1/2 & 1/2 & 1/2 \\ 1/2 & -1/2 & 1/2 \end{bmatrix}$$

25. We wish to show that $A(B - C) = AB - AC$. By Part (*d*) of Theorem 1.4.1, we have $A(B - C) = A(B + (-C)) = AB + A(-C)$. Finally by Part (*m*), we have $A(-C) = -AC$ and the desired result can be obtained by substituting this result in the above equation.

27. (a) We have

$$A^r A^s = \underbrace{(AA\cdots A)}_{\substack{r \\ \text{factors}}}\underbrace{(AA\cdots A)}_{\substack{s \\ \text{factors}}}$$

$$= \underbrace{AA\cdots A}_{\substack{r+s \\ \text{factors}}} = A^{r+s}$$

On the other hand,

$$(A^r)^s = \underbrace{\underbrace{(AA\cdots A)}_{\substack{r \\ \text{factors}}}\underbrace{(AA\cdots A)}_{\substack{r \\ \text{factors}}}\cdots\underbrace{(AA\cdots A)}_{\substack{r \\ \text{factors}}}}_{\substack{s \\ \text{factors}}}$$

$$= \underbrace{AA\cdots A}_{\substack{rs \\ \text{factors}}}$$

(b) Suppose that $r < 0$ and $s < 0$; let $\rho = -r$ and $\sigma = -s$, so that

$A^r A^s = A^{-\rho} A^{-\sigma}$

$= (A^{-1})^{\rho} (A^{-1})^{\sigma}$ (by the definition)

$= (A^{-1})^{\rho+\sigma}$ (by Part (a))

$= A^{-(\rho+\sigma)}$ (by the definition)

$= A^{-\rho-\sigma}$

$= A^{r+s}$

Also

$$
\begin{aligned}
(A^r)^s &= (A^{-\rho})^{-\sigma} \\
&= [(A^{-1})^{\rho}]^{-\sigma} && \text{(by the definition)} \\
&= \left([(A^{-1})^{\rho}]^{-1}\right)^{\sigma} && \text{(by the definition)} \\
&= \left([(A^{-1})^{-1})]^{\rho}\right)^{\sigma} && \text{(by Theorem 1.4.8b)} \\
&= ([A]^{\rho})^{\sigma} && \text{(by Theorem 1.4.8a)} \\
&= A^{\rho\sigma} && \text{(by Part (a))} \\
&= A^{(-\rho)(-\sigma)} \\
&= A^{rs}
\end{aligned}
$$

29. **(a)** If $AB = AC$, then

$$A^{-1}(AB) = A^{-1}(AC)$$

or

$$(A^{-1}A)B = (A^{-1}A)C$$

or

$$B = C$$

(b) The matrix A in Example 3 is not invertible.

31. **(a)** Any pair of matrices that do not commute will work. For example, if we let

$$A = \begin{bmatrix} 1 & 0 \\ 0 & 0 \end{bmatrix} \qquad B = \begin{bmatrix} 0 & 1 \\ 0 & 1 \end{bmatrix}$$

then

$$(A + B)^2 = \begin{bmatrix} 1 & 1 \\ 0 & 1 \end{bmatrix}^2 = \begin{bmatrix} 1 & 2 \\ 0 & 1 \end{bmatrix}$$

whereas

$$A^2 + 2AB + B^2 = \begin{bmatrix} 1 & 3 \\ 0 & 1 \end{bmatrix}$$

(b) In general,

$$(A + B)^2 = (A + B)(A + B) = A^2 + AB + BA + B^2$$

33. If

$$A = \begin{bmatrix} a_{11} & 0 & 0 \\ 0 & a_{22} & 0 \\ 0 & 0 & a_{33} \end{bmatrix} \text{ then } A^2 = \begin{bmatrix} a_{11}^2 & 0 & 0 \\ 0 & a_{22}^2 & 0 \\ 0 & 0 & a_{33}^2 \end{bmatrix}$$

Thus, $A^2 = I$ if and only if $a_{11}^2 = a_{22}^2 = a_{33}^2 = 1$, or $a_{11} = \pm 1$, $a_{22} = \pm 1$, and $a_{33} = \pm 1$. There are exactly eight possibilities:

$$\begin{bmatrix} 1 & 0 & 0 \\ 0 & 1 & 0 \\ 0 & 0 & 1 \end{bmatrix} \begin{bmatrix} 1 & 0 & 0 \\ 0 & 1 & 0 \\ 0 & 0 & -1 \end{bmatrix} \begin{bmatrix} 1 & 0 & 0 \\ 0 & -1 & 0 \\ 0 & 0 & 1 \end{bmatrix} \begin{bmatrix} 1 & 0 & 0 \\ 0 & -1 & 0 \\ 0 & 0 & -1 \end{bmatrix}$$

$$\begin{bmatrix} -1 & 0 & 0 \\ 0 & 1 & 0 \\ 0 & 0 & 1 \end{bmatrix} \begin{bmatrix} -1 & 0 & 0 \\ 0 & 1 & 0 \\ 0 & 0 & -1 \end{bmatrix} \begin{bmatrix} -1 & 0 & 0 \\ 0 & -1 & 0 \\ 0 & 0 & 1 \end{bmatrix} \begin{bmatrix} -1 & 0 & 0 \\ 0 & -1 & 0 \\ 0 & 0 & -1 \end{bmatrix}$$

35. **(b)** The statement is true, since $(A - B)^2 = (-(B - A))^2 = (B - A)^2$.

(c) The statement is true only if A^{-1} and B^{-1} exist, in which case

$$(AB^{-1})(BA^{-1}) = A(B^{-1}B)A^{-1} = AI_nA^{-1} = AA^{-1} = I_n$$

EXERCISE SET 1.5

1. **(a)** The matrix may be obtained from I_2 by adding –5 times Row 1 to Row 2. Thus, it is elementary.

(c) The matrix may be obtained from I_2 by multiplying Row 2 of I_2 by $\sqrt{3}$. Thus it is elementary.

(e) This is not an elementary matrix because it is not invertible.

(g) The matrix may be obtained from I_4 only by performing two elementary row operations such as replacing Row 1 of I_4 by Row 1 plus Row 4, and then multiplying Row 1 by 2. Thus it is not an elementary matrix.

3. **(a)** If we interchange Rows 1 and 3 of A, then we obtain B. Therefore, E_1 must be the matrix obtained from I_3 by interchanging Rows 1 and 3 of I_3, i.e.,

$$E_1 = \begin{bmatrix} 0 & 0 & 1 \\ 0 & 1 & 0 \\ 1 & 0 & 0 \end{bmatrix}$$

(c) If we multiply Row 1 of A by –2 and add it to Row 3, then we obtain C. Therefore, E_3 must be the matrix obtained from I_3 by replacing its third row by –2 times Row 1 plus Row 3, i.e.,

$$E_3 = \begin{bmatrix} 1 & 0 & 0 \\ 0 & 1 & 0 \\ -2 & 0 & 1 \end{bmatrix}$$

5. **(a)** $R_1 \leftrightarrow R_{2,}$ Row 1 and Row 2 are swapped

(b) $R_1 \to 2R_1$

$R_2 \to -3R_2$

(c) $R_2 \to -2R_1 + R_2$

7. (a)

$$\left[\begin{array}{rrr|rrr} 3 & 4 & -1 & 1 & 0 & 0 \\ 1 & 0 & 3 & 0 & 1 & 0 \\ 2 & 5 & -4 & 0 & 0 & 1 \end{array}\right]$$

$$\left[\begin{array}{rrr|rrr} 1 & 0 & -3 & 0 & 1 & 0 \\ 3 & 4 & -1 & 1 & 0 & 0 \\ 2 & 5 & -4 & 0 & 0 & 1 \end{array}\right]$$

Interchange Rows 1 and 2.

$$\left[\begin{array}{rrr|rrr} 1 & 0 & 3 & 0 & 1 & 0 \\ 0 & 4 & -10 & 1 & -3 & 0 \\ 0 & 5 & -10 & 0 & -2 & 1 \end{array}\right]$$

Add –3 times Row 1 to Row 2 and –2 times Row 1 to 3.

$$\left[\begin{array}{rrr|rrr} 1 & 0 & 3 & 0 & 1 & 0 \\ 0 & 4 & -10 & 1 & -3 & 0 \\ 0 & 1 & 0 & -1 & 1 & 1 \end{array}\right]$$

Add –1 times Row 2 to Row 3.

$$\left[\begin{array}{rrr|rrr} 1 & 0 & 3 & 0 & 1 & 0 \\ 0 & 1 & 0 & -1 & 1 & 1 \\ 0 & 0 & -10 & 5 & -7 & -4 \end{array}\right]$$

Add –4 times Row 3 to Row 2 and interchange Rows 2 and 3.

$$\left[\begin{array}{rrr|rrr} 1 & 0 & 0 & \frac{3}{2} & -\frac{11}{10} & -\frac{6}{5} \\ 0 & 1 & 0 & -1 & 1 & 1 \\ 0 & 0 & 1 & -\frac{1}{2} & \frac{7}{10} & \frac{2}{5} \end{array}\right]$$

Multiply Row 3 by –1/10. Then add –3 times Row 3 to Row 1.

Thus, the desired inverse is

$$\begin{bmatrix} \frac{3}{2} & -\frac{11}{10} & -\frac{6}{5} \\ -1 & 1 & 1 \\ -\frac{1}{2} & \frac{7}{10} & \frac{2}{5} \end{bmatrix}$$

7. (c)

$$\left[\begin{array}{ccc|ccc} 1 & 0 & 1 & 1 & 0 & 0 \\ 0 & 1 & 1 & 0 & 1 & 0 \\ 1 & 1 & 0 & 0 & 0 & 1 \end{array}\right]$$

$$\left[\begin{array}{ccc|ccc} 1 & 0 & 1 & 1 & 0 & 0 \\ 0 & 1 & 1 & 0 & 1 & 0 \\ 0 & 1 & -1 & -1 & 0 & 1 \end{array}\right]$$

Subtract Row 1 from Row 3.

$$\left[\begin{array}{ccc|ccc} 1 & 0 & 1 & 1 & 0 & 0 \\ 0 & 1 & 1 & 0 & 1 & 0 \\ 0 & 0 & 1 & \frac{1}{2} & \frac{1}{2} & -\frac{1}{2} \end{array}\right]$$

Subtract Row 2 from Row 3 and multiply Row 3 by –1/2.

$$\left[\begin{array}{ccc|ccc} 1 & 0 & 0 & \frac{1}{2} & -\frac{1}{2} & \frac{1}{2} \\ 0 & 1 & 0 & -\frac{1}{2} & \frac{1}{2} & \frac{1}{2} \\ 1 & 1 & 0 & \frac{1}{2} & \frac{1}{2} & -\frac{1}{2} \end{array}\right]$$

Subtract Row 3 from Rows 1 and 2.

Thus

$$\begin{bmatrix} 1 & 0 & 1 \\ 0 & 1 & 1 \\ 1 & 1 & 0 \end{bmatrix}^{-1} = \begin{bmatrix} \frac{1}{2} & -\frac{1}{2} & \frac{1}{2} \\ -\frac{1}{2} & \frac{1}{2} & \frac{1}{2} \\ \frac{1}{2} & \frac{1}{2} & -\frac{1}{2} \end{bmatrix}$$

(e)

$$\left[\begin{array}{ccc|ccc} 1 & 0 & 1 & 1 & 0 & 0 \\ -1 & 1 & 1 & 0 & 1 & 0 \\ 0 & 1 & 0 & 0 & 0 & 1 \end{array}\right]$$

$$\left[\begin{array}{ccc|ccc} 1 & 0 & 1 & 1 & 0 & 0 \\ 0 & 1 & 2 & 1 & 1 & 0 \\ 0 & 0 & -2 & -1 & -1 & 1 \end{array}\right]$$

Add Row 1 to Row 2 and subtract the new Row 2 from Row 3.

$$\left[\begin{array}{ccc|ccc} 1 & 0 & 1 & 1 & 0 & 0 \\ 0 & 1 & 0 & 0 & 0 & 1 \\ 0 & 0 & 1 & \frac{1}{2} & \frac{1}{2} & -\frac{1}{2} \end{array}\right]$$

Add Row 3 to Row 2 and then multiply Row 3 by -1/2.

$$\left[\begin{array}{ccc|ccc} 1 & 0 & 0 & \frac{1}{2} & -\frac{1}{2} & \frac{1}{2} \\ 0 & 1 & 0 & 0 & 0 & 1 \\ 0 & 0 & 1 & \frac{1}{2} & \frac{1}{2} & -\frac{1}{2} \end{array}\right]$$

Subtract Row 3 from Row 1.

Thus

$$\begin{bmatrix} 1 & 0 & 1 \\ -1 & 1 & 1 \\ 0 & 1 & 0 \end{bmatrix}^{-1} = \begin{bmatrix} \frac{1}{2} & -\frac{1}{2} & \frac{1}{2} \\ 0 & 0 & 1 \\ \frac{1}{2} & \frac{1}{2} & -\frac{1}{2} \end{bmatrix}$$

9. **(b)** Multiplying Row i of

$$\left[\begin{array}{cccc|cccc} 0 & 0 & 0 & k_1 & 1 & 0 & 0 & 0 \\ 0 & 0 & k_2 & 0 & 0 & 1 & 0 & 0 \\ 0 & k_3 & 0 & 0 & 0 & 0 & 1 & 0 \\ k_4 & 0 & 0 & 0 & 0 & 0 & 0 & 1 \end{array}\right]$$

by $1/k_i$ for $i = 1, 2, 3, 4$ and then reversing the order of the rows yields I_4 on the left and the desired inverse

$$\begin{bmatrix} 0 & 0 & 0 & 1/k_4 \\ 0 & 0 & 1/k_3 & 0 \\ 0 & 1/k_2 & 0 & 0 \\ 1/k_1 & 0 & 0 & 0 \end{bmatrix}$$

on the right.

(c) To reduce

$$\left[\begin{array}{cccc|cccc} k & 0 & 0 & 0 & 1 & 0 & 0 & 0 \\ 1 & k & 0 & 0 & 0 & 1 & 0 & 0 \\ 0 & 1 & k & 0 & 0 & 0 & 1 & 0 \\ 0 & 0 & 1 & k & 0 & 0 & 0 & 1 \end{array}\right]$$

we multiply Row i by $1/k$ and then subtract Row i from Row $(i + 1)$ for $i = 1, 2, 3$. Then multiply Row 4 by $1/k$. This produces I_4 on the left and the inverse,

$$\begin{bmatrix} 1/k & 0 & 0 & 0 \\ -1/k^2 & 1/k & 0 & 0 \\ 1/k^3 & -1/k^2 & 1/k & 0 \\ 1/k^4 & 1/k^3 & -1/k^2 & 1/k \end{bmatrix}$$

on the right.

13. **(a)** $E_3E_2E_1A =$

$$\begin{bmatrix} 1 & 0 & 0 \\ 0 & 1/4 & 0 \\ 0 & 0 & 1 \end{bmatrix}\begin{bmatrix} 1 & 0 & 0 \\ 0 & 1 & -3 \\ 0 & 0 & 1 \end{bmatrix}\begin{bmatrix} 1 & 0 & 2 \\ 0 & 1 & 0 \\ 0 & 0 & 1 \end{bmatrix}\begin{bmatrix} 1 & 0 & -2 \\ 0 & 4 & 3 \\ 0 & 0 & 1 \end{bmatrix} = I_3$$

(b) $A = (E_3E_2E_1)^{-1} = E_1^{-1}E_2^{-1}E_3^{-1}$

$$= \begin{bmatrix} 1 & 0 & -2 \\ 0 & 1 & 0 \\ 0 & 0 & 1 \end{bmatrix} \begin{bmatrix} 1 & 0 & 0 \\ 0 & 1 & 3 \\ 0 & 0 & 1 \end{bmatrix} \begin{bmatrix} 1 & 0 & 0 \\ 0 & 4 & 0 \\ 0 & 0 & 1 \end{bmatrix}$$

15. If A is an elementary matrix, then it can be obtained from the identity matrix I by a single elementary row operation. If we start with I and multiply Row 3 by a nonzero constant, then $a = b = 0$. If we interchange Row 1 or Row 2 with Row 3, then $c = 0$. If we add a nonzero multiple of Row 1 or Row 2 to Row 3, then either $b = 0$ or $a = 0$. Finally, if we operate only on the first two rows, then $a = b = 0$. Thus at least one entry in Row 3 must equal zero.

17. Every $m \times n$ matrix A can be transformed into reduced row-echelon form B by a sequence of row operations. From Theorem 1.5.1,

$$B = E_kE_{k-1} \cdots E_1A$$

where $E_1, E_2, \cdots, E_k$ are the elementary matrices corresponding to the row operations. If we take $C = E_kE_{k-1} \cdots E_1$, then C is invertible by Theorem 1.5.2 and the rule following Theorem 1.4.6.

19. **(a)** First suppose that A and B are row equivalent. Then there are elementary matrices $E_1, \cdots, E_p$ such that $A = E_1 \cdots E_pB$. There are also elementary matrices $E_{p+1}, \cdots, E_{p+q}$ such that $E_{p+1} \cdots E_{p+q}A$ is in reduced row-echelon form. Therefore, the matrix $E_{p+1} \cdots E_{p+q}E_1 \cdots E_pB$ is also in (the same) reduced row-echelon form. Hence we have found, via elementary matrices, a sequence of elementary row operations which will put B in the same reduced row-echelon form as A.

Now suppose that A and B have the same reduced row-echelon form. Then there are elementary matrices $E_1, \cdots, E_p$ and $E_{p+1}, \cdots, E_{p+q}$ such that $E_1 \cdots E_pA = E_{p+1} \cdots E_{p+q}B$. Since elementary matrices are invertible, this equation implies that $A = E_p^{-1} \cdots E_1^{-1}E_{p+1} \cdots E_{p+q}B$. Since the inverse of an elementary matrix is also an elementary matrix, we have that A and B are row equivalent.

21. The matrix A, by hypothesis, can be reduced to the identity matrix via a sequence of elementary row operations. We can therefore find elementary matrices $E_1, E_2, \cdots E_k$ such that

$$E_k \cdots E_2 \cdot E_1 \cdot A = I_n$$

Since every elementary matrix is invertible, it follows that

$$A = E_1^{-1}E_2^{-1} \cdots E_k^{-1}I_n$$

23. **(a)** True. Suppose we reduce A to its reduced row-echelon form via a sequence of elementary row operations. The resulting matrix must have at least one row of zeros, since otherwise we would obtain the identity matrix and A would be invertible. Thus at least one of the variables in $\mathbf{x}$ must be arbitrary and the system of equations will have infinitely many solutions.

(b) See Part (a).

(d) False. If $B = EA$ for any elementary matrix E, then $A = E^{-1}B$. Thus, if B were invertible, then A would also be invertible, contrary to hypothesis.

EXERCISE SET 1.6

1. This system of equations is of the form $Ax = \mathbf{b}$, where

$$A = \begin{bmatrix} 1 & 1 \\ 5 & 6 \end{bmatrix} \quad x = \begin{bmatrix} x_1 \\ x_2 \end{bmatrix} \quad \text{and} \quad \mathbf{b} = \begin{bmatrix} 2 \\ 9 \end{bmatrix}$$

By Theorem 1.4.5,

$$A^{-1} = \begin{bmatrix} 6 & -1 \\ -5 & 1 \end{bmatrix}$$

Thus

$$x = A^{-1}\mathbf{b} = \begin{bmatrix} 6 & -1 \\ -5 & 1 \end{bmatrix}\begin{bmatrix} 2 \\ 9 \end{bmatrix} = \begin{bmatrix} 3 \\ -1 \end{bmatrix}$$

That is,

$$x_1 = 3 \quad \text{and} \quad x_2 = -1$$

3. This system is of the form $Ax = \mathbf{b}$, where

$$A = \begin{bmatrix} 1 & 3 & 1 \\ 2 & 2 & 1 \\ 2 & 3 & 1 \end{bmatrix} \quad \mathbf{x} = \begin{bmatrix} x_1 \\ x_2 \\ x_3 \end{bmatrix} \quad \text{and} \quad \mathbf{b} = \begin{bmatrix} 4 \\ -1 \\ 3 \end{bmatrix}$$

By direct computation we obtain

$$A^{-1} = \begin{bmatrix} -1 & 0 & 1 \\ 0 & -1 & 1 \\ 2 & 3 & -4 \end{bmatrix}$$

so that

$$\mathrm{x} = A^{-1}\mathbf{b} = \begin{bmatrix} -1 \\ 4 \\ -7 \end{bmatrix}$$

That is,

$$x_1 = -1,\ x_2 = 4, \text{ and } x_3 = -7$$

5. The system is of the form Ax = **b**, where

$$A = \begin{bmatrix} 1 & 1 & 1 \\ 1 & 1 & -4 \\ -4 & 1 & 1 \end{bmatrix} \quad \mathrm{x} = \begin{bmatrix} x_1 \\ x_2 \\ x_3 \end{bmatrix} \quad \text{and} \quad \mathbf{b} = \begin{bmatrix} 5 \\ 10 \\ 0 \end{bmatrix}$$

By direct computation, we obtain

$$A^{-1} = \left(\frac{1}{5}\right)\begin{bmatrix} 1 & 0 & -1 \\ 3 & 1 & 1 \\ 1 & -1 & 0 \end{bmatrix}$$

Thus,

$$\mathrm{x} = A^{-1}\mathbf{b} = \begin{bmatrix} 1 \\ 5 \\ -1 \end{bmatrix}$$

That is, $x_1 = 1$, $x_2 = 5$, and $x_3 = -1$.

7. The system is of the form $A\text{x} = \mathbf{b}$ where

$$A = \begin{bmatrix} 3 & 5 \\ 1 & 2 \end{bmatrix} \quad \text{x} = \begin{bmatrix} x_1 \\ x_2 \end{bmatrix} \quad \text{and} \quad \mathbf{b} = \begin{bmatrix} b_1 \\ b_2 \end{bmatrix}$$

By Theorem 1.4.5, we have

$$A^{-1} = \begin{bmatrix} 2 & -5 \\ -1 & 3 \end{bmatrix}$$

Thus

$$\mathbf{x} = A^{-1}\mathbf{b} = \begin{bmatrix} 2b_1 - 5b_2 \\ -b_1 + 3b_2 \end{bmatrix}$$

That is,

$$x_1 = 2b_1 - 5b_2 \quad \text{and} \quad x_2 = -b_1 + 3b_2$$

9. The system is of the form $A\text{x} = \mathbf{b}$, where

$$A = \begin{bmatrix} 1 & 2 & 1 \\ 1 & -1 & 1 \\ 1 & 1 & 0 \end{bmatrix} \quad \mathbf{x} = \begin{bmatrix} x_1 \\ x_2 \\ x_3 \end{bmatrix} \quad \text{and} \quad \mathbf{b} = \begin{bmatrix} b_1 \\ b_2 \\ b_3 \end{bmatrix}$$

We compute

$$A^{-1} = \begin{bmatrix} -1/3 & 1/3 & 1 \\ 1/3 & -1/3 & 0 \\ 2/3 & 1/3 & -1 \end{bmatrix}$$

so that

$$\mathbf{x} = A^{-1}\mathbf{b} \begin{bmatrix} -(1/3)b_1 + (1/3)b_2 + b_3 \\ (1/3)b_1 - (1/3)b_2 \\ (2/3)b_1 + (1/3)b_2 - b_3 \end{bmatrix}$$

9. **(a)** In this case, we let

$$\mathbf{b} = \begin{bmatrix} -1 \\ 3 \\ 4 \end{bmatrix}$$

Then

$$x = A^{-1}\mathbf{b} \begin{bmatrix} 16/3 \\ -4/3 \\ -11/3 \end{bmatrix}$$

That is, $x_1 = 16/3$, $x_2 = -4/3$, and $x_3 = -11/3$.

(c) In this case, we let

$$\mathbf{b} = \begin{bmatrix} -1 \\ -1 \\ 3 \end{bmatrix}$$

Then

$$x = A^{-1}\mathbf{b} \begin{bmatrix} 3 \\ 0 \\ -4 \end{bmatrix}$$

That is, $x_1 = 3$, $x_2 = 0$, and $x_3 = -4$.

11. The coefficient matrix, augmented by the two **b** matrices, yields

$$\left[\begin{array}{cc|c|c} 1 & -5 & 1 & -2 \\ 3 & 2 & 4 & 5 \end{array}\right]$$

This reduces to

$$\left[\begin{array}{cc|c|c} 1 & -5 & 1 & -2 \\ 0 & 17 & 1 & 11 \end{array}\right]$$

Add –3 times Row 1 to Row 2.

or

$$\left[\begin{array}{cc|c|c} 1 & 0 & 22/17 & 21/17 \\ 0 & 1 & 1/17 & 11/17 \end{array}\right]$$

Divide Row 2 by 17 and add 5 times Row 2 to Row 1.

Thus the solution to Part (a) is $x_1 = 22/17$, $x_2 = 1/17$, and to Part (b) is $x_1 = 21/17$, $x_2 = 11/17$.

15. As above, we set up the matrix

$$\left[\begin{array}{ccc|c|c} 1 & -2 & 1 & -2 & 1 \\ 2 & -5 & 1 & 1 & -1 \\ 3 & -7 & 2 & -1 & 0 \end{array}\right]$$

This reduces to

$$\left[\begin{array}{ccc|c|c} 1 & -2 & 1 & -2 & 1 \\ 0 & -1 & -1 & 5 & -3 \\ 0 & -1 & -1 & 5 & -3 \end{array}\right]$$

Add appropriate multiples of Row 1 to Rows 2 and 3.

or

$$\left[\begin{array}{ccc|c|c} 1 & -2 & 1 & -2 & 1 \\ 0 & 1 & 1 & -5 & 3 \\ 0 & 0 & 0 & 0 & 0 \end{array}\right]$$

Add –1 times Row 2 to Row 3 and multiply Row 2 by –1.

or

$$\left[\begin{array}{ccc|c|c} 1 & 0 & 3 & -12 & 7 \\ 0 & 1 & 1 & -5 & 3 \\ 0 & 0 & 0 & 0 & 0 \end{array}\right]$$

Add twice Row 2 to Row 3.

Thus if we let $x_3 = t$, we have for Part (a) $x_1 = -12 - 3t$ and $x_2 = -5 - t$, while for Part (b) $x_1 = 7 - 3t$ and $x_2 = 3 - t$.

17. The augmented matrix for this system of equations is

$$\begin{bmatrix} 1 & -2 & 5 & b_1 \\ 4 & -5 & 8 & b_2 \\ -3 & 3 & -3 & b_3 \end{bmatrix}$$

If we reduce this matrix to row-echelon form, we obtain

$$\begin{bmatrix} 1 & -2 & 5 & b_1 \\ 0 & 1 & -4 & \frac{1}{3}(b_2 - 4b_1) \\ 0 & 0 & 0 & -b_1 + b_2 + b_3 \end{bmatrix}$$

The third row implies that $b_3 = b_1 - b_2$. Thus, $A\mathrm{x} = \mathbf{b}$ is consistent if and only if $\mathbf{b}$ has the form

$$\mathbf{b} = \begin{bmatrix} b_1 \\ b_2 \\ b_1 - b_2 \end{bmatrix}$$

23. Since $A\mathrm{x} =$ has only $\mathrm{x} = \mathbf{0}$ as a solution, Theorem 1.6.4 guarantees that A is invertible. By Theorem 1.4.8 (b), A^k is also invertible. In fact,

$$(A^k)^{-1} = (A^{-1})^k$$

Since the proof of Theorem 1.4.8 (b) was omitted, we note that

$$\underbrace{A^{-1}A^{-1}\cdots A^{-1}}_{\substack{k\\ \text{factors}}}\ \underbrace{AA\cdots A}_{\substack{k\\ \text{factors}}}=I$$

Because A^k is invertible, Theorem 1.6.4 allows us to conclude that $A^k\mathrm{x} = \mathbf{0}$ has only the trivial solution.

25. Suppose that x_1 is a fixed matrix which satisfies the equation $A\mathrm{x}_1 = \mathbf{b}$. Further, let x be any matrix whatsoever which satisfies the equation $A\mathrm{x} = \mathbf{b}$. We must then show that there is a matrix x_0 which satisfies both of the equations $\mathrm{x} = \mathrm{x}_1 + \mathrm{x}_0$ and $A\mathrm{x}_0 = \mathbf{0}$.

Clearly, the first equation implies that

$$\mathrm{x}_0 = \mathrm{x} - \mathrm{x}_1$$

This candidate for x_0 will satisfy the second equation because

$$A\mathrm{x}_0 = A(\mathrm{x} - \mathrm{x}_1) = A\mathrm{x} - A\mathrm{x}_1 = \mathbf{b} - \mathbf{b} = \mathbf{0}$$

We must also show that if both $A\mathrm{x}_1 = \mathbf{b}$ and $A\mathrm{x}_0 = \mathbf{0}$, then $A(\mathrm{x}_1 + \mathrm{x}_0) = \mathbf{b}$. But

$$A(\mathrm{x}_1 + \mathrm{x}_0) = A\mathrm{x}_1 + A\mathrm{x}_0 = \mathbf{b} + \mathbf{0} = \mathbf{b}$$

27. **(a)** $\mathrm{x} \neq 0$ and $\mathrm{x} \neq y$

(b) $\mathrm{x} \neq 0$ and $y \neq 0$

(c) $\mathrm{x} \neq y$ and $\mathrm{x} \neq -y$

Gaussian elimination has to be performed on $(A \mid I)$ to find A^{-1}. Then the product $A^{-1}B$ is performed, to find x. Instead, use Gaussian elimination on $(A \mid B)$ to find x. There are fewer steps in the Gaussian elimination, since $(A \mid B)$ is a $m \times (n+1)$ matrix in general, or $n \times (n+1)$ where A is square $(n \times n)$. Compare this with $(A \mid I)$ which is $n \times (2n)$ in the inversion approach. Also, the inversion approach only works for A $n \times n$ and invertible.

29. No. The system of equations $A\mathbf{x} = \mathbf{x}$ is equivalent to the system $(A - I)\mathbf{x} = \mathbf{0}$. For this system to have a unique solution, $A - I$ must be invertible. If, for instance, $A = I$, then any vector $\mathbf{x}$ will be a solution to the system of equations $A\mathbf{x} = \mathbf{x}$.

Note that if $\mathbf{x} \neq \mathbf{0}$ is a solution to the equation $A\mathbf{x} = \mathbf{x}$, then so is $k\mathbf{x}$ for any real number k. A unique solution can only exist if $A - I$ is invertible, in which case, $\mathbf{x} = \mathbf{0}$.

31. Let A and B be square matrices of the same size. If either A or B is singular, then AB is singular.

EXERCISE SET 1.7

7. The matrix A fails to be invertible if and only if $a + b - 1 = 0$ and the matrix B fails to be invertible if and only if $2a - 3b - 7 = 0$. For both of these conditions to hold, we must have $a = 2$ and $b = -1$.

9. We know that A and B will commute if and only if

$$AB = \begin{bmatrix} 2 & 1 \\ 1 & -5 \end{bmatrix}\begin{bmatrix} a & b \\ b & d \end{bmatrix} = \begin{bmatrix} 2a+b & 2b+d \\ a-5b & b-5d \end{bmatrix}$$

is symmetric. So $2b + d = a - 5b$, from which it follows that $a - d = 7b$.

11. **(b)** Clearly

$$A = \begin{bmatrix} ka_{11} & ka_{12} & ka_{13} \\ ka_{21} & ka_{22} & ka_{23} \\ ka_{31} & ka_{32} & ka_{33} \end{bmatrix}\begin{bmatrix} 3/k & 0 & 0 \\ 0 & 5/k & 0 \\ 0 & 0 & 7/k \end{bmatrix}$$

for any real number $k = 0$.

13. We verify the result for the matrix A by finding its inverse.

$$\left[\begin{array}{ccc|ccc} -1 & 2 & 5 & 1 & 0 & 0 \\ 0 & 1 & 3 & 0 & 1 & 0 \\ 0 & 0 & -4 & 0 & 0 & 1 \end{array}\right]$$

$$\left[\begin{array}{ccc|ccc} 1 & -2 & -5 & -1 & 0 & 0 \\ 0 & 1 & 3 & 0 & 1 & 0 \\ 0 & 0 & 1 & 0 & 0 & -1/4 \end{array}\right]$$

Multiply Row 1 by –1 and Row 3 by –1/4.

$$\left[\begin{array}{ccc|ccc} 1 & 0 & 1 & -1 & 2 & 0 \\ 0 & 1 & 0 & 0 & 1 & 3/4 \\ 0 & 0 & 1 & 0 & 0 & -1/4 \end{array}\right]$$

Add 2 times Row 2 to Row 1 and –3 times Row 3 to Row 2.

$$\left[\begin{array}{ccc|ccc} 1 & 0 & 0 & -1 & 2 & 1/4 \\ 0 & 1 & 0 & 0 & 1 & 3/4 \\ 0 & 0 & 1 & 0 & 0 & -1/4 \end{array}\right]$$

Add –1 times Row 3 to Row 1.

Thus A^{-1} is indeed upper triangular.

15. **(a)** If A is symmetric, then $A^T = A$. Then $(A^2)^T = (AA)^T = A^TA^T = A \cdot A = A^2$, so A^2 is symmetric.

(b) We have from part (a) that

$$(2A^2 - 3A + I)^T = 2(A^2)^T - 3A^T + I^T = 2A^2 - 3A + I$$

17. From Theorem 1.7.1(b), we have if A is an $n \times n$ upper triangular matrix, so is A^2. By induction, if A is an $n \times n$ upper triangular matrix, so is A^k, $k = 1, 2, 3, \ldots$ We note that the identity matrix $I_n = A^0$ is also upper triangular. Next, if A is $n \times n$ upper triangular, and K is any (real) scalar, then KA is upper triangular. Also, if A and B are $n \times n$ upper triangular matrices, then so is $A+B$. These facts allow us to conclude if $p(x)$ is any (real) polynomial, and A is $n \times n$ upper triangular, then $P(A)$ is an $n \times n$ upper triangular matrix.

19. Let

$$A = \begin{bmatrix} x & 0 & 0 \\ 0 & y & 0 \\ 0 & 0 & z \end{bmatrix}$$

Then if $A^2 - 3A - 4I = O$, we have

$$\begin{bmatrix} x^2 & 0 & 0 \\ 0 & y^2 & 0 \\ 0 & 0 & z^2 \end{bmatrix} - 3\begin{bmatrix} x & 0 & 0 \\ 0 & y & 0 \\ 0 & 0 & z \end{bmatrix} - 4\begin{bmatrix} 1 & 0 & 0 \\ 0 & 1 & 0 \\ 0 & 0 & 1 \end{bmatrix} = O$$

This leads to the system of equations

$$x^2 - 3x - 4 = 0$$

$$y^2 - 3y - 4 = 0$$

$$z^2 - 3z - 4 = 0$$

which has the solutions $x = 4, -1$, $y = 4, -1$, $z = 4, -1$. Hence, there are 8 possible choices for x, y, and z, respectively, namely (4, 4, 4), (4, 4, –1), (4, –1, 4), (4, –1, –1), (–1, 4, 4), (–1, 4, –1), (–1, –1, 4), and (–1, –1, –1).

23. The matrix

$$A = \begin{bmatrix} 0 & 1 \\ -1 & 0 \end{bmatrix}$$

is skew-symmetric but

$$AA = A^2 = \begin{bmatrix} -1 & 0 \\ 0 & -1 \end{bmatrix}$$

is not skew-symmetric. Therefore, the result does not hold.

In general, suppose that A and B are commuting skew-symmetric matrices. Then $(AB)^T = (BA)^T = A^T B^T = (-A)(-B) = AB$, so that AB is symmetric rather than skew-symmetric. [We note that if A and B are skew-symmetric and their product is symmetric, then $AB = (AB)^T = B^T A^T = (-B)(-A) = BA$, so the matrices commute and thus skew-symmetric matrices, too, commute if and only if their product is symmetric.]

25. Let

$$A = \begin{bmatrix} x & y \\ 0 & z \end{bmatrix}$$

Then

$$A^3 = \begin{bmatrix} x^3 & y\left(x^2 + xz + z^2\right) \\ 0 & z^3 \end{bmatrix} = \begin{bmatrix} 1 & 30 \\ 0 & -8 \end{bmatrix}$$

Hence, $x^3 = 1$ which implies that $x = 1$, and $z^3 = -8$ which implies that $z = -2$. Therefore, $3y = 30$ and thus $y = 10$.

27. To multiply two diagonal matrices, multiply their corresponding diagonal elements to obtain a new diagonal matrix. Thus, if D_1 and D_2 are diagonal matrices with diagonal elements $d_1, \ldots, d_n$ and $e_1, \ldots, e_n$ respectively, then D_1D_2 is a diagonal matrix with diagonal elements $d_1e_1, \ldots, d_ne_n$. The proof follows directly from the definition of matrix multiplication.

29. In general, let $A = [a_{ij}]_{n \times n}$ denote a lower triangular matrix with no zeros on or below the diagonal and let $A\mathrm{x} = \mathbf{b}$ denote the system of equations where $\mathbf{b} = [b_1, b_2, \ldots, b_n]^T$. Since A is lower triangular, the first row of A yields the equation $a_{11}x_1 = b_1$. Since $a_{11} \neq 0$, we can solve for x_1. Next, the second row of A yields the equation $a_{21}x_1 + a_{22}x_2 = b_2$. Since we know x_1 and since $a_{22} \neq 0$, we can solve for x_2. Continuing in this way, we can solve for successive values of x_i by back substituting all of the previously found values $x_1, x_2, \ldots, x_{i-1}$.

SUPPLEMENTARY EXERCISES 1

1.

$$\begin{bmatrix} \frac{3}{5} & -\frac{4}{5} & x \\ \frac{4}{5} & \frac{3}{5} & y \end{bmatrix}$$

$$\begin{bmatrix} 1 & -\frac{4}{3} & \frac{5}{3}x \\ \frac{4}{5} & \frac{3}{5} & y \end{bmatrix}$$

Multiply Row 1 by 5/3.

$$\begin{bmatrix} 1 & -\frac{4}{3} & \frac{5}{3}x \\ 0 & \frac{5}{3} & -\frac{4}{3}x+y \end{bmatrix}$$

Add –4/5 times Row 1 to Row 2.

$$\begin{bmatrix} 1 & -\frac{4}{3} & \frac{5}{3}x \\ 0 & 1 & -\frac{4}{5}x+\frac{3}{5}y \end{bmatrix}$$

Multiply Row 2 by 3/5.

$$\begin{bmatrix} 1 & 0 & \frac{3}{5}x+\frac{4}{5}y \\ 0 & 1 & -\frac{4}{5}x+\frac{3}{5}y \end{bmatrix}$$

Add –4/3 times Row 2 to Row 1.

Thus,

$$x' = \frac{3}{5}x + \frac{4}{5}y$$

$$y' = -\frac{4}{5}x + \frac{3}{5}y$$

3. We denote the system of equations by

$$a_{11}x_1 + a_{12}x_2 + a_{13}x_3 + a_{14}x_4 = 0$$

$$a_{21}x_1 + a_{22}x_2 + a_{23}x_3 + a_{24}x_4 = 0$$

If we substitute both sets of values for x_1, x_2, x_3, and x_4 into the first equation, we obtain

$$\begin{aligned} a_{11} - a_{12} + a_{13} + 2a_{14} &= 0 \\ 2a_{11} \qquad + 3a_{13} - 2a_{14} &= 0 \end{aligned}$$

where a_{11}, a_{12}, a_{13}, and a_{14} are variables. If we substitute both sets of values for x_1, x_2, x_3, and x_4 into the second equation, we obtain

$$\begin{aligned} a_{21} - a_{22} + a_{23} + 2a_{24} &= 0 \\ 2a_{21} \qquad + 3a_{23} - a_{24} &= 0 \end{aligned}$$

where a_{21}, a_{22}, a_{23}, and a_{24} are again variables. The two systems above both yield the matrix

$$\begin{bmatrix} 1 & -1 & 1 & 2 & 0 \\ 2 & 0 & 3 & -1 & 0 \end{bmatrix}$$

which reduces to

$$\begin{bmatrix} 1 & 0 & 3/2 & -1/2 & 0 \\ 0 & 1 & 1/2 & -5/2 & 0 \end{bmatrix}$$

This implies that

$$a_{11} = -(3/2)a_{13} + (1/2)a_{14}$$

$$a_{12} = -(1/2)a_{13} + (5/2)a_{14}$$

and similarly,

$$a_{21} = (-3/2)a_{23} + (1/2)a_{24}$$

$$a_{22} = (-1/2)a_{23} + (5/2)a_{24}$$

As long as our choice of values for the numbers a_{ij} is consistent with the above, then the system will have a solution. For simplicity, and to insure that neither equation is a multiple of the other, we let $a_{13} = a_{14} = -1$ and $a_{23} = 0$, $a_{24} = 2$. This means that $a_{11} = 1$, $a_{12} = -2$, $a_{21} = 1$, and $a_{22} = 5$, so that the system becomes

$$\begin{aligned} x_1 - 2x_2 - x_3 - x_4 &= 0 \\ x_1 + 5x_2 + \quad 2x_4 &= 0 \end{aligned}$$

Of course, this is just one of infinitely many possibilities.

5. As in Exercise 4, we reduce the system to the equations

$$x = \frac{1+5z}{4}$$

$$y = \frac{35-9z}{4}$$

Since x, y, and z must all be positive integers, we have $z > 0$ and $35 - 9z > 0$ or $4 > z$. Thus we need only check the three values $z = 1, 2, 3$ to see whether or not they produce integer solutions for x and y. This yields the unique solution $x = 4$, $y = 2$, $z = 3$.

9. Note that K must be a 2×2 matrix. Let

$$K = \begin{bmatrix} a & b \\ c & d \end{bmatrix}$$

Then

$$\begin{bmatrix} 1 & 4 \\ -2 & 3 \\ 1 & -2 \end{bmatrix} \begin{bmatrix} a & b \\ c & d \end{bmatrix} \begin{bmatrix} 2 & 0 & 0 \\ 0 & 1 & -1 \end{bmatrix} = \begin{bmatrix} 8 & 6 & -6 \\ 6 & -1 & 1 \\ -4 & 0 & 0 \end{bmatrix}$$

or

$$\begin{bmatrix} 1 & 4 \\ -2 & 3 \\ 1 & -2 \end{bmatrix}\begin{bmatrix} 2a & b & -b \\ 2c & d & -d \end{bmatrix} = \begin{bmatrix} 8 & 6 & -6 \\ 6 & -1 & 1 \\ -4 & 0 & 0 \end{bmatrix}$$

or

$$\begin{bmatrix} 2a+8c & b+4d & -b-4d \\ -4a+6c & -2b+3d & 2b-3d \\ 2a-4c & b-2d & -b+2d \end{bmatrix} = \begin{bmatrix} 8 & 6 & -6 \\ 6 & -1 & 1 \\ -4 & 0 & 0 \end{bmatrix}$$

Thus

$$\begin{aligned} 2a \quad + 8c \quad &= 8 \\ b \quad + 4d &= 6 \\ -4a \quad + 6c \quad &= 6 \\ -2b \quad + 3d &= -1 \\ 2a \quad - 4c \quad &= -4 \\ b \quad - 2d &= 0 \end{aligned}$$

Note that we have omitted the 3 equations obtained by equating elements of the last columns of these matrices because the information so obtained would be just a repeat of that gained by equating elements of the second columns. The augmented matrix of the above system is

$$\begin{bmatrix} 2 & 0 & 8 & 0 & 8 \\ 0 & 1 & 0 & 4 & 6 \\ -4 & 0 & 6 & 0 & 6 \\ 0 & -2 & 0 & 3 & -1 \\ 2 & 0 & -4 & 0 & -4 \\ 0 & 1 & 0 & -2 & 0 \end{bmatrix}$$

The reduced row-echelon form of this matrix is

$$\begin{bmatrix} 1 & 0 & 0 & 0 & 0 \\ 0 & 1 & 0 & 0 & 2 \\ 0 & 0 & 1 & 0 & 1 \\ 0 & 0 & 0 & 1 & 1 \\ 0 & 0 & 0 & 0 & 0 \\ 0 & 0 & 0 & 0 & 0 \end{bmatrix}$$

Thus $a = 0$, $b = 2$, $c = 1$, and $d = 1$.

11. The matrix X in Part (a) must be 2×3 for the operations to make sense. The matrices in Parts (b) and (c) must be 2×2.

(b) Let $X = \begin{bmatrix} x & y \\ z & w \end{bmatrix}$. Then

$$X\begin{bmatrix} 1 & -1 & 2 \\ 3 & 0 & 1 \end{bmatrix} = \begin{bmatrix} x+3y & -x & 2x+y \\ z+3w & -z & 2z+w \end{bmatrix}$$

If we equate matrix entries, this gives us the equations

$$x + 3y = -5 \qquad x + 3w = 6$$

$$-x = -1 \qquad -z = -3$$

$$2x + y = 0 \qquad 2z + w = 7$$

Thus $x = 1$ and $z = 3$, so that the top two equations give $y = -2$ and $w = 1$. Since these values are consistent with the bottom two equations, we have that

$$X = \begin{bmatrix} 1 & -2 \\ 3 & 1 \end{bmatrix}$$

11. (c) As above, let $X = \begin{bmatrix} x & y \\ z & w \end{bmatrix}$, so that the matrix equation becomes

$$\begin{bmatrix} 3x+z & 3y+w \\ -x+2z & -y+2w \end{bmatrix} - \begin{bmatrix} x+2y & 4x \\ z+2w & 4z \end{bmatrix} = \begin{bmatrix} 2 & -2 \\ 5 & 4 \end{bmatrix}$$

This yields the system of equations

$$\begin{aligned} 2x - 2y + z \qquad\quad &= 2 \\ -4x + 3y \qquad + w &= -2 \\ -x \qquad + z - 2w &= 5 \\ -y - 4z + 2w &= 4 \end{aligned}$$

with matrix

$$\begin{bmatrix} 2 & -2 & 1 & 0 & 2 \\ -4 & 3 & 0 & 1 & -2 \\ -1 & 0 & 1 & -2 & 5 \\ 0 & -1 & -4 & 2 & 4 \end{bmatrix}$$

which reduces to

$$\begin{bmatrix} 1 & 0 & 0 & 0 & -113/37 \\ 0 & 1 & 0 & 0 & -160/37 \\ 0 & 0 & 1 & 0 & -20/37 \\ 0 & 0 & 0 & 1 & -46/37 \end{bmatrix}$$

Hence, $x = -113/37$, $y = -160/37$, $z = -20/37$, and $w = -46/37$.

15. Since the coordinates of the given points must satisfy the polynomial, we have

$$\begin{aligned} p(1) = 2 \quad &\Rightarrow \quad a + b + c = 2 \\ p(-1) = 6 \quad &\Rightarrow \quad a - b + c = 6 \\ p(2) = 3 \quad &\Rightarrow \quad 4a + 2b + c = 3 \end{aligned}$$

The reduced row-echelon form of the augmented matrix of this system of equations is

$$\begin{bmatrix} 1 & 0 & 0 & 1 \\ 0 & 1 & 0 & -2 \\ 0 & 0 & 1 & 3 \end{bmatrix}$$

Thus, $a = 1$, $b = -2$, and $c = 3$.

17. We must show that $(I - J_n)\,(I - \frac{1}{n-1} J_n) = I$ or that $(I - \frac{1}{n-1} J_n)\,(I - J_n) = I$. (By virtue of Theorem 1.6.3, we need only demonstrate one of these equalities.) We have

$$\left(I - J_n\right)\left(I - \frac{1}{n-1} J_n\right) = I^2 - \frac{1}{n-1} IJ_n - J_n I + \frac{1}{n-1} J_n^2$$
$$= I - \frac{n}{n-1} J_n + \frac{1}{n-1} J_n^2$$

But $J_n^2 = nJ_n$ (think about actually squaring J_n), so that the right-hand side of the above equation is just I, as desired.

19. First suppose that $AB^{-1} = B^{-1}A$. Note that all matrices must be square and of the same size. Therefore

$$(AB^{-1})B = (B^{-1}A)B$$

or

$$A = B^{-1}AB$$

so that

$$BA = B(B^{-1}AB) = (BB^{-1})(AB) = AB$$

It remains to show that if $AB = BA$ then $AB^{-1} = B^{-1}A$. An argument similar to the one given above will serve, and we leave the details to you.

21. **(b)** Let the ij^{th} entry of A be a_{ij}. Then $\operatorname{tr}(A) = a_{11} + a_{22} + \cdots + a_{nn}$, so that

$$\operatorname{tr}(kA) = ka_{11} + ka_{22} + \cdots + ka_{nn}$$
$$= k\,(a_{11} + a_{22} + \cdots + a_{nn})$$
$$= k\operatorname{tr}(A)$$

(d) Let the ij^{th} entries of A and B be a_{ij} and b_{ij}, respectively. Then

$$\operatorname{tr}(AB) = a_{11}b_{11} + a_{12}b_{21} + \cdots + a_{1n}b_{n1}$$
$$+ a_{21}b_{12} + a_{22}b_{22} + \cdots + a_{2n}b_{n2}$$
$$+ \cdots$$
$$+ a_{n1}b_{1n} + a_{n2}b_{2n} + \cdots + a_{nn}b_{nn}$$

and

$$\begin{aligned}\operatorname{tr}(BA) = {} & b_{11}a_{11} + b_{12}a_{21} + \cdots + b_{1n}a_{n1} \\ & + b_{21}a_{12} + b_{22}a_{22} + \cdots + b_{2n}a_{n2} \\ & + \cdots \\ & + b_{n1}a_{1n} + b_{n2}a_{2n} + \cdots + b_{nn}a_{nn}\end{aligned}$$

If we rewrite each of the terms $b_{ij}a_{ji}$ in the above expression as $a_{ji}b_{ij}$ and list the terms in the order indicated by the arrows below,

$$\begin{aligned}\operatorname{tr}(BA) = {} & a_{11}b_{11} + a_{21}b_{12} + \cdots + a_{n1}b_{1n} \\ & + a_{12}b_{21} + a_{22}b_{22} + \cdots + a_{n2}b_{2n} \\ & + \cdots \\ & + a_{1n}b_{n1} + a_{2n}b_{n2} + \cdots + a_{nn}b_{nn}\end{aligned}$$

then we have $\operatorname{tr}(AB) = \operatorname{tr}(BA)$.

25. Suppose that A is a square matrix whose entries are differentiable functions of x. Suppose also that A has an inverse, A^{-1}. Then we shall show that A^{-1} also has entries which are differentiable functions of x and that

$$\frac{dA^{-1}}{dx} = -A^{-1}\frac{dA}{dx}A^{-1}$$

Since we can find A^{-1} by the method used in Chapter 1, its entries are functions of x which are obtained from the entries of A by using only addition together with multiplication and division by constants or entries of A. Since sums, products, and quotients of differentiable functions are differentiable wherever they are defined, the resulting entries in the inverse will be differentiable functions except, perhaps, for values of x where their denominators are zero. (Note that we never have to divide by a function which is identically zero.) That is, the entries of A^{-1} are differentiable wherever they are defined. But since we are assuming that A^{-1} is defined, its entries must be differentiable. Moreover,

$$\frac{d}{dx}(AA^{-1}) = \frac{d}{dx}(I) = 0$$

or

$$\frac{dA}{dx}A^{-1} + A\frac{dA^{-1}}{dx} = 0$$

Therefore

$$A\frac{dA^{-1}}{dx} = \frac{dA}{dx}A^{-1}$$

so that

$$\frac{dA^{-1}}{dx} = -A^{-1}\frac{dA}{dx}A^{-1}$$

27. **(b)** Let H be a Householder matrix, so that $H = I - 2PP^T$ where P is an $n \times 1$ matrix. Then using Theorem 1.4.9,

$$\begin{aligned} H^T &= (I - 2PP^T)^T \\ &= I^T - (2PP^T)^T \\ &= I - 2(P^T)^T P^T \\ &= I - 2\,PP^T \\ &= H \end{aligned}$$

and (using Theorem 1.4.1)

$$\begin{aligned} H^T H &= H^2 && \text{(by the above result)} \\ &= (I - 2PP^T)^2 \\ &= I^2 - 2PP^T - 2PP^T + (-2PP^T)^2 \\ &= I - 4PP^T + 4PP^T PP^T \\ &= I - 4PP^T + 4PP^T && \text{(because } P^T P = I\text{)} \\ &= I \end{aligned}$$

29. **(b)** A bit of experimenting and an application of Part (a) indicates that

$$A^n = \begin{bmatrix} a^n & 0 & 0 \\ 0 & b^n & 0 \\ d & 0 & c^n \end{bmatrix}$$

where

$$d = a^{n-1} + a^{n-2}c + \cdots + ac^{n-2} + c^{n-1} = \frac{a^n - c^n}{a - c} \quad \text{if } a \neq c$$

If $a = c$, then $d = na^{n-1}$. We prove this by induction. Observe that the result holds when $n = 1$. Suppose that it holds when $n = N$. Then

$$A^{N+1} = AA^N = A\begin{bmatrix} a^N & 0 & 0 \\ 0 & b^N & 0 \\ d & 0 & c^N \end{bmatrix} = \begin{bmatrix} a^{N+1} & 0 & 0 \\ 0 & b^{N+1} & 0 \\ a^N + cd & 0 & c^{N+1} \end{bmatrix}$$

Here

$$a^N + cd = \begin{cases} a^N + c\dfrac{a^N - c^N}{a - c} = \dfrac{a^{N+1} - a^N c + a^N c - c^{N+1}}{a - c} = \dfrac{a^{N+1} - c^{N+1}}{a - c} & \text{if } a \neq c \\ a^N + a\left(Na^{N-1}\right) = (N+1)a^N & \text{if } a = c \end{cases}$$

Thus the result holds when $n = N + 1$ and so must hold for all values of n.

EXERCISE SET 2.1

1. **(a)** $M_{11} = 7 \bullet 4-(-1) \bullet 1 = 29$, $M_{12} = 21$, $M_{13} = 27$, $M_{21} = -11$, $M_{22} = 13$, $M_{23} = -5$, $M_{31} = -19$, $M_{32} = -19$, $M_{33} = 19$

(b) $C_{11} = 29$, $C_{12} = -21$, $C_{13} = 27$, $C_{21} = 11$, $C_{22} = 13$, $C_{23} = 5$, $C_{31} = -19$, $C_{32} = 19$, $C_{33} = 19$

3. **(a)**

$$|A| = 1 \cdot \begin{vmatrix} 7 & -1 \\ 1 & 4 \end{vmatrix} + 2 \cdot \begin{vmatrix} 6 & -1 \\ -3 & 4 \end{vmatrix} + 3 \cdot \begin{vmatrix} 6 & 7 \\ -3 & 1 \end{vmatrix} = 29 + 42 + 81 = 152$$

(b) $|A| = 1 \bullet M_{11} - 6 \bullet M_{21} - 3 \bullet M_{23} = 152$

(c) $|A = 6 \bullet M_{21} + 7 \bullet M_{22} + 1 \bullet M_{23} = 152$

(d) $|A| = 2 \bullet M_{12} + 7 \bullet M_{22} + 1 \bullet M_{32} = 152$

(e) $|A| = -3 \bullet M_{31} - 1 \bullet M_{32} + 4 \bullet M_{33} = 152$

(f) $|A| = 3 \bullet M_{13} + 1 \bullet M_{23} + 4 \bullet M_{33} = 152$

5. Second column:

$$|A| = 5 \cdot \begin{vmatrix} -3 & 7 \\ -1 & 5 \end{vmatrix} = 5 \cdot -8 = -40$$

7. First column:

$$|A| = 1 \cdot \begin{vmatrix} k & k^2 \\ k & k^2 \end{vmatrix} - 1 \cdot \begin{vmatrix} k & k^2 \\ k & k^2 \end{vmatrix} + 1 \cdot \begin{vmatrix} k & k^2 \\ k & k^2 \end{vmatrix} = 0$$

$$|A| = -(k-1) \cdot \begin{vmatrix} 2 & 4 \\ 5 & k \end{vmatrix} + (k-3) \cdot \begin{vmatrix} k+1 & 5 \\ 7 & k \end{vmatrix} - (k+1) \cdot \begin{vmatrix} k+1 & 7 \\ 2 & 4 \end{vmatrix}$$

9. Third column:

$$|A| = -3 \cdot \begin{vmatrix} 3 & 3 & 5 \\ 2 & 2 & -2 \\ 2 & 10 & 2 \end{vmatrix} - 3 \cdot \begin{vmatrix} 3 & 3 & 5 \\ 2 & 2 & -2 \\ 4 & 1 & 0 \end{vmatrix} = -240$$

11.

$$\text{adj}(A) = \begin{pmatrix} -3 & 5 & 5 \\ 3 & -4 & -5 \\ -2 & 2 & 3 \end{pmatrix}; A = -1; A^{-1} = \begin{pmatrix} 3 & -5 & -5 \\ -3 & 4 & 5 \\ 2 & -2 & -3 \end{pmatrix}$$

13.

$$\text{adj}(A) = \begin{pmatrix} 2 & 6 & 4 \\ 0 & 4 & 6 \\ 0 & 0 & 2 \end{pmatrix}; |A| = 4; A^{-1} = \begin{pmatrix} 1/2 & 3/2 & 1 \\ 0 & 1 & 3/2 \\ 0 & 0 & 1/2 \end{pmatrix}$$

15. **(a)**

$$A^{-1} = \begin{pmatrix} -4 & 3 & 0 & -1 \\ 2 & -1 & 0 & 0 \\ -7 & 0 & -1 & 8 \\ 6 & 0 & 1 & -7 \end{pmatrix}$$

(b) Same as (a).

(c) Gaussian elimination is significantly more efficient for finding inverses.

17.

$$A = \begin{pmatrix} 4 & 5 & 0 \\ 11 & 1 & 2 \\ 1 & 5 & 2 \end{pmatrix}, \mathbf{b} = \begin{pmatrix} 2 \\ 3 \\ 1 \end{pmatrix}; |A| = -132$$

$|A_1| = -36$, $|A_2| = -24$, $|A_3| = -12$

$x_1 = -36/-132 = 3/11$, $x_2 = -24/-132 = 2/11$, $x_3 = 12/-132 = -1/11$

19.

$$A = \begin{pmatrix} 1 & -3 & 1 \\ 2 & -1 & 0 \\ 4 & 0 & -3 \end{pmatrix}, \mathbf{b} = \begin{pmatrix} 4 \\ -2 \\ 0 \end{pmatrix}; |A| = -11$$

$|A_1| = 30$, $|A_2| = 38$, $|A_3| = -40$

$x_1 = 30/-11 = -30/11$, $x_2 = 38/-11 = -38/11$, $x_3 = 40/-11 = -40/11$

21.

$$A = \begin{pmatrix} 3 & -1 & 1 \\ -1 & 7 & 2 \\ 2 & 6 & -1 \end{pmatrix}, \mathbf{b} = \begin{pmatrix} 4 \\ 1 \\ 5 \end{pmatrix}; |A| = 0$$

The method is not applicable to this problem because the determinant of the coefficient matrix is zero.

23.

$$A = \begin{pmatrix} 4 & 1 & 1 & 1 \\ 3 & 7 & -1 & 1 \\ 7 & 3 & -5 & 8 \\ 1 & 1 & 1 & 2 \end{pmatrix}, \mathbf{b} = \begin{pmatrix} 6 \\ 1 \\ -3 \\ 3 \end{pmatrix}; |A| = -424$$

$$A_2 = \begin{pmatrix} 4 & 6 & 1 & 1 \\ 3 & 1 & -1 & 1 \\ 7 & -3 & -5 & 8 \\ 1 & 3 & 1 & 2 \end{pmatrix}; |A_2| = 0$$

$$y = 0/-424 = 0$$

25. This follows from Theorem 2.1.2 and the fact that the cofactors of A are integers if A has only integer entries.

27. Let A be an upper (not lower) triangular matrix. Consider $AX = I$; the solution X of this equation is the inverse of A. To solve for column 1 of X, we could use Cramer's Rule. Note that if we do so then $A_2, \ldots, A_n$ are each upper triangular matrices with a zero on the main diagonal; hence their determinants are all zero, and so $x_{2,1}, \ldots, x_{n,1}$ are all zero. In a similar way, when solving for column 2 of X we find that $x_{3,2}, \ldots, x_{n,2}$ are all zero, and so on. Hence, X is upper triangular; the inverse of an invertible upper triangular matrix is itself upper triangular. Now apply Theorem 1.4.10 to obtain the corresponding result for lower triangular matrices.

29. Expanding the determinant gives $x(b_1 - b_2) - y(a_1 - a_2) + a_1b_2 - a_2b_1 = 0$

$$x(b_1 - b_2) - y(a_1 - a_2) + a_1b_1 - a_2b_1 = 0$$
$$y = \frac{b_1 - b_2}{a_1 - a_2}x + \frac{a_1b_2 - a_2b_1}{a_1 - a_2}$$

which is the slope-intercept form of the line through these two points, assuming that $a_1 \neq a_2$.

31. **(a)** $|A| = A_{11}| \bullet |A_{22}| = (2 \bullet 3 - 4 \bullet -1) \bullet (1 \bullet 2 - 3 \bullet -10 + \bullet -28) = -1080$

(b.) Expand along the first column; $|A| = -1080$.

33. From I_4 we see that such a matrix can have at least 12 zero entries (i.e., 4 nonzero entries). If a 4×4 matrix has only 3 nonzero entries, some row has only zero entries. Expanding along that row shows that its determinant is necessarily zero.

35. **(a)** True (see the proof of Theorem 2.1.2).

(b) False (requires an invertible, and hence in particular square, coefficient matrix).

(c) True (Theorem 2.1.2).

(d) True (a row of all zeroes will appear in every minor's submatrix).

EXERCISE SET 2.2

1. **(b)** We have

$$\det(A) = \begin{vmatrix} 2 & -1 & 3 \\ 1 & 2 & 4 \\ 5 & -3 & 6 \end{vmatrix} = \begin{vmatrix} 0 & -5 & -5 \\ 1 & 2 & 4 \\ 0 & -13 & -14 \end{vmatrix}$$

Add –2 times Row 2 to Row 1 and –5 times Row 2 to Row 3.

$$= (-1)(-5)\begin{vmatrix} 1 & 2 & 4 \\ 0 & 1 & 1 \\ 0 & -13 & -14 \end{vmatrix}$$

Factor –5 from Row 1 and interchange Row 1 and Row 2.

$$= (-1)(-5)\begin{vmatrix} 1 & 2 & 4 \\ 0 & 1 & 1 \\ 0 & 0 & -1 \end{vmatrix}$$

Add 13 times Row 2 to Row 3.

$$= (-1)(-5)(-1) = -5$$

By Theorem 2.2.2.

$$\det(A^T) = \begin{vmatrix} 2 & 1 & 5 \\ -1 & 2 & -3 \\ 3 & 4 & 6 \end{vmatrix} = \begin{vmatrix} 0 & 5 & -1 \\ -1 & 2 & -3 \\ 0 & 10 & -3 \end{vmatrix}$$

Add 2 times Row 2 to Row 1 and 3 times Row 2 to Row 3.

$$= (-1)\begin{vmatrix} -1 & 2 & -3 \\ 0 & 5 & -1 \\ 0 & 0 & -1 \end{vmatrix}$$

Add –2 times Row 1 to Row 3, and interchange Row 1 and Row 2.

$$= (-1)(-1)(5)(-1) = -5$$

By Theorem 2.2.2.

3. **(b)** Since this matrix is just I_4 with Row 2 and Row 3 interchanged, its determinant is –1.

5.

$$\det(A) = \begin{vmatrix} 0 & 3 & 1 \\ 1 & 1 & 2 \\ 3 & 2 & 4 \end{vmatrix} = (-1)\begin{vmatrix} 1 & 1 & 2 \\ 0 & 3 & 1 \\ 3 & 2 & 4 \end{vmatrix}$$

Interchange Row 1 and Row 2.

$$= (-1)\begin{vmatrix} 1 & 1 & 2 \\ 0 & 3 & 1 \\ 0 & -1 & -2 \end{vmatrix}$$

Add –3 times Row 1 to Row 3.

$$= (-1)(3)\begin{vmatrix} 1 & 1 & 2 \\ 0 & 1 & 1/3 \\ 0 & -1 & -2 \end{vmatrix}$$

Factor 3 from Row 2.

$$= -3\begin{vmatrix} 1 & 1 & 2 \\ 0 & 1 & 1/3 \\ 0 & 0 & -5/3 \end{vmatrix}$$

Add Row 2 to Row 3.

If we factor –5/3 from Row 3 and apply Theorem 2.2.2 we find that

$$\det(A) = -3(-5/3)(1) = 5$$

7.

$$\det(A) = \begin{vmatrix} 3 & -6 & 9 \\ -2 & 7 & -2 \\ 0 & 1 & 5 \end{vmatrix} = 3\begin{vmatrix} 1 & -2 & 3 \\ 0 & 3 & 4 \\ 0 & 1 & 5 \end{vmatrix}$$

Factor 3 from Row 1 and Add twice Row 1 to Row 2.

$$= (3)(3)\begin{vmatrix} 1 & -2 & 3 \\ 0 & 1 & 4/3 \\ 0 & 0 & 11/3 \end{vmatrix}$$

Factor 3 from Row 2 and subtract Row 2 from Row 3.

$$= 9\left(\frac{11}{3}\right)\begin{vmatrix} 1 & -2 & 3 \\ 0 & 1 & 4/3 \\ 0 & 0 & 1 \end{vmatrix}$$

Factor 11/3 from Row 3.

$$= 9(11/3)(1) = 33$$

9.

$$\det(A)=\begin{vmatrix} 2 & 1 & 3 & 1 \\ 1 & 0 & 1 & 1 \\ 0 & 2 & 1 & 0 \\ 0 & 1 & 2 & 3 \end{vmatrix}=(-1)\begin{vmatrix} 1 & 0 & 1 & 1 \\ 2 & 1 & 3 & 1 \\ 0 & 2 & 1 & 0 \\ 0 & 1 & 2 & 3 \end{vmatrix}$$

Interchange Row 1 and Row 2.

$$=(-1)\begin{vmatrix} 1 & 0 & 1 & 1 \\ 0 & 1 & 1 & -1 \\ 0 & 2 & 1 & 0 \\ 0 & 1 & 2 & 3 \end{vmatrix}$$

Add –2 times Row 1 to Row 2.

$$=(-1)\begin{vmatrix} 1 & 0 & 1 & 1 \\ 0 & 1 & 1 & -1 \\ 0 & 0 & -1 & 2 \\ 0 & 0 & 1 & 4 \end{vmatrix}$$

Add –2 times Row 2 to Row 3; subtract Row 2 from Row 4.

$$=(-1)\begin{vmatrix} 1 & 0 & 1 & 1 \\ 0 & 1 & 1 & -1 \\ 0 & 0 & -1 & 2 \\ 0 & 0 & 0 & 6 \end{vmatrix}$$

Add Row 3 to Row 4.

$$=(-1)(-1)(6)(1)=6$$

11.

$$\det(A)=\begin{vmatrix} 1 & 3 & 1 & 5 & 3 \\ -2 & -7 & 0 & -4 & 2 \\ 0 & 0 & 1 & 0 & 1 \\ 0 & 0 & 2 & 1 & 1 \\ 0 & 0 & 0 & 1 & 1 \end{vmatrix}$$

$$=\begin{vmatrix} 1 & 3 & 1 & 5 & 3 \\ 0 & -1 & 2 & 6 & 8 \\ 0 & 0 & 1 & 0 & 1 \\ 0 & 0 & 0 & 1 & -1 \\ 0 & 0 & 0 & 1 & 1 \end{vmatrix}$$

Add 2 times Row 1 to Row 2; add –2 times Row 3 to Row 4.

$$=\begin{vmatrix} 1 & 3 & 1 & 5 & 3 \\ 0 & -1 & 2 & 6 & 8 \\ 0 & 0 & 1 & 0 & 1 \\ 0 & 0 & 0 & 1 & -1 \\ 0 & 0 & 0 & 0 & 2 \end{vmatrix}$$

Add –1 times Row 4 to Row 5.

Hence, $\det(A) = (-1)(2)(1) = -2$.

13.

$$\det(A) = \begin{vmatrix} 1 & 1 & 1 \\ a & b & c \\ a^2 & b^2 & c^2 \end{vmatrix}$$

$$= \begin{vmatrix} 1 & 1 & 1 \\ 0 & b-a & c-a \\ 0 & b^2-a^2 & c^2-a^2 \end{vmatrix}$$

Add $-a$ times Row 1 to Row 2; add $-a^2$ times Row 1 to Row 3.

Since $b^2 - a^2 = (b - a)(b + a)$, we add $-(b + a)$ times Row 2 to Row 3 to obtain

$$\det(A) = \begin{vmatrix} 1 & 1 & 1 \\ 0 & b-a & c-a \\ 0 & 0 & (c^2-a^2)-(c-a)(b+a) \end{vmatrix}$$

$$= (b - a)[(c^2 - a^2) - (c - a)(b + a)]$$

$$= (b - a)(c - a)[(c + a) - (b + a)]$$

$$= (b - a)(c - a)(c - b)$$

15. In each case, d will denote the determinant on the left and, as usual, $\det(A) = \sum \pm a_{1j_1}a_{2j_2}a_{3j_3}$, where $\sum$ denotes the sum of all such elementary products.

(a) $d = \sum \pm (ka_{1j_1})a_{2j_2}a_{3j_3} = k \sum \pm a_{1j_1}a_{2j_2}a_{3j_3} = k\det(A)$

(b) $d = \sum \pm a_{2j_1}a_{1j_2}a_{3j_3} = \sum \pm a_{1j_2}a_{2j_1}a_{3j_3}$

$$\begin{vmatrix} a_{11}+ka_{21} & a_{12}+ka_{22} & a_{13}+ka_{23} \\ a_{21} & a_{22} & a_{23} \\ a_{31} & a_{32} & a_{33} \end{vmatrix}$$

$$= (a_{11} + ka_{21})(a_{22})(a_{33}) + (a_{12} + ka_{22})(a_{23})(a_{31})$$

$$+ (a_{13} + ka_{23})(a_{21})(a_{32}) - (a_{13} + ka_{23})(a_{22})(a_{31})$$

$$- (a_{12} + ka_{22})(a_{21})(a_{33}) - (a_{11} + ka_{21})(a_{23})(a_{32})$$

$$= a_{11}\, a_{22}\, a_{33} + a_{12}\, a_{23}\, a_{31} + a_{13}\, a_{21}\, a_{32} - a_{13}\, a_{22}\, a_{31}$$

$$- a_{12}\, a_{22}\, a_{33} - a_{11}\, a_{23}\, a_{32}$$

$$+ ka_{21}\, a_{22}\, a_{33} + ka_{22}\, a_{23}\, a_{31} + ka_{23}\, a_{21}\, a_{32}$$

$$- ka_{23}\, a_{22}\, a_{31} - ka_{22}\, a_{21}\, a_{33} - ka_{21}\, a_{23}\, a_{32}$$

$$= \begin{vmatrix} a_{11} & a_{12} & a_{13} \\ a_{21} & a_{22} & a_{23} \\ a_{31} & a_{32} & a_{33} \end{vmatrix}$$

17. (8)

$$\begin{vmatrix} 1 & -2 & 3 & 1 \\ 5 & -9 & 6 & 3 \\ -1 & 2 & -6 & -2 \\ 2 & 8 & 6 & 1 \end{vmatrix} = \begin{vmatrix} 1 & -2 & 3 & 1 \\ 3 & -5 & 0 & 1 \\ 1 & -2 & 0 & 0 \\ 0 & 12 & 0 & -1 \end{vmatrix}$$

$R_2 \to R_2 - 2R_1$

$R_3 \to R_3 + 2R_1$

$R_4 \to R_4 - 2R_1$

$$= 3\begin{vmatrix} 3 & -5 & 1 \\ 1 & -2 & 0 \\ 0 & 12 & -1 \end{vmatrix} = 3\begin{vmatrix} 3 & -5 & 1 \\ 1 & -2 & 0 \\ 3 & 7 & 0 \end{vmatrix} = 3\begin{vmatrix} 1 & -2 \\ 3 & 7 \end{vmatrix} = 39$$

$R_3 \to R_3 + R_1$

(9)

$$\begin{vmatrix} 2 & 1 & 3 & 1 \\ 1 & 0 & 1 & 1 \\ 0 & 2 & 1 & 0 \\ 0 & 1 & 2 & 3 \end{vmatrix} = \begin{vmatrix} 0 & 1 & 1 & -1 \\ 1 & 0 & 1 & 1 \\ 0 & 2 & 1 & 0 \\ 0 & 1 & 2 & 3 \end{vmatrix} = (-1)\begin{vmatrix} 1 & 1 & -1 \\ 2 & 1 & 0 \\ 1 & 2 & 3 \end{vmatrix}$$

$R_1 \to R_1 - 2R_2$ $R_3 \to R_3 + 3R_1$

$$= (-1)\begin{vmatrix} 1 & 1 & -1 \\ 2 & 1 & 0 \\ 4 & 5 & 0 \end{vmatrix} = (-1)(-1)\begin{vmatrix} 2 & 1 \\ 4 & 5 \end{vmatrix} = 6$$

(10)

$$\begin{vmatrix} 0 & 1 & 1 & 1 \\ 1/2 & 1/2 & 1 & 1/2 \\ 2/3 & 1/3 & 1/3 & 0 \\ -1/3 & 2/3 & 0 & 0 \end{vmatrix} = \begin{vmatrix} -1 & 0 & -1 & 0 \\ 1/2 & 1/2 & 1 & 1/2 \\ 2/3 & 1/3 & 1/3 & 0 \\ -1/3 & 2/3 & 0 & 0 \end{vmatrix}$$

$R_1 \to R_1 - 2R_2$

$$= \frac{1}{2}\begin{vmatrix} -1 & 0 & -1 \\ 2/3 & 1/3 & 1/3 \\ -1/3 & 2/3 & 0 \end{vmatrix} = \frac{1}{2}\begin{vmatrix} 1 & 1 & 0 \\ 2/3 & 1/3 & 1/3 \\ -1/3 & 2/3 & 0 \end{vmatrix} = (1/2)(-1/3)\begin{vmatrix} 1 & 1 \\ -1/3 & 2/3 \end{vmatrix}$$

$R_1 \to R_1 + 3R_2$

$$= \frac{-1}{6}\left(\frac{2}{3} + \frac{1}{3}\right) = \frac{-1}{6}$$

(11)

$$\begin{vmatrix} 1 & 3 & 1 & 5 & 3 \\ -2 & -7 & 0 & -4 & 2 \\ 0 & 0 & 1 & 0 & 1 \\ 0 & 0 & 2 & 1 & 1 \\ 0 & 0 & 0 & 1 & 1 \end{vmatrix} = \begin{vmatrix} 1 & 3 & 1 & 5 & 3 \\ 0 & -1 & 2 & 6 & 8 \\ 0 & 0 & 1 & 0 & 1 \\ 0 & 0 & 2 & 1 & 1 \\ 0 & 0 & 0 & 1 & 1 \end{vmatrix} = \begin{vmatrix} -1 & 2 & 6 & 8 \\ 0 & 1 & 0 & 1 \\ 0 & 2 & 1 & 1 \\ 0 & 0 & 1 & 1 \end{vmatrix}$$

$R_2 \to R_2 + 2R_1$

$$= (-1)\begin{vmatrix} 1 & 0 & 1 \\ 2 & 1 & 1 \\ 0 & 1 & 1 \end{vmatrix} = (-1)\begin{vmatrix} 1 & 0 & 1 \\ 2 & 0 & 0 \\ 0 & 1 & 1 \end{vmatrix} = (-1)(-2)\begin{vmatrix} 0 & 1 \\ 1 & 1 \end{vmatrix} = -2$$

$R_2 \to R_2 - R_3$

19. Since the given matrix is upper triangular, its determinant is the product of the diagonal elements. That is, the determinant is $x(x+1)(2x-1)$. This product is zero if and only if $x = 0$, $x = -1$, or $x = 1/2$.

EXERCISE SET 2.3

1. **(a)** We have

$$\det(A) = \begin{vmatrix} -1 & 2 \\ 3 & 4 \end{vmatrix} = -4 - 6 = -10$$

and

$$\det(2A) = \begin{vmatrix} -2 & 4 \\ 6 & 8 \end{vmatrix} = (-2)(8) - (4)(6) = -40 = 2^2(-10)$$

5. **(a)** By Equation (1),

$$\det(3A) = 3^3 \det(A) = (27)(-7) = -189$$

(c) Again, by Equation (1), $\det(2A^{-1}) = 2^3 \det(A^{-1})$. By Theorem 2.3.5, we have

$$\det(2A^{-1}) = \frac{8}{\det(A)} = -\frac{8}{7}$$

(d) Again, by Equation (1), $\det(2A) = 2^3 \det(A) = -56$. By Theorem 2.3.5, we have

$$\det[(2A)^{-1}] = \frac{1}{\det(2A)} = \frac{-1}{56}$$

(e)

$$\begin{vmatrix} a & g & d \\ b & h & e \\ c & i & f \end{vmatrix} = -\begin{vmatrix} a & d & g \\ b & e & h \\ c & f & i \end{vmatrix}$$

Interchange Columns 2 and 3.

$$= -\begin{vmatrix} a & b & c \\ d & e & f \\ g & h & i \end{vmatrix}.$$

Take the transpose of the matrix.

$$= 7$$

7. If we replace Row 1 by Row 1 plus Row 2, we obtain

$$\begin{vmatrix} b+c & c+a & b+a \\ a & b & c \\ 1 & 1 & f \end{vmatrix} = \begin{vmatrix} a+b+c & b+c+a & c+b+a \\ a & b & c \\ 1 & 1 & f \end{vmatrix} = 0$$

because the first and third rows are proportional.

13. By adding Row 1 to Row 2 and using the identity $\sin^2 x + \cos^2 x = 1$, we see that the determinant of the given matrix can be written as

$$\begin{vmatrix} \sin^2\alpha & \sin^2\beta & \sin^2\gamma \\ 1 & 1 & 1 \\ 1 & 1 & 1 \end{vmatrix}$$

But this is zero because two of its rows are identical. Therefore the matrix is not invertible.

15. We work with the system from Part (b).

(i) Here

$$\det(\lambda I - A) = \begin{bmatrix} \lambda-2 & 3 \\ -4 & \lambda-3 \end{bmatrix} = (\lambda-2)(\lambda-3)-12 = \lambda^2 - 5\lambda - 6$$

so the characteristic equation is $\lambda^2 - 5\lambda - 6 = 0$.

(ii) The eigenvalues are just the solutions to this equation, or $\lambda = 6$ and $\lambda = -1$.

(iii) If $\lambda = 6$, then the corresponding eigenvectors are the nonzero solutions $\mathbf{x} = \begin{bmatrix} x_1 \\ x_2 \end{bmatrix}$ to the equation

$$\begin{bmatrix} 6-2 & -3 \\ -4 & 6-3 \end{bmatrix}\begin{bmatrix} x_1 \\ x_2 \end{bmatrix} = \begin{bmatrix} 4 & -3 \\ -4 & 3 \end{bmatrix}\begin{bmatrix} x_1 \\ x_2 \end{bmatrix} = \begin{bmatrix} 0 \\ 0 \end{bmatrix}$$

The solution to this system is $x_1 = (3/4)t$, $x_2 = t$, so $\mathbf{x} = \begin{bmatrix} (3/4)t \\ t \end{bmatrix}$ is an eigenvector whenever $t \neq 0$.

If $\lambda = -1$, then the corresponding eigenvectors are the nonzero solutions

$\mathbf{x} = \begin{bmatrix} x_1 \\ x_2 \end{bmatrix}$ to the equation

$$\begin{bmatrix} -3 & -3 \\ -4 & -4 \end{bmatrix}\begin{bmatrix} x_1 \\ x_2 \end{bmatrix} = \begin{bmatrix} 0 \\ 0 \end{bmatrix}$$

If we let $x_1 = t$, then $x_2 = -t$, so $\mathbf{x} = \begin{bmatrix} t \\ -t \end{bmatrix}$ is an eigenvector whenever $t \neq 0$.

It is easy to check that these eigenvalues and their corresponding eigenvectors satisfy the original system of equations by substituting for x_1, x_2, and λ. The solution is valid for all values of t.

17. **(a)** We have, for instance,

$$\begin{vmatrix} a_1+b_1 & c_1+d_1 \\ a_2+b_2 & c_2+d_2 \end{vmatrix} = \begin{vmatrix} a_1+b_1 & c_1+d_1 \\ a_2 & c_2 \end{vmatrix} + \begin{vmatrix} a_1+b_1 & c_1+d_1 \\ b_2 & d_2 \end{vmatrix}$$

$$= \begin{vmatrix} a_1 & c_1 \\ a_2 & c_2 \end{vmatrix} + \begin{vmatrix} b_1 & d_1 \\ a_2 & c_2 \end{vmatrix} + \begin{vmatrix} a_1 & c_1 \\ b_2 & d_2 \end{vmatrix} + \begin{vmatrix} b_1 & d_1 \\ b_2 & d_2 \end{vmatrix}$$

The answer is clearly not unique.

19. Let B be an $n \times n$ matrix and E be an $n \times n$ elementary matrix.

Case 2: Let E be obtained by interchanging two rows of I_n. Then $\det(E) = -1$ and EA is just A with (the same) two rows interchanged. By Theorem 2.2.3, $\det(EA) = -\det(A) = \det(E)\det(A)$.

Case 3: Let E be obtained by adding a multiple of one row of I_n to another. Then $\det(E) = 1$ and $\det(EA) = \det(A)$. Hence $\det(EA) = \det(A) = \det(E)\det(A)$.

21. If either A or B is singular, then either $\det(A)$ or $\det(B)$ is zero. Hence, $\det(AB) = \det(A)\det(B) = 0$. Thus AB is also singular.

23. **(a)** False. If $\det(A) = 0$, then A cannot be expressed as the product of elementary matrices. If it could, then it would be invertible as the product of invertible matrices.

(b) True. The reduced row echelon form of A is the product of A and elementary matrices, all of which are invertible. Thus for the reduced row echelon form to have a row of zeros and hence zero determinant, we must also have $\det(A) = 0$.

(c) False. Consider the 2×2 identity matrix. In general, reversing the order of the columns may change the sign of the determinant.

(d) True. Since $\det(AA^T) = \det(A)\det(A^T) = [\det(A)]^2$, $\det(AA^T)$ cannot be negative.

EXERCISE SET 2.4

1. **(a)** The number of inversions in (4,1,3,5,2) is 3 + 0 + 1 + 1 = 5.

(d) The number of inversions in (5,4,3,2,1) is 4 + 3 + 2 + 1 = 10.

3. $$\begin{vmatrix} 3 & 5 \\ -2 & 4 \end{vmatrix} = 12-(-10) = 22$$

5. $$\begin{vmatrix} -5 & 6 \\ -7 & -2 \end{vmatrix} = (-5)(-2)-(-7)(6) = 52$$

7. $$\begin{vmatrix} a-3 & 5 \\ -3 & a-2 \end{vmatrix} = (a-3)(a-2)-(-3)(5) = a^2-5a+21$$

9. $$\begin{vmatrix} -2 & 1 & 4 \\ 3 & 5 & -7 \\ 1 & 6 & 2 \end{vmatrix} = (-20-7+72)-(20+84+6) = -65$$

11. $$\begin{vmatrix} 3 & 0 & 0 \\ 2 & -1 & 5 \\ 1 & 9 & -4 \end{vmatrix} = (12+0+0)-(0+135+0) = -123$$

13. (a)

$$\det(A) = \begin{vmatrix} \lambda-2 & 1 \\ -5 & \lambda+4 \end{vmatrix} = (\lambda-2)(\lambda+4)+5$$

$$= \lambda^2 + 2\lambda - 3 = (\lambda - 1)(\lambda + 3)$$

Hence, $\det(A) = 0$ if and only if $\lambda = 1$ or $\lambda = -3$.

15. If A is a 4×4 matrix, then

$$\det(A) = \sum (-1)^p a_{1i_1} a_{2i_2} a_{3i_3} a_{4i_4}$$

where $p = 1$ if (i_1, i_2, i_3, i_4) is an odd permutation of $\{1,2,3,4\}$ and $p = 2$ otherwise. There are 24 terms in this sum.

17. (a) The only nonzero product in the expansion of the determinant is

$$a_{15}a_{24}a_{33}a_{42}a_{51} = (-3)(-4)(-1)(2)(5) = -120$$

Since (5,4,3,2,1) is even, $\det(A) = -120$.

(b) The only nonzero product in the expansion of the determinant is

$$a_{11}a_{25}a_{33}a_{44}a_{52} = (5)(-4)(3)(1)(-2) = 120$$

Since (1,5,3,4,2) is odd, $\det(A) = -120$.

19. The value of the determinant is

$$\sin^2 \theta - (-\cos^2 \theta) = \sin^2 \theta + \cos^2 \theta = 1$$

The identity $\sin^2 \theta + \cos^2 \theta = 1$ holds for all values of θ.

21. Since the product of integers is always an integer, each elementary product is an integer. The result then follows from the fact that the sum of integers is always an integer.

23. (a) Since each elementary product in the expansion of the determinant contains a factor from each row, each elementary product must contain a factor from the row of zeros. Thus, each signed elementary product is zero and $\det(A) = 0$.

25. Let $U = [a_{ij}]$ be an n by n upper triangular matrix. That is, suppose that $a_{ij} = 0$ whenever $i > j$. Now consider any elementary product $a_{1j_1}a_{2j_2} \cdots a_{nj_n}$. If $k > j_k$ for any factor a_{kj_k} in this product, then the product will be zero. But if $k \leq j_k$ for all $k = 1, 2, \ldots, n$, then $k = j_k$ for all k because $j_1, j_2, \ldots, j_n$ is just a permutation of the integers $1, 2, \ldots, n$. Hence, $a_{11}a_{22} \cdots a_{nn}$ is the only elementary product which is not guaranteed to be zero. Since the column indices in this product are in natural order, the product appears with a plus sign. Thus, the determinant of U is the product of its diagonal elements. A similar argument works for lower triangular matrices.

SUPPLEMENTARY EXERCISES 2

1.

$$x' = \frac{\begin{vmatrix} x & -\frac{4}{5} \\ y & \frac{3}{5} \end{vmatrix}}{\begin{vmatrix} \frac{3}{5} & -\frac{4}{5} \\ \frac{4}{5} & \frac{3}{5} \end{vmatrix}} = \frac{\frac{3}{5}x + \frac{4}{5}y}{\frac{9}{25} + \frac{16}{25}} = \frac{3}{5}x + \frac{4}{5}y$$

$$y' = \frac{\begin{vmatrix} \frac{3}{5} & x \\ \frac{4}{5} & y \end{vmatrix}}{\begin{vmatrix} \frac{3}{5} & -\frac{4}{5} \\ \frac{4}{5} & \frac{3}{5} \end{vmatrix}} = \frac{\frac{3}{5}y - \frac{4}{5}x}{1} = -\frac{4}{5}x + \frac{3}{5}y$$

3. The determinant of the coefficient matrix is

$$\begin{vmatrix} 1 & 1 & \alpha \\ 1 & 1 & \beta \\ \alpha & \beta & 1 \end{vmatrix} = \begin{vmatrix} 1 & 1 & \alpha \\ 0 & 0 & \beta - \alpha \\ \alpha & \beta & 1 \end{vmatrix} = -(\beta - a)\begin{vmatrix} 1 & 1 \\ \alpha & \beta \end{vmatrix} = -(\beta - a)(\beta - a)$$

The system of equations has a nontrivial solution if and only if this determinant is zero; that is, if and only if $\alpha = \beta$. (See Theorem 2.3.6.)

5. **(a)** If the perpendicular from the vertex of angle α to side a meets side a between angles β and γ, then we have the following picture:

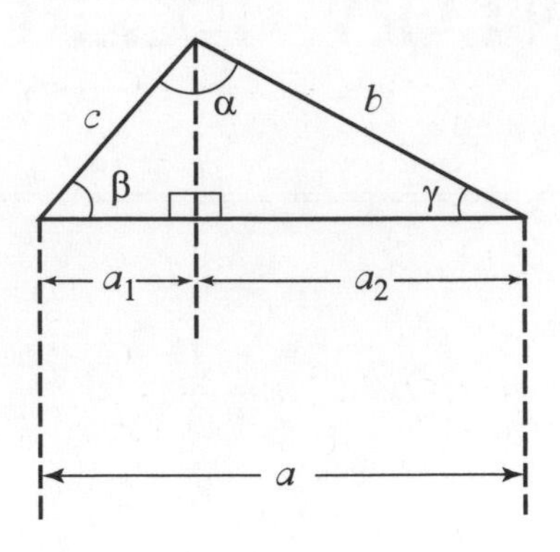

Thus $\cos\beta = \frac{a_1}{c}$ and $\cos\gamma = \frac{a_2}{b}$ and hence

$$a = a_1 + a_2 = c\cos\beta + b\cos\gamma$$

This is the first equation which you are asked to derive. If the perpendicular intersects side a outside of the triangle, the argument must be modified slightly, but the same result holds. Since there is nothing sacred about starting at angle α, the same argument starting at angles β and γ will yield the second and third equations.

Cramer's Rule applied to this system of equations yields the following results:

(b.)

$$\cos\alpha = \frac{\begin{vmatrix} a & c & b \\ b & 0 & a \\ c & a & 0 \end{vmatrix}}{\begin{vmatrix} 0 & c & b \\ c & 0 & a \\ b & a & 0 \end{vmatrix}} = \frac{a(-a^2+b^2+c^2)}{2abc} = \frac{b^2+c^2-a^2}{2bc}$$

$$\cos\beta = \frac{\begin{vmatrix} 0 & a & b \\ c & b & a \\ b & c & a \end{vmatrix}}{2abc} = \frac{b(a^2-b^2+c^2)}{2abc} = \frac{a^2+c^2-b^2}{2ac}$$

$$\cos\gamma = \frac{\begin{vmatrix} 0 & c & a \\ c & 0 & b \\ b & a & c \end{vmatrix}}{2abc} = \frac{c(a^2+b^2-c^2)}{2abc} = \frac{a^2+b^2-c^2}{2ab}$$

7. If A is invertible, then $A^{-1} = \dfrac{1}{\det(A)}\,\text{adj}(A)$, or $\text{adj}(A) = [\det(A)]A^{-1}$. Thus

$$\text{adj}(A) = \frac{A}{\det(A)} = I$$

That is, adj(A) is invertible and

$$[\text{adj}(A)]^{-1} = \frac{1}{\det(A)}A$$

It remains only to prove that $A = \det(A)\text{adj}(A^{-1})$. This follows from Theorem 2.4.2 and Theorem 2.3.5 as shown:

$$A = [A^{-1}]^{-1} = \frac{1}{\det(A^{-1})}\text{adj}(A^{-1}) = \det(A)\text{adj}(A^{-1})$$

9. We simply expand W. That is,

$$\frac{dW}{dx} = \frac{d}{dx}\begin{vmatrix} f_1(x) & f_2(x) \\ g_1(x) & g_2(x) \end{vmatrix}$$

$$= \frac{d}{dx}(f_1(x)g_2(x) - f_2(x)g_1(x))$$

$$= f_1'(x)g_2(x) + f_1(x)g_2'(x) - f_2'(x)g_1(x) - f_2(x)g_1'(x)$$

$$= [f_1'(x)g_2(x) - f_2'(x)g_1(x)] + [f_1(x)g_2'(x) - f_2(x)g_1'(x)]$$

$$= \begin{vmatrix} f_1'(x) & f_2'(x) \\ g_1(x) & g_2(x) \end{vmatrix} + \begin{vmatrix} f_1(x) & f_2(x) \\ g_1'(x) & g_2'(x) \end{vmatrix}$$

11. Let A be an $n \times n$ matrix for which the entries in each row add up to zero and let x be the $n \times 1$ matrix each of whose entries is one. Then all of the entries in the $n \times 1$ matrix Ax are zero since each of its entries is the sum of the entries of one of the rows of A. That is, the homogeneous system of linear equations

$$A\text{x} = \begin{bmatrix} 0 \\ \vdots \\ 0 \end{bmatrix}$$

has a nontrivial solution. Hence $\det(A) = 0$. (See Theorem 2.3.6.)

13. **(a)** If we interchange the ith and j^{th} rows of A, then we claim that we must interchange the ith and j^{th} columns of A^{-1}. To see this, let

$$A = \begin{bmatrix} \text{Row 1} \\ \text{Row 2} \\ \vdots \\ \text{Row n} \end{bmatrix} \quad \text{and } A^{-1} = [\text{Col. 1, Col. 2}, \cdots, \text{Col. n}]$$

where $AA^{-1} = I$. Thus, the sum of the products of corresponding entries from Row s in A and from Column r in A^{-1} must be 0 unless $s = r$, in which case it is 1. That is, if Rows i and j are interchanged in A, then Columns i and j must be interchanged in A^{-1} in order to insure that only 1's will appear on the diagonal of the product AA^{-1}.

(b) If we multiply the ith row of A by a nonzero scalar c, then we must divide the ith column of A^{-1} by c. This will insure that the sum of the products of corresponding entries from the ith row of A and the ith column of A^{-1} will remain equal to 1.

(c) Suppose we add c times the ith row of A to the jth row of A. Call that matrix B. Now suppose that we add $-c$ times the jth column of A^{-1} to the ith column of A^{-1}. Call that matrix C. We claim that $C = B^{-1}$. To see that this is so, consider what happens when

$$\text{Row } j \to \text{Row } j + c \text{ Row } i \qquad [\text{in } A]$$

$$\text{Column } i \to \text{Column } i - c \text{ Column } j \qquad [\text{in } A^{-1}]$$

The sum of the products of corresponding entries from the jth row of B and any k^{th} column of C will clearly be 0 unless $k = i$ or $k = j$. If $k = i$, then the result will be $c - c = 0$. If $k = j$, then the result will be 1. The sum of the products of corresponding entries from any other row of B—say the rth row—and any column of C—say the kth column—will be 1 if $r = k$ and 0 otherwise. This follows because there have been no changes unless $k = i$. In case $k = i$, the result is easily checked.

15. **(a)** We have

$$\det(\lambda I - A) = \begin{vmatrix} \lambda - a_{11} & -a_{12} & -a_{13} \\ -a_{21} & \lambda - a_{22} & -a_{23} \\ -a_{31} & -a_{32} & \lambda - a_{33} \end{vmatrix}$$

If we calculate this determinant by any method, we find that

$$\begin{aligned} \det(\lambda I - A) &= (\lambda - a_{11})(\lambda - a_{22})(\lambda - a_{33}) - a_{23}a_{32}(\lambda - a_{11}) \\ &\quad -a_{13}a_{31}(\lambda - a_{22}) - a_{12}a_{21}(\lambda - a_{33}) \\ &\quad -a_{13}a_{21}a_{32} - a_{12}a_{23}a_{31} \\ &= \lambda^3 + (-a_{11} - a_{22} - a_{33})\lambda^2 \\ &\quad + (a_{11}a_{22} + a_{11}a_{33} + a_{22}a_{33} - a_{12}a_{21} - a_{13}a_{31} - a_{23}a_{32})\lambda \\ &\quad + (a_{11}a_{23}a_{32} + a_{12}a_{21}a_{33} + a_{13}a_{22}a_{31} \\ &\qquad -a_{11}a_{22}a_{33} - a_{12}a_{23}a_{31} - a_{13}a_{21}a_{32}) \end{aligned}$$

(b) From Part (a) we see that $b = -tr(A)$ and $d = -\det(A)$. (It is less obvious that c is the trace of the matrix of minors of the entries of A; that is, the sum of the minors of the diagonal entries of A.)

17. If we multiply Column 1 by 10^4, Column 2 by 10^3, Column 3 by 10^2, Column 4 by 10, and add the results to Column 5, we obtain a new Column 5 whose entries are just the 5 numbers listed in the problem. Since each is divisible by 19, so is the resulting determinant.

TECHNOLOGY EXERCISES 2

T3. Let $y = ax^3 + bx^3 + cx + d$ be the polynomial of degree three to pass through the four given points. Substitution of the x and y coordinates of these points into the equation of the polynomial yields the system

$$\begin{aligned} 7 &= 27a + 9b + 3c + d \\ -1 &= 8a + 4b + 2c + d \\ -1 &= a + b + c + d \\ 1 &= 0a + 0b + 0c + d \end{aligned}$$

Using Cramer's Rule,

$$a = \frac{\begin{vmatrix} 7 & 9 & 3 & 1 \\ -1 & 4 & 2 & 1 \\ -1 & 1 & 1 & 1 \\ 1 & 0 & 0 & 1 \end{vmatrix}}{\begin{vmatrix} 27 & 9 & 3 & 1 \\ 8 & 4 & 2 & 1 \\ 1 & 1 & 1 & 1 \\ 0 & 0 & 0 & 1 \end{vmatrix}} = \frac{12}{12} = 1, \quad b = \frac{\begin{vmatrix} 27 & 7 & 3 & 1 \\ 8 & -1 & 2 & 1 \\ 1 & -1 & 1 & 1 \\ 0 & 1 & 0 & 1 \end{vmatrix}}{12} = \frac{-24}{12} = -2$$

$$c = \frac{\begin{vmatrix} 27 & 9 & 7 & 1 \\ 8 & 4 & -1 & 1 \\ 1 & 1 & -1 & 1 \\ 0 & 0 & 1 & 1 \end{vmatrix}}{12} = \frac{-12}{12} = -1, \quad d = \frac{\begin{vmatrix} 27 & 9 & 3 & 7 \\ 8 & 4 & 2 & -1 \\ 1 & 1 & 1 & -1 \\ 0 & 0 & 0 & 1 \end{vmatrix}}{12} = \frac{12}{12} = 1$$

Plot. $y = x^3 - 2x^2 - x + 1$

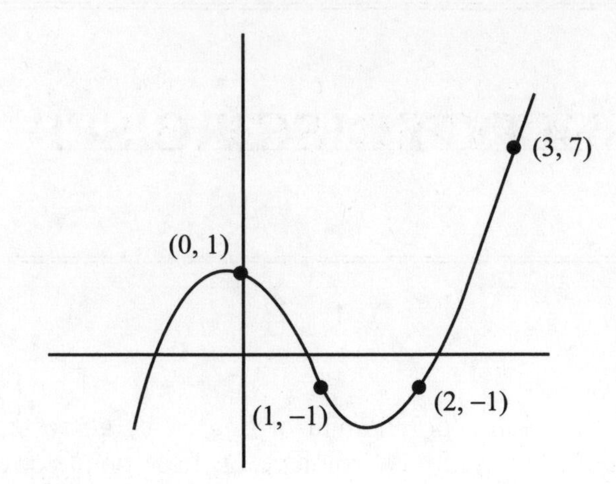
(3, 7)
(0, 1)
(1, −1)
(2, −1)

EXERCISE SET 3.1

1. **(a)**

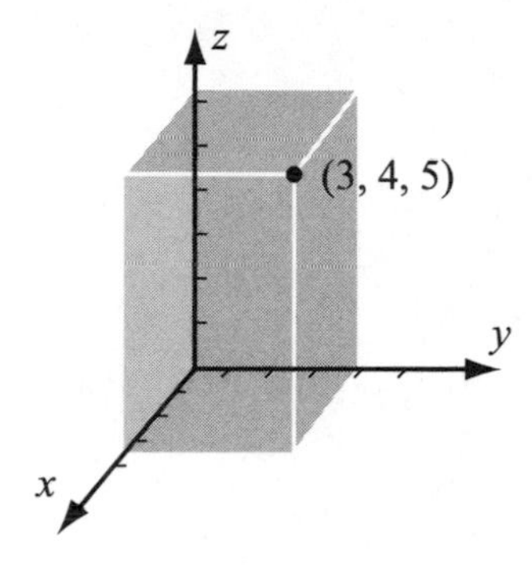

(c)

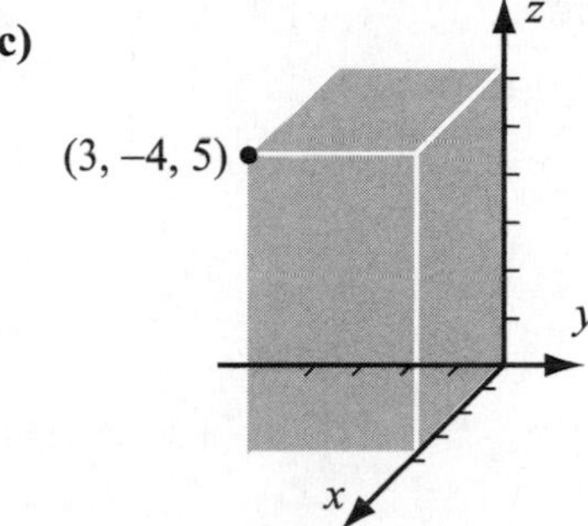

(e)

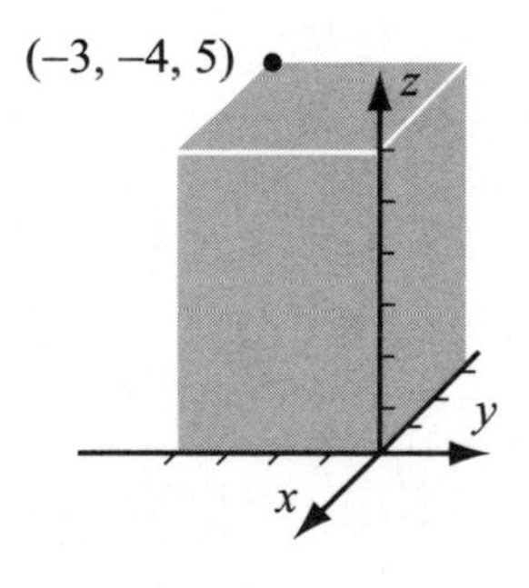

(j)

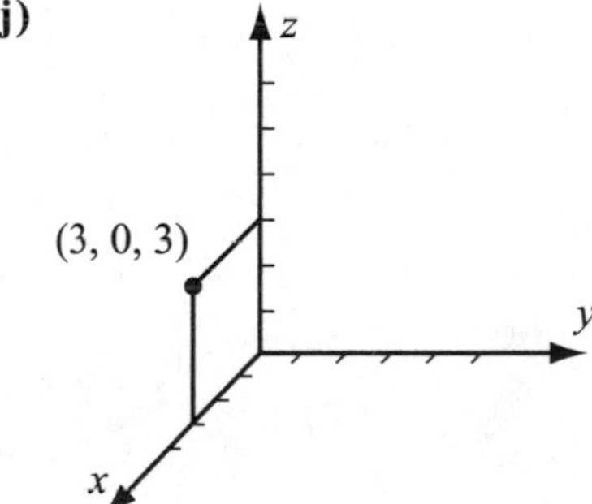

3. **(a)** $\overrightarrow{P_1P_2} = (3 - 4, 7 - 8) = (-1, -1)$

(e) $\overrightarrow{P_1P_2} = (-2 - 3, 5 + 7, -4 -2) = (-5, 12, -6)$

5. **(a)** Let $P = (x, y, z)$ be the initial point of the desired vector and assume that this vector has the same length as **v**. Since $\overrightarrow{PQ}$ has the same direction as $\mathbf{v} = (4, -2, -1)$, we have the equation

$$\overrightarrow{PQ} = (3 - x, 0 - y, -5 - z) = (4, -2, -1)$$

If we equate components in the above equation, we obtain

$$x = -1, \ y = 2, \ \text{and} \ z = -4$$

Thus, we have found a vector $\overrightarrow{PQ}$ which satisfies the given conditions. Any positive multiple $k\ \overrightarrow{PQ}$ will also work provided the terminal point remains fixed at Q. Thus, P could be any point $(3 - 4k, 2k, k - 5)$ where $k > 0$.

(b) Let $P = (x, y, z)$ be the initial point of the desired vector and assume that this vector has the same length as **v**. Since $\overrightarrow{PQ}$ is oppositely directed to $\mathbf{v} = (4, -2, -1)$, we have the equation

$$\overrightarrow{PQ} = (3 - x, 0 - y, -5 - z) = (-4, 2, 1)$$

If we equate components in the above equation, we obtain

$$x = 7, y = -2, \text{and} \ z = -6$$

Thus, we have found a vector $\overrightarrow{PQ}$ which satisfies the given conditions. Any positive multiple $k\ \overrightarrow{PQ}$ will also work, provided the terminal point remains fixed at Q. Thus, P could be any point $(3 + 4k, -2k, -k - 5)$ where $k > 0$.

7. Let $\mathbf{x} = (x_1, x_2, x_3)$. Then

$$\begin{aligned} 2\mathbf{u} - \mathbf{v} + \mathbf{x} &= (-6, 2, 4) - (4, 0, -8) + (x_1, x_2, x_3) \\ &= (-10 + x_1, 2 + x_2, 12 + x_3) \end{aligned}$$

On the other hand,

$$\begin{aligned} 7\mathbf{x} + \mathbf{w} &= 7(x_1, x_2, x_3) + (6, -1, -4) \\ &= (7x_1 + 6, 7x_2 - 1, 7x_3 - 4) \end{aligned}$$

If we equate the components of these two vectors, we obtain

$$7x_1 + 6 = x_1 - 10$$

$$7x_2 - 1 = x_2 + 2$$

$$7x_3 - 4 = x_3 + 12$$

Hence, $\mathbf{x} = (-8/3, 1/2, 8/3)$.

9. Suppose there are scalars c_1, c_2, and c_3 which satisfy the given equation. If we equate components on both sides, we obtain the following system of equations:

$$-2c_1 - 3c_2 + c_3 = 0$$

$$9c_1 + 2c_2 + 7c_3 = 5$$

$$6c_1 + c_2 + 5c_3 = 4$$

The augmented matrix of this system of equations can be reduced to

$$\begin{bmatrix} 2 & 3 & -1 & 0 \\ 0 & 2 & -2 & -1 \\ 0 & 0 & 0 & -1 \end{bmatrix}$$

The third row of the above matrix implies that $0c_1 + 0c_2 + 0c_3 = -1$. Clearly, there do not exist scalars c_1, c_2, and c_3 which satisfy the above equation, and hence the system is inconsistent.

11. We work in the plane determined by the three points $O = (0, 0, 0)$, $P = (2, 3, -2)$, and $Q = (7, -4, 1)$. Let X be a point on the line through P and Q and let $t\,\overrightarrow{PQ}$ (where t is a positive, real number) be the vector with initial point P and terminal point X. Note that the length of $t\,\overrightarrow{PQ}$ is t times the length of $\overrightarrow{PQ}$. Referring to the figure below, we see that

$$\overrightarrow{OP} + t\,\overrightarrow{PQ} = \overrightarrow{OX}$$

and

$$\overrightarrow{OP} + \overrightarrow{PQ} = \overrightarrow{OQ}$$

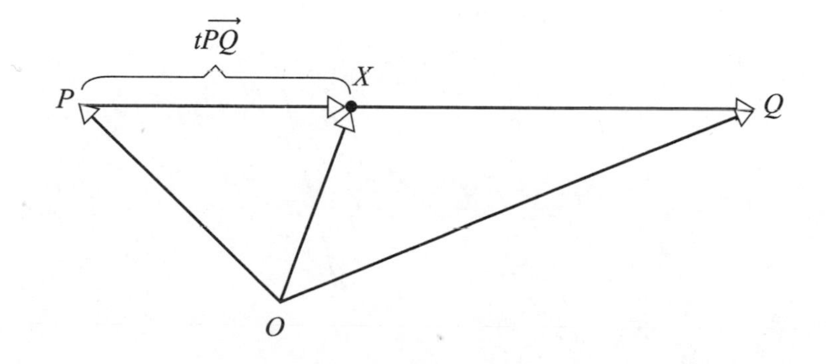

Therefore,

$$\overrightarrow{OX} = \overrightarrow{OP} + t(\overrightarrow{OQ} - \overrightarrow{OP})$$
$$= (1-t)\,\overrightarrow{OP} + t\,\overrightarrow{OQ}$$

(a) To obtain the midpoint of the line segment connecting P and Q, we set $t = 1/2$. This gives

$$\overrightarrow{OX} = \frac{1}{2}\overrightarrow{OP} + \frac{1}{2}\overrightarrow{OQ}$$
$$= \frac{1}{2}(2,3,-2) + \frac{1}{2}(7,-4,1)$$
$$= \left(\frac{9}{2}, -\frac{1}{2}, -\frac{1}{2}\right)$$

(b) Now set $t = 3/4$. This gives

$$\overrightarrow{OX} = \frac{1}{4}(2,3,-2) + \frac{3}{4}(7,-4,1) = \left(\frac{23}{4}, -\frac{9}{4}, \frac{1}{4}\right)$$

13. $Q = (7, -3, -19)$

17. The vector **u** has terminal point Q which is the midpoint of the line segment connecting P_1 and P_2.

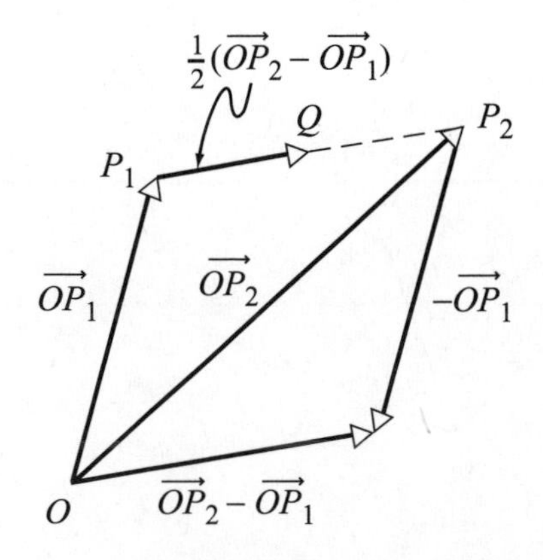

19. Geometrically, given 4 nonzero vectors attach the "tail" of one to the "head" of another and continue until all 4 have been strung together.The vector from the "tail" of the first vector to the "head" of the last one will be their sum.

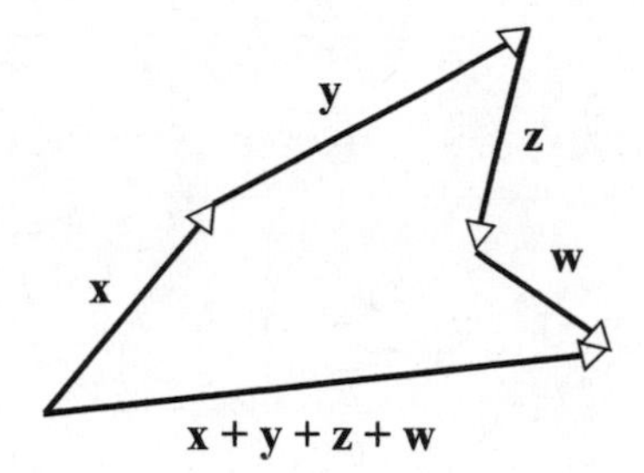

EXERCISE SET 3.2

1. **(a)** $\|\mathbf{v}\| = \left(4^2 + (-3)^2\right)^{1/2} = 5$

(c) $\|\mathbf{v}\| = \left[(-5)^2 + 0^2\right]^{1/2} = 5$

(e) $\|\mathbf{v}\| = \left[(-7)^2 + 2^2 + (-1)^2\right]^{1/2} = \sqrt{54}$

3. **(a)** Since $\mathbf{u} + \mathbf{v} = (3, -5, 7)$, then

$$\|\mathbf{u} + \mathbf{v}\| = \left[3^2 + (-5)^2 + 7^2\right]^{1/2} = \sqrt{83}$$

(c) Since

$$\| -2\mathbf{u}\| = \left[(-4)^2 + 4^2 + (-6)^2\right]^{1/2} = 2\sqrt{17}$$

and

$$2\|\mathbf{u}\| = 2\left[2^2 + (-2)^2 + 3^2\right]^{1/2} = 2\sqrt{17}$$

then

$$\| -2\mathbf{u}\| + 2\|\mathbf{u}\| = 4\sqrt{17}$$

(e) Since $\|\mathbf{w}\| = \left[3^2 + 6^2 + (-4)^2\right]^{1/2} = \sqrt{61}$, then

$$\frac{1}{\|\mathbf{w}\|}\mathbf{w} = \left(\frac{3}{\sqrt{61}}, \frac{6}{\sqrt{61}}, \frac{-4}{\sqrt{61}}\right)$$

5. **(a)** $k = 1, l = 3$

(b) no possible solution

7. Since $k\mathbf{v} = (-k, 2k, 5k)$, then

$$\|k\mathbf{v}\| = \left[k^2 + 4k^2 + 25k^2\right]^{1/2} = |k|\sqrt{30}$$

If $\|k\mathbf{v}\| = 4$, it follows that $|k|\sqrt{30} = 4$ or $k = \pm 4/\sqrt{30}$.

9. **(b)** From Part (a), we know that the norm of $\mathbf{v}/\|\mathbf{v}\|$ is 1. But if $\mathbf{v} = (3, 4)$, then $\|\mathbf{v}\| = 5$. Hence $\mathbf{u} = \mathbf{v}/\|\mathbf{v}\| = (3/5, 4/5)$ has norm 1 and has the same direction as $\mathbf{v}$.

11. Note that $\|\mathbf{p} - \mathbf{p}_0\| = 1$ if and only if $\|\mathbf{p} - \mathbf{p}_0\|^2 = 1$. Thus

$$(x - x_0)^2 + (y - y_0)^2 + (z - z_0)^2 = 1$$

The points (x, y, z) which satisfy these equations are just the points on the sphere of radius 1 with center (x_0, y_0, z_0); that is, they are all the points whose distance from (x_0, y_0, z_0) is 1.

13. These proofs are for vectors in 3-space. To obtain proofs in 2-space, just delete the 3rd component. Let $\mathbf{u} = (u_1, u_2, u_3)$ and $\mathbf{v} = (v_1, v_2, v_3)$. Then

(a) $$\begin{aligned}\mathbf{u} + \mathbf{v} &= (u_1 + v_1, u_2 + v_2, u_3 + v_3) \\ &= (v_1 + u_1, v_2 + u_2, v_3 + u_3) = \mathbf{v} + \mathbf{u}\end{aligned}$$

(c) $$\begin{aligned}\mathbf{u} + \mathbf{0} &= (u_1 + 0, u_2 + 0, u_3 + 0) \\ &= (0 + u_1, 0 + u_2, 0 + u_3) \\ &= (u_1, u_2, u_3) = 0 + u = u\end{aligned}$$

(e) $$k(l\mathbf{u}) = k(lu_1, lu_2, lu_3) = (klu_1, klu_2, klu_3) = (kl)\mathbf{u}$$

15. See Exercise 9. Equality occurs only when $\mathbf{u}$ and $\mathbf{v}$ have the same direction or when one is the zero vector.

17. **(a)** If $\|\mathbf{x}\| < 1$, then the point $\mathbf{x}$ lies inside the circle or sphere of radius one with center at the origin.

(b) Such points $\mathbf{x}$ must satisfy the inequality $\|\mathbf{x} - \mathbf{x}_0\| > 1$.

EXERCISE SET 3.3

1. **(a)** $\mathbf{u} \cdot \mathbf{v} = (2)(5) + (3)(-7) = -11$

(c) $\mathbf{u} \cdot \mathbf{v} = (1)(3) + (-5)(3) + (4)(3) = 0$

3. **(a)** $\mathbf{u} \cdot \mathbf{v} = (6)(2) + (1)(0) + (4)(-3) = 0$. Thus the vectors are orthogonal.

(b) $\mathbf{u} \cdot \mathbf{v} = -1 < 0$. Thus θ is obtuse.

5. **(a)** From Problem 4(a), we have

$$\mathbf{w}_2 = \mathbf{u} - \mathbf{w}_1 = \mathbf{u} = (6, 2)$$

(c) From Problem 4(c), we have

$$\mathbf{w}_2 = (3, 1, -7) - (-16/13, 0, -80/13) = (55/13, 1, -11/13)$$

13. Let $\mathbf{w} = (x, y, z)$ be orthogonal to both $\mathbf{u}$ and $\mathbf{v}$. Then $\mathbf{u} \cdot \mathbf{w} = 0$ implies that $x + z = 0$ and $\mathbf{v} \cdot \mathbf{w} = 0$ implies that $y + z = 0$. That is $\mathbf{w} = (x, x, -x)$. To transform into a unit vector, we divide each component by $\|\mathbf{w}\| = \sqrt{3x^2}$. Thus either $(1/\sqrt{3}, 1/\sqrt{3}, -1/\sqrt{3})$ or $(-1/\sqrt{3}, -1/\sqrt{3}, 1/\sqrt{3})$ will work.

The minus sign in the above equation is extraneous because it yields an angle of $2\pi/3$.

17. **(b)** Here

$$D = \frac{|4(2) + 1(-5) - 2|}{\sqrt{(4)^2 + (1)^2}} = \frac{1}{\sqrt{17}}$$

19. If we subtract Equation (**) from Equation (*) in the solution to Problem 18, we obtain

$$\|\mathbf{u}+\mathbf{v}\|^2 - \|\mathbf{u}-\mathbf{v}\|^2 = 4(\mathbf{u}\cdot\mathbf{v})$$

If we then divide both sides by 4, we obtain the desired result.

21. **(a)** Let $\mathbf{i} = (1, 0, 0)$, $\mathbf{j} = (0, 1, 0)$, and $\mathbf{k} = (0, 0, 1)$ denote the unit vectors along the x, y, and z axes, respectively. If $\mathbf{v}$ is the arbitrary vector (a, b, c), then we can write $\mathbf{v} = a\mathbf{i} + b\mathbf{j} + c\mathbf{k}$. Hence, the angle α between $\mathbf{v}$ and $\mathbf{i}$ is given by

$$\cos\alpha = \frac{\mathbf{v}\cdot\mathbf{i}}{\|\mathbf{v}\|\,\|\mathbf{i}\|} = \frac{a}{\sqrt{a^2+b^2+c^2}} = \frac{a}{\|\mathbf{v}\|}$$

since $\|\mathbf{i}\| = 1$ and $\mathbf{i}\cdot\mathbf{j} = \mathbf{i}\cdot\mathbf{k} = 0$.

23. By the results of Exercise 21, we have that if $\mathbf{v}_i = (a_i, b_i, c_i)$ for $i = 1$ and 2, then $\cos\alpha_i = \frac{a_i}{\|\mathbf{v}_i\|}$, $\cos\beta_i = \frac{b_i}{\|\mathbf{v}_i\|}$, and $\cos\gamma_i = \frac{c_i}{\|\mathbf{v}_i\|}$. Now

$$\begin{aligned} \mathrm{v}_1 \text{ and } \mathrm{v}_2 \text{ are orthogonal} &\Leftrightarrow \mathbf{v}_1\cdot\mathbf{v}_2 = 0 \\ &\Leftrightarrow a_1a_2 + b_1b_2 + c_1c_2 = 0 \\ &\Leftrightarrow \frac{a_1a_2}{\|\mathbf{v}_1\|\,\|\mathbf{v}_2\|} + \frac{b_1b_2}{\|\mathbf{v}_1\|\,\|\mathbf{v}_2\|} + \frac{c_1c_2}{\|\mathbf{v}_1\|\,\|\mathbf{v}_2\|} = 0 \\ &\Leftrightarrow \cos\alpha_1\cos\alpha_2 + \cos\beta_1\cos\beta_2 + \cos\gamma_1\cos\gamma_2 = 0 \end{aligned}$$

25. Note that

$$\mathbf{v}\cdot(k_1\mathbf{w}_1 + k_2\mathbf{w}_2) = k_1(\mathbf{v}\cdot\mathbf{w}_1) + k_2(\mathbf{v}\cdot\mathbf{w}_2) = 0$$

because, by hypothesis, $\mathbf{v}\cdot\mathbf{w}_1 = \mathbf{v}\cdot\mathbf{w}_2 = 0$. Therefore $\mathbf{v}$ is orthogonal to $k_1\mathbf{w}_1 + k_2\mathbf{w}_2$ for any scalars k_1 and k_2.

27. **(a)** The inner product $\mathbf{x}\cdot\mathbf{y}$ is defined only if both $\mathbf{x}$ and $\mathbf{y}$ are vectors, but here $\mathbf{v}\cdot\mathbf{w}$ is a scalar.

(b) We can add two vectors or two scalars, but not one of each.

(c) The norm of $\mathbf{x}$ is defined only for $\mathbf{x}$ a vector, but $\mathbf{u}\cdot\mathbf{v}$ is a scalar.

(d) Again, the dot product of a scalar and a vector is undefined.

29. If, for instance, $\mathbf{u} = (1, 0, 0)$, $\mathbf{v} = (0, 1, 0)$ and $\mathbf{w} = (0, 0, 1)$, we have $\mathbf{u} \cdot \mathbf{v} = \mathbf{u} \cdot \mathbf{w} = 0$, but $\mathbf{v} \neq \mathbf{w}$.

31. This is just the Pythagorean Theorem.

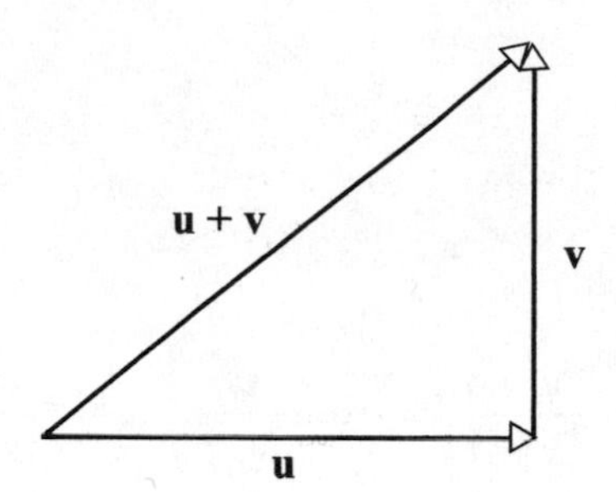

EXERCISE SET 3.4

1. **(a)** $\mathbf{v} \times \mathbf{w} = \left(\begin{vmatrix} 2 & -3 \\ 6 & 7 \end{vmatrix}, -\begin{vmatrix} 0 & -3 \\ 2 & 7 \end{vmatrix}, \begin{vmatrix} 0 & 2 \\ 2 & 6 \end{vmatrix} \right) = (32, -6, -4)$

(c) Since

$$\mathbf{u} \times \mathbf{v} = \left(\begin{vmatrix} 2 & -1 \\ 2 & -3 \end{vmatrix}, -\begin{vmatrix} 3 & -1 \\ 0 & -3 \end{vmatrix}, \begin{vmatrix} 3 & 2 \\ 0 & 2 \end{vmatrix} \right) = (-4, 9, 6)$$

we have

$$(\mathbf{u} \times \mathbf{v}) \times \mathbf{w} = \left(\begin{vmatrix} 9 & 6 \\ 6 & 7 \end{vmatrix}, -\begin{vmatrix} -4 & 6 \\ 2 & 7 \end{vmatrix}, \begin{vmatrix} -4 & 9 \\ 2 & 6 \end{vmatrix} \right) = (27, 40, -42)$$

(e) Since

$$\mathbf{v} - 2\mathbf{w} = (0, 2, -3) - (4, 12, 14) = (-4, -10, -17)$$

we have

$$\mathbf{u} \times (\mathbf{v} - 2\mathbf{w}) = \left(\begin{vmatrix} 2 & -1 \\ -10 & -17 \end{vmatrix}, -\begin{vmatrix} 3 & -1 \\ -4 & -17 \end{vmatrix}, \begin{vmatrix} 3 & 2 \\ -4 & -10 \end{vmatrix} \right) = (-44, 55, -22)$$

3. **(a)** Since $\mathbf{u} \times \mathbf{v} = (-7, -1, 3)$, the area of the parallelogram is $\|\mathbf{u} \times \mathbf{v}\| = \sqrt{59}$.

(c) Since **u** and **v** are proportional, they lie on the same line and hence the area of the parallelogram they determine is zero, which is, of course, $\|\mathbf{u} \times \mathbf{v}\|$.

7. Choose any nonzero vector $\mathbf{w}$ which is not parallel to $\mathbf{u}$. For instance, let $\mathbf{w} = (1, 0, 0)$ or $(0, 1, 0)$. Then $\mathbf{v} = \mathbf{u} \times \mathbf{w}$ will be orthogonal to $\mathbf{u}$. Note that if $\mathbf{u}$ and $\mathbf{w}$ were parallel, then $\mathbf{v} = \mathbf{u} \times \mathbf{w}$ would be the zero vector.

Alternatively, let $\mathbf{w} = (x, y, z)$. Then $\mathbf{w}$ orthogonal to $\mathbf{u}$ implies $2x - 3y + 5z = 0$. Now assign nonzero values to any two of the variables x, y, and z and solve for the remaining variable.

9. **(e)** Since $(\mathbf{u} \times \mathbf{w}) \cdot \mathbf{v} = \mathbf{v} \cdot (\mathbf{u} \times \mathbf{w})$ is a determinant whose rows are the components of $\mathbf{v}$, $\mathbf{u}$, and $\mathbf{w}$, respectively, we interchange Rows 1 and 2 to obtain the determinant which represents $\mathbf{u} \cdot (\mathbf{v} \times \mathbf{w})$. Since the value of this determinant is 3, we have $(\mathbf{u} \times \mathbf{w}) \cdot \mathbf{v} = -3$.

11. **(a)** Since the determinant

$$\begin{vmatrix} -1 & -2 & 1 \\ 3 & 0 & -2 \\ 5 & -4 & 0 \end{vmatrix} = 16 \neq 0$$

the vectors do not lie in the same plane.

15. By Theorem 3.4.2, we have

$$\begin{aligned}(\mathbf{u} + \mathbf{v}) \times (\mathbf{u} - \mathbf{v}) &= \mathbf{u} \times (\mathbf{u} - \mathbf{v}) + \mathbf{v} \times (\mathbf{u} - \mathbf{v}) \\ &= (\mathbf{u} \times \mathbf{u}) + (\mathbf{u} \times (-\mathbf{v})) + (\mathbf{v} \times \mathbf{u}) + (\mathbf{v} \times (-\mathbf{v})) \\ &= \mathbf{0} - (\mathbf{u} \times \mathbf{v}) - (\mathbf{u} \times \mathbf{v}) - (\mathbf{v} - \mathbf{v}) \\ &= -2(\mathbf{u} \times \mathbf{v})\end{aligned}$$

17. **(a)** The area of the triangle with sides $\overrightarrow{AB}$ and $\overrightarrow{AC}$ is the same as the area of the triangle with sides $(-1, 2, 2)$ and $(1, 1, -1)$ where we have "moved" A to the origin and translated B and C accordingly. This area is $\frac{1}{2}\|(-1, 2, 2) \times (1, 1, -1)\| = \frac{1}{2}\|(-4, 1, -3)\| = \sqrt{26}/2$.

19. **(a)** Let $\mathbf{u} = \overrightarrow{AP} = (-4, 0, 2)$ and $\mathbf{v} = \overrightarrow{AB} = (-3, 2, -4)$. Then the distance we want is

$$\|(-4, 0, 2) \times (-3, 2, -4)\|/\|(-3, 2, -4)\| = \|(-4, -22, -8)\|/\sqrt{29}. = 2\sqrt{141}/\sqrt{29}$$

21. **(b)** One vector **n** which is perpendicular to the plane containing **v** and **w** is given by

$$\mathbf{n} = \mathbf{w} \times \mathbf{v} = (1, 3, 3) \times (1, 1, 2) = (3, 1, -2)$$

Therefore the angle ϕ between **u** and **n** is given by

$$\phi = \cos^{-1}\left(\frac{\mathbf{u} \cdot \mathbf{n}}{\|\mathbf{u}\|\|\mathbf{n}\|}\right) = \cos^{-1}\left(\frac{9}{14}\right)$$
$$\approx 0.8726 \text{ radians (or } 49.99^\circ)$$

Hence the angle θ between **u** and the plane is given by

$$\theta = \frac{\pi}{2} - \phi \approx .6982 \text{ radians (or } 40^\circ 19')$$

If we had interchanged the roles of **v** and **w** in the formula for **n** so that $\mathbf{n} = \mathbf{v} \times \mathbf{w} = (-3, -1, 2)$, then we would have obtained $\phi = \cos^{-1}\left(-\frac{9}{14}\right) \approx 2.269$ radians or 130.0052°. In this case, $\theta = \phi \approx -\frac{\pi}{2}$.

In either case, note that θ may be computed using the formula

$$\theta = \left|\cos^{-1}\left(\frac{\mathbf{u} \cdot \mathbf{n}}{\|\mathbf{u}\|\|\mathbf{n}\|}\right)\right|.$$

25. **(a)** By Theorem 3.4.1, we know that the vector $\mathbf{v} \times \mathbf{w}$ is perpendicular to both **v** and **w**. Hence $\mathbf{v} \times \mathbf{w}$ is perpendicular to every vector in the plane determined by **v** and **w**; moreover the only vectors perpendicular to $\mathbf{v} \times \mathbf{w}$ which share its initial point must be in this plane. But also by Theorem 3.4.1, $\mathbf{u} \times (\mathbf{v} \times \mathbf{w})$ is perpendicular to $\mathbf{v} \times \mathbf{w}$ for any vector $\mathbf{u} \neq \mathbf{0}$ and hence must lie in the plane determined by **v** and **w**.

(b) The argument is completely similar to Part (a), above.

29. If **a**, **b**, **c**, and **d** lie in the same plane, then $(\mathbf{a} \times \mathbf{b})$ and $(\mathbf{c} \times \mathbf{d})$ are both perpendicular to this plane, and are therefore parallel. Hence, their cross-product is zero.

31. **(a)** The required volume is

$$\frac{1}{6}\left(\,|(-1-3, 2+2, 0-3) \cdot ((2-3, 1+2, -3-3) \times (1-3, 0+2, 1-3))|\,\right)$$

$$= \frac{1}{6}\left(\,|(-4,4,-3) \cdot (6,10,4)|\,\right)$$

$$= 2/3$$

33. Let $\mathbf{u} = (u_1, u_2, u_3)$, $\mathbf{v} = (v_1, v_2, v_3)$, and $\mathbf{w} = (w_1, w_2, w_3)$.

For Part (c), we have

$$\mathbf{u} \times \mathbf{w} = (u_2w_3 - u_3w_2, u_3w_1 - u_1w_3, u_1w_2 - u_2w_1)$$

and

$$\mathbf{v} \times \mathbf{w} = (v_2w_3 - v_3w_2, v_3w_1 - v_1w_3, v_1w_2 - v_2w_1)$$

Thus

$$(\mathbf{u} \times \mathbf{w}) + (\mathbf{v} \times \mathbf{w})$$

$$= ([u_2 + v_2]w_3 - [u_3 + v_3]w_2, [u_3 + v_3]w_1 - [u_1 + v_1]w_3, [u_1 + v_1]w_2 - [u_2 + v_2]w_1)$$

But, by definition, this is just $(\mathbf{u} + \mathbf{v}) \times \mathbf{w}$.

For Part (d), we have

$$k(\mathbf{u} \times \mathbf{v}) = (k[u_2v_3 - u_3v_2], k[u_3v_1 - u_1v_3], k[u_1v_2 - u_2v_1])$$

and

$$(k\mathbf{u}) \times \mathbf{v} = (ku_2v_3 - ku_3v_3, ku_3v_1 - ku_1v_3, ku_1v_2 - ku_2v_1)$$

Thus, $k(\mathbf{u} \times \mathbf{v}) = (k\mathbf{u}) \times \mathbf{v}$. The identity $k(\mathbf{u} \times \mathbf{v}) = \mathbf{u} \times (k\mathbf{v})$ may be proved in an analogous way.

35. **(a)** Observe that $\mathbf{u} \times \mathbf{v}$ is perpendicular to both $\mathbf{u}$ and $\mathbf{v}$, and hence to all vectors in the plane which they determine. Similarly, $\mathbf{w} = \mathbf{v} \times (\mathbf{u} \times \mathbf{v})$ is perpendicular to both $\mathbf{v}$ and to $\mathbf{u} \times \mathbf{v}$. Hence, it must lie on the line through the origin perpendicular to $\mathbf{v}$ and in the plane determined by $\mathbf{u}$ and $\mathbf{v}$.

(b) From the above, $\mathbf{v} \cdot \mathbf{w} = \mathbf{0}$. Applying Part (*d*) of Theorem 3.7.1, we have

$$\mathbf{w} = \mathbf{v} \times (\mathbf{u} \times \mathbf{v}) = (\mathbf{v} \cdot \mathbf{v})\mathbf{u} - (\mathbf{v} \cdot \mathbf{u})\mathbf{v}$$

so that

$$\begin{aligned} \mathbf{u} \cdot \mathbf{w} &= (\mathbf{v} \cdot \mathbf{v})(\mathbf{u} \cdot \mathbf{u}) - (\mathbf{v} \cdot \mathbf{u})(\mathbf{u} \cdot \mathbf{v}) \\ &= \|\mathbf{v}\|^2\|\mathbf{u}\|^2 - (\mathbf{u} \cdot \mathbf{v})^2 \end{aligned}$$

37. The expression $\mathbf{u} \cdot (\mathbf{v} \times \mathbf{w})$ is clearly well-defined.

Since the cross product is not associative, the expression $\mathbf{u} \times \mathbf{v} \times \mathbf{w}$ is not well-defined because the result is dependent upon the order in which we compute the cross products, i.e., upon the way in which we insert the parentheses. For example, $(\mathbf{i} \times \mathbf{j}) \times \mathbf{j} = \mathbf{k} \times \mathbf{j} = -\mathbf{i}$ but $\mathbf{i} \times (\mathbf{j} \times \mathbf{j}) = \mathbf{i} \times \mathbf{0} = \mathbf{0}$.

The expression $\mathbf{u} \cdot \mathbf{v} \times \mathbf{w}$ may be deemed to be acceptable because there is only one meaningful way to insert parenthesis, namely, $\mathbf{u} \cdot (\mathbf{v} \times \mathbf{w})$. The alternative, $(\mathbf{u} \cdot \mathbf{v}) \times \mathbf{w}$, does not make sense because it is the cross product of a scalar with a vector.

EXERCISE SET 3.5

5. **(a)** Normal vectors for the planes are $(4, -1, 2)$ and $(7, -3, 4)$. Since these vectors are not multiples of one another, the planes are not parallel.

(b) Normal vectors are $(1, -4, -3)$ and $(3, -12, -9)$. Since one vector is three times the other, the planes are parallel.

7. **(a)** Normal vectors for the planes are $(3, -1, 1)$ and $(1, 0, 2)$. Since the inner product of these two vectors is not zero, the planes are not perpendicular.

11. **(a)** As in Example 6, we solve the two equations simultaneously. If we eliminate y, we have $x + 7z + 12 = 0$. Let, say, $z = t$, so that $x = -12 - 7t$, and substitute these values into the equation for either plane to get $y = -41 - 23t$.

Alternatively, recall that a direction vector for the line is just the cross-product of the normal vectors for the two planes, i.e.,

$$(7, -2, 3) \times (-3, 1, 2) = (-7, -23, 1)$$

Thus if we can find a point which lies on the line (that is, any point whose coordinates satisfy the equations for both planes), we are done. If we set $z = 0$ and solve the two equations simultaneously, we get $x = -12$ and $y = -41$, so that $x = -12 - 7t$, $y = -41 - 23t$, $z = 0 + t$ is one set of equations for the line (see above).

13. **(a)** Since the normal vectors $(-1, 2, 4)$ and $(2, -4, -8)$ are parallel, so are the planes.

(b) Since the normal vectors $(3, 0, -1)$ and $(-1, 0, 3)$ are not parallel, neither are the planes.

17. Since the plane is perpendicular to a line with direction $(2, 3, -5)$, we can use that vector as a normal to the plane. The point-normal form then yields the equation $2(x + 2) + 3(y - 1) - 5(z - 7) = 0$, or $2x + 3y - 5z + 36 = 0$.

19. **(a)** Since the vector $(0, 0, 1)$ is perpendicular to the xy-plane, we can use this as the normal for the plane. The point-normal form then yields the equation $z - z_0 = 0$. This equation could just as well have been derived by inspection, since it represents the set of all points with fixed z and x and y arbitrary.

21. A normal to the plane is $\mathbf{n} = (5, -2, 1)$ and the point $(3, -6, 7)$ is in the desired plane. Hence, an equation for the plane is $5(x - 3) - 2(y + 6) + (z - 7) = 0$ or $5x - 2y + z - 34 = 0$.

25. Call the points A, B, C, and D, respectively. Since the vectors $\overrightarrow{AB} = (-1, 2, 4)$ and $\overrightarrow{BC} = (-2, -1, -2)$ are not parallel, then the points A, B, and C do determine a plane (and not just a line). A normal to this plane is $\overrightarrow{AB} \times \overrightarrow{BC} = (0, -10, 5)$. Therefore an equation for the plane is

$$2y - z + 1 = 0$$

Since the coordinates of the point D satisfy this equation, all four points must lie in the same plane.

Alternatively, it would suffice to show that (for instance) $\overrightarrow{AB} \times \overrightarrow{BC}$ and $\overrightarrow{AD} \times \overrightarrow{DC}$ are parallel, so that the planes determined by A, B, and C and A, D, and C are parallel. Since they have points in common, they must coincide.

27. Normals to the two planes are $(4, -2, 2)$ and $(3, 3, -6)$ or, simplifying, $\mathbf{n}_1 = (2, -1, 1)$ and $\mathbf{n}_2 = (1, 1, -2)$. A normal $\mathbf{n}$ to a plane which is perpendicular to both of the given planes must be perpendicular to both $\mathbf{n}_1$ and $\mathbf{n}_2$. That is, $\mathbf{n} = \mathbf{n}_1 \times \mathbf{n}_2 = (1, 5, 3)$. The plane with this normal which passes through the point $(-2, 1, 5)$ has the equation

$$(x + 2) + 5(y - 1) + 3(z - 5) = 0$$

or

$$x + 5y + 3z - 18 = 0$$

31. If, for instance, we set $t = 0$ and $t = -1$ in the line equation, we obtain the points $(0, 1, -3)$ and $(-1, 0, -5)$. These, together with the given point and the methods of Example 2, will yield an equation for the desired plane.

33. The plane we are looking for is just the set of all points $P = (x, y, z)$ such that the distances from P to the two fixed points are equal. If we equate the squares of these distances, we have

$$(x + 1)^2 + (y + 4)^2 + (z + 2)^2 = (x - 0)^2 + (y + 2)^2 + (z - 2)^2$$

or

$$2x + 1 + 8y + 16 + 4z + 4 = 4y + 4 - 4z + 4$$

or

$$2x + 4y + 8z + 13 = 0$$

35. We change the parameter in the equations for the second line from t to s. The two lines will then intersect if we can find values of s and t such that the x, y, and z coordinates for the two lines are equal; that is, if there are values for s and t such that

$$\begin{aligned} 4t + 3 &= 12s - 1 \\ t + 4 &= 6s + 7 \\ 1 &= 3s + 5 \end{aligned}$$

This system of equations has the solution $t = -5$ and $s = -4/3$. If we then substitute $t = -5$ into the equations for the first line or $s = -4/3$ into the equations for the second line, we find that $x = -17$, $y = -1$, and $z = 1$ is the point of intersection.

37. **(a)** If we set $z = t$ and solve for x and y in terms of z, then we find that

$$x = \frac{11}{23} + \frac{7}{23}t, \quad y = -\frac{41}{23} - \frac{1}{23}t, \quad z = t$$

39. **(b)** By Theorem 3.5.2, the distance is

$$D = \frac{|2(-1)+3(2)-4(1)-1|}{\sqrt{2^2+3^2+(-4)^2}} = \frac{1}{\sqrt{29}}$$

41. **(a)** $d = \sqrt{\frac{30}{11}}$

(b) $d = \sqrt{\frac{382}{11}}$

(c) $d = 0$ since point is on the line

$P_0 = (-1, 2, 0)$
$n = (3, -1, 1)$
$3(x+1) - 1(y-2) + 1(z) = 0$
$3x + 3 - y + 2 + z = 0$
$3x - y + z + 5 = 0$
$\frac{|3(0) - 1(0) + 0 + 5|}{\sqrt{3^2+1^2+1^2}} = \frac{5}{\sqrt{11}}$

45. **(a)** Normals to the two planes are (1, 0, 0) and (2, –1, 1). The angle between them is given by

$$\cos\theta = \frac{(1,0,0)\cdot(2,-1,1)}{\sqrt{1}\,\sqrt{4+1+1}} = \frac{2}{\sqrt{6}}$$

Thus $\theta = \cos^{-1}(2/\sqrt{6}) \approx 35°15'52''$.

47. If we substitute any value of the parameter—say t_0—into $\mathbf{r} = \mathbf{r}_0 + t\mathbf{v}$ and $-t_0$ into $\mathbf{r} = \mathbf{r}_0 - t\mathbf{v}$, we clearly obtain the same point. Hence, the two lines coincide. They both pass through the point $\mathbf{r}_0$ and both are parallel to $\mathbf{v}$.

49. The equation $\mathbf{r} = (1 - t)\mathbf{r}_1 + t\mathbf{r}_2$ can be rewritten as $\mathbf{r} = \mathbf{r}_1 + t(\mathbf{r}_2 - \mathbf{r}_1)$. This represents a line through the point P_1 with direction $\mathbf{r}_2 - \mathbf{r}_1$. If $t = 0$, we have the point P_1. If $t = 1$, we have the point P_2. If $0 < t < 1$, we have a point on the line segment connecting P_1 and P_2. Hence the given equation represents this line segment.

EXERCISE SET 4.1

3. We must find numbers c_1, c_2, c_3, and c_4 such that

$$c_1(-1, 3, 2, 0) + c_2(2, 0, 4, -1) + c_3(7, 1, 1, 4) + c_4(6, 3, 1, 2) = (0, 5, 6, -3)$$

If we equate vector components, we obtain the following system of equations:

$$\begin{array}{rcrcrcrcl} -c_1 & + & 2c_2 & + & 7c_3 & + & 6c_4 & = & 0 \\ 3c_1 & & & + & c_3 & + & 3c_4 & = & 5 \\ 2c_1 & + & 4c_2 & + & c_3 & + & c_4 & = & 6 \\ & & -c_2 & + & 4c_3 & + & 2c_4 & = & 3 \end{array}$$

The augmented matrix of this system is

$$\begin{bmatrix} -1 & 2 & 7 & 6 & 0 \\ 3 & 0 & 1 & 3 & 5 \\ 2 & 4 & 1 & 1 & 6 \\ 0 & -1 & 4 & 2 & -3 \end{bmatrix}$$

The reduced row-echelon form of this matrix is

$$\begin{bmatrix} 1 & 0 & 0 & 0 & 1 \\ 0 & 1 & 0 & 0 & 1 \\ 0 & 0 & 1 & 0 & -1 \\ 0 & 0 & 0 & 1 & 1 \end{bmatrix}$$

Thus $c_1 = 1$, $c_2 = 1$, $c_3 = -1$, and $c_4 = 1$.

5. **(c)** $\|\mathbf{v}\| = [3^2 + 4^2 + 0^2 + (-12)^2]^{1/2} = \sqrt{169} = 13$

9. **(a)** $(2,5) \bullet (-4,3) = (2)(-4) + (5)(3) = 7$

(c) $(3, 1, 4, -5) \bullet (2, 2, -4, -3) = 6 + 2 - 16 + 15 = 7$

11. **(a)** $d(\mathbf{u},\mathbf{v}) = [(1-2)^2 + (-2-1)^2]^{1/2} = \sqrt{10}$

(c) $d(\mathbf{u}, \mathbf{v}) = [(0+3)^2 + (-2-2)^2 + (-1-4)^2 + (1-4)^2]^{1/2} = \sqrt{59}$

15. **(a)** We look for values of k such that

$$\mathbf{u} \bullet \mathbf{v} = 2 + 7 + 3k = 0$$

Clearly $k = -3$ is the only possiblity.

17. **(a)** We have $|\mathbf{u} \bullet \mathbf{v}| = |3(4) + 2(-1)| = 10$, while

$$\|\mathbf{u}\|\,\|\mathbf{v}\| = [3^2 + 2^2]^{1/2}[4^2 + (-1)^2]^{1/2} = \sqrt{221}.$$

(d) Here $|\mathbf{u} \bullet \mathbf{v}| = 0 + 2 + 2 + 1 = 5$, while

$$\|\mathbf{u}\|\,\|\mathbf{v}\| = [0^2 + (-2)^2 + 2^2 + 1^2]^{1/2}[(-1)^2 + (-1)^2 + 1^2 + 1^2]^{1/2} = 6.$$

23. We must see if the system

$$\begin{aligned} 3 + 4t &= s \\ 2 + 6t &= 3 - 3s \\ 3 + 4t &= 5 - 4s \\ -1 - 2t &= 4 - 2s \end{aligned}$$

is consistent. Solving the first two equations yield $t = -4/9$, $s = 11/9$. Substituting into the 3rd equation yields $5/3 = 1/9$. Thus the system is *inconsistent*, so the lines are *skew*.

25. This is just the Cauchy-Schwarz inequality applied to the vectors $\mathbf{v}^T A^T$ and $\mathbf{u}^T A^T$ with both sides of the inequality squared. Why?

27. Let $\mathbf{u} = (u_1, \ldots, u_n)$, $\mathbf{v} = (\mathbf{v}_1, \ldots, \mathbf{v}_n)$, and $\mathbf{w} = (w_1, \ldots, w_n)$.

(a) $\mathbf{u} \cdot (k\mathbf{v}) = (u_1, \ldots, u_n) \cdot (kv_1, \ldots, kv_n)$

$$= u_1 kv_1 + \cdots + u_n kv_n$$

$$= k(u_1v_1 + \cdots + u_nv_n)$$

$$= k(\mathbf{u} \cdot \mathbf{v})$$

(b) $\mathbf{u} \cdot (\mathbf{v} + \mathbf{w}) = (u_1, \ldots, u_n) \cdot (v_1 + w_1, \ldots, v_n + w_n)$

$$= u_1(v_1 + w_1) + \cdots + u_n (v_n + w_n)$$

$$= (u_1v_1 + \cdots + u_nv_n) + (u_1w_1 + \cdots + u_nw_n)$$

$$= \mathbf{u} \cdot \mathbf{v} + \mathbf{u} \cdot \mathbf{w}$$

35. **(a)** By theorem 4.1.7, we have $d(\mathbf{u}, \mathbf{v}) = \|\mathbf{u} - \mathbf{v}\| = \sqrt{\|\mathbf{u}\|^2 + \|\mathbf{v}\|^2} = \sqrt{2}$.

37. **(a)** True. In general, we know that

$$\|\mathbf{u} + \mathbf{v}\|^2 = \|\mathbf{u}\|^2 + \|\mathbf{v}\|^2 + 2(\mathbf{u} \cdot \mathbf{v})$$

So in this case $\mathbf{u} \cdot \mathbf{v} = 0$ and the vectors are orthogonal.

(b) True. We are given that $\mathbf{u} \cdot \mathbf{v} = \mathbf{u} \cdot \mathbf{w} = 0$. But since $\mathbf{u} \cdot (\mathbf{v} + \mathbf{w}) = \mathbf{u} \cdot \mathbf{v} + \mathbf{u} \cdot \mathbf{w}$, it follows that $\mathbf{u}$ is orthogonal to $\mathbf{v} + \mathbf{w}$.

(c) False. To obtain a counterexample, let $\mathbf{u} = (1, 0, 0)$, $\mathbf{v} = (1, 1, 0)$, and $\mathbf{w} = (-1, 1, 0)$.

EXERCISE SET 4.2

1. **(b)** Since the transformation maps (x_1, x_2) to (w_1, w_2, w_3), the domain is R^2 and the codomain is R^3. The transformation is not linear because of the terms $2x_1x_2$ and $3x_1x_2$.

3. The standard matrix is A, where

$$\mathbf{w} = A\mathbf{x} = \begin{bmatrix} 3 & 5 & -1 \\ 4 & -1 & 1 \\ 3 & 2 & -1 \end{bmatrix} \begin{bmatrix} x_1 \\ x_2 \\ x_3 \end{bmatrix}$$

so that

$$T(-1, 2, 4) = \begin{bmatrix} 3 & 5 & -1 \\ 4 & -1 & 1 \\ 3 & 2 & -1 \end{bmatrix} \begin{bmatrix} -1 \\ 2 \\ 4 \end{bmatrix} = \begin{bmatrix} 3 \\ -2 \\ -3 \end{bmatrix}$$

5. **(a)** The standard matrix is

$$\begin{bmatrix} 0 & 1 \\ -1 & 0 \\ 1 & 3 \\ 1 & -1 \end{bmatrix}$$

Note that $T(1, 0) = (0, -1, 1, 1)$ and $T(0, 1) = (1, 0, 3, -1)$.

7. **(b)** Here

$$T(2,1-3)=\begin{bmatrix} 2 & -1 & 1 \\ 0 & 1 & 1 \\ 0 & 0 & 0 \end{bmatrix}\begin{bmatrix} 2 \\ 1 \\ -3 \end{bmatrix}=\begin{bmatrix} 0 \\ -2 \\ 0 \end{bmatrix}$$

9. **(a)** In this case,

$$T(2,-5,3)=\begin{bmatrix} 1 & 0 & 0 \\ 0 & 1 & 0 \\ 0 & 0 & -1 \end{bmatrix}\begin{bmatrix} 2 \\ -5 \\ 3 \end{bmatrix}=\begin{bmatrix} 2 \\ -5 \\ -3 \end{bmatrix}$$

so the reflection of (2, –5, 3) is (2, –5, –3).

13. **(b)** The image of (–2, 1, 2) is $(0, 1, 2\sqrt{2})$, since

$$\begin{bmatrix} \cos(45°) & 0 & \sin(45°) \\ 0 & 1 & 0 \\ -\sin(45°) & 0 & \cos(45°) \end{bmatrix}\begin{bmatrix} -2 \\ 1 \\ 2 \end{bmatrix}=\begin{bmatrix} \frac{1}{\sqrt{2}} & 0 & \frac{1}{\sqrt{2}} \\ 0 & 1 & 0 \\ -\frac{1}{\sqrt{2}} & 0 & \frac{1}{\sqrt{2}} \end{bmatrix}\begin{bmatrix} -2 \\ 1 \\ 2 \end{bmatrix}=\begin{bmatrix} 0 \\ 1 \\ 2\sqrt{2} \end{bmatrix}$$

15. **(b)** The image of (–2, 1, 2) is $(0, 1, 2\sqrt{2})$, since

$$\begin{bmatrix} \cos(-45°) & 0 & -\sin(-45°) \\ 0 & 1 & 0 \\ \sin(-45°) & 0 & \cos(-45°) \end{bmatrix}\begin{bmatrix} -2 \\ 1 \\ 2 \end{bmatrix}$$

$$=\begin{bmatrix} \frac{1}{\sqrt{2}} & 0 & \frac{1}{\sqrt{2}} \\ 0 & 1 & 0 \\ -\frac{1}{\sqrt{2}} & 0 & \frac{1}{\sqrt{2}} \end{bmatrix}\begin{bmatrix} -2 \\ 1 \\ 2 \end{bmatrix}=\begin{bmatrix} 0 \\ 1 \\ 2\sqrt{2} \end{bmatrix}$$

17. **(a)** The standard matrix is

$$\begin{bmatrix} 0 & 1 \\ 1 & 0 \end{bmatrix}\begin{bmatrix} 1 & 0 \\ 0 & 0 \end{bmatrix}\begin{bmatrix} \frac{1}{2} & -\frac{\sqrt{3}}{2} \\ \frac{\sqrt{3}}{2} & \frac{1}{2} \end{bmatrix} = \begin{bmatrix} 0 & 0 \\ \frac{1}{2} & -\frac{\sqrt{3}}{2} \end{bmatrix}$$

(c) The standard matrix for a counterclockwise rotation of $15° + 105° + 60° = 180°$ is

$$\begin{bmatrix} \cos(180°) & -\sin(180°) \\ \sin(180°) & \cos(180°) \end{bmatrix} = \begin{bmatrix} -1 & 0 \\ 0 & -1 \end{bmatrix}$$

19. **(c)** The standard matrix is

$$\begin{bmatrix} \cos(180°) & -\sin(180°) & 0 \\ \sin(180°) & \cos(180°) & 0 \\ 0 & 0 & 1 \end{bmatrix}\begin{bmatrix} \cos(90°) & 0 & \sin(90°) \\ 0 & 1 & 0 \\ -\sin(90°) & 0 & \cos(90°) \end{bmatrix}\begin{bmatrix} 1 & 0 & 0 \\ 0 & \cos(270°) & -\sin(270°) \\ 0 & \sin(270°) & \cos(270°) \end{bmatrix}$$

$$= \begin{bmatrix} -1 & 0 & 0 \\ 0 & -1 & 0 \\ 0 & 0 & 1 \end{bmatrix}\begin{bmatrix} 0 & 0 & 1 \\ 0 & 1 & 0 \\ -1 & 0 & 0 \end{bmatrix}\begin{bmatrix} 1 & 0 & 0 \\ 0 & 0 & 1 \\ 0 & -1 & 0 \end{bmatrix}$$

$$= \begin{bmatrix} 0 & 1 & 0 \\ 0 & 0 & -1 \\ -1 & 0 & 0 \end{bmatrix}$$

21. **(a)** Geometrically, it doesn't make any difference whether we rotate and then dilate or whether we dilate and then rotate. In matrix terms, a dilation or contraction is represented by a scalar multiple of the identity matrix. Since such a matrix commutes with any square matrix of the appropriate size, the transformations commute.

23. Set (a, b, c) equal to (1, 0, 0), (0, 1, 0), and (0, 0, 1) in turn.

25. **(a)** Since $T_2(T_1(x_1, x_2)) = (3(x_1 + x_2), 2(x_1 + x_2) + 4(x_1 - x_2)) = (3x_1 + 3x_2, 6x_1 - 2x_2)$, we have

$$[T_2 \circ T_1] = \begin{bmatrix} 3 & 3 \\ 6 & -2 \end{bmatrix}$$

We also have

$$[T_2]\,[T_1] = \begin{bmatrix} 3 & 0 \\ 2 & 4 \end{bmatrix} \begin{bmatrix} 1 & 1 \\ 1 & -1 \end{bmatrix} = \begin{bmatrix} 3 & 3 \\ 6 & -2 \end{bmatrix}$$

27. Compute the trace of the matrix given in Formula (17) and use the fact that (a, b, c) is a unit vector.

29. **(a)** This is an orthogonal projection on the x-axis and a dilation by a factor of 2.

(b) This is a reflection about the x-axis and a dilation by a factor of 2.

31. Since $\cos(2\theta) = \cos^2\theta - \sin^2\theta$ and $\sin(2\theta) = 2\sin\theta\cos\theta$, this represents a rotation through an angle of 2θ.

EXERCISE SET 4.3

1. **(a)** Projections are not one-to-one since two distinct vectors can have the same image vector.

(b) Since a reflection is its own inverse, it is a one-to-one mapping of R^2 or R^3 onto itself.

3. If we reduce the system of equations to row-echelon form, we find that $w_1 = 2w_2$, so that any vector in the range must be of the form $(2w, w)$. Thus $(3, 1)$, for example, is not in the range.

5. **(a)** Since the determinant of the matrix

$$[T] = \begin{bmatrix} 1 & 2 \\ -1 & 1 \end{bmatrix}$$

is 3, the transformation T is one-to-one with

$$[T]^{-1} = \begin{bmatrix} \frac{1}{3} & -\frac{2}{3} \\ \frac{1}{3} & \frac{1}{3} \end{bmatrix}$$

Thus $T^{-1}(w_1, w_2) = \left(\frac{1}{3}w_1 - \frac{2}{3}w_2, \frac{1}{3}w_1 + \frac{1}{3}w_2\right)$.

(b) Since the determinant of the matrix

$$[T] = \begin{bmatrix} 4 & -6 \\ -2 & 3 \end{bmatrix}$$

is zero, T is not one-to-one.

9. **(a)** T is linear since

$$\begin{aligned} T((x_1, y_1) + (x_2, y_2)) &= (2(x_1 + x_2) + (y_1 + y_2), (x_1 + x_2) - (y_1 + y_2)) \\ &= (2x_1 + y_1, x_1 - y_1) + (2x_2 + y_2, x_2 - y_2) \\ &= T(x_1, y_1) + T(x_2, y_2) \end{aligned}$$

and

$$\begin{aligned} T(k(x, y)) &= (2kx + ky, kx - ky) \\ &= k(2x + y, x - y) = kT(x, y) \end{aligned}$$

(b) Since

$$\begin{aligned} T((x_1, y_1) + (x_2, y_2)) &= (x_1 + x_2 + 1, y_1 + y_2) \\ &= (x_1 + 1, y_1) + (x_2, y_2) \\ &\neq T(x_1, y_1) + T(x_2, y_2) \end{aligned}$$

and $T(k(x, y)) = (kx + 1, ky) \neq kT(x, y)$ unless $k = 1$, T is nonlinear.

13. **(a)** The projection sends $\mathbf{e}_1$ to itself and the reflection sends $\mathbf{e}_1$ to $-\mathbf{e}_1$, while the projection sends $\mathbf{e}_2$ to the zero vector, which remains fixed under the reflection. Therefore

$$T(\mathbf{e}_1) = (-1, 0) \text{ and } T(\mathbf{e}_2) = (0, 0), \text{ so that } [T] = \begin{bmatrix} -1 & 0 \\ 0 & 0 \end{bmatrix}$$

(b) We have $\mathbf{e}_1 = (1, 0) \to (0, 1) \to (0, -1) = 0\mathbf{e}_1 - \mathbf{e}_2$ while $\mathbf{e}_2 = (0, 1) \to (1, 0) \to (1, 0)$ $= \mathbf{e}_1 + 0\mathbf{e}_2$. Hence $[T] = \begin{bmatrix} 0 & 1 \\ -1 & 0 \end{bmatrix}$

(c) Here $\mathbf{e}_1 = (1, 0) \to (3, 0) \to (0, 3) \to (0, 3) = 0\mathbf{e}_1 + 3\mathbf{e}_2$ and $\mathbf{e}_2 = (0, 1) \to (0, 3) \to$ $(3, 0) \to (0, 0) = 0\mathbf{e}_1 + 0\mathbf{e}_2$. Therefore $[T] = \begin{bmatrix} 0 & 0 \\ 3 & 0 \end{bmatrix}$

17. **(a)** By the result of Example 5,

$$T\left(\begin{bmatrix}-1\\2\end{bmatrix}\right)=\begin{bmatrix}(1/\sqrt{2})^2 & (1/\sqrt{2})(1/\sqrt{2})\\(1/\sqrt{2})(1/\sqrt{2}) & (1/\sqrt{2})^2\end{bmatrix}\begin{bmatrix}-1\\2\end{bmatrix}=\begin{bmatrix}1/2\\1/2\end{bmatrix}$$

or $T(-1, 2) = (1/2, 1/2)$

19. **(a)** $A=\begin{pmatrix}-1 & 0 & 0\\0 & 1 & 0\\0 & 0 & 1\end{pmatrix}$

Eigenvalue $\lambda_1 = -1$, eigenvector $\xi_1 = \begin{pmatrix}1\\0\\0\end{pmatrix}$

Eigenvaluc $\lambda_2 = 1$, eigenvector $\xi_{21} = \begin{pmatrix}0\\1\\0\end{pmatrix}$, $\xi_{22} = \begin{pmatrix}0\\0\\1\end{pmatrix}$

(b) $A=\begin{pmatrix}1 & 0 & 0\\0 & 0 & 0\\0 & 0 & 1\end{pmatrix}$

$\lambda_1 = 0,\quad \xi_1 = \begin{pmatrix}0\\1\\0\end{pmatrix}$

$\lambda_2 = 1,\quad \xi_{21} = \begin{pmatrix}1\\0\\0\end{pmatrix}$, $\xi_{22} = \begin{pmatrix}0\\0\\1\end{pmatrix}$, or in general $\begin{pmatrix}s\\0\\t\end{pmatrix}$

(c) This transformation doubles the length of each vector while leaving its direction unchanged. Therefore $\lambda = 2$ is the only eigenvalue and every nonzero vector in R^3 is a corresponding eigenvector. To verify this, observe that the characteristic equation is

$$\begin{vmatrix}\lambda-2 & 0 & 0\\0 & \lambda-2 & 0\\0 & 0 & \lambda-2\end{vmatrix}=0$$

or $(\lambda - 2)^3 = 0$. Thus the only eigenvalue is $\lambda = 2$. If (x, y, z) is a corresponding eigenvector, then

$$\begin{bmatrix} 0 & 0 & 0 \\ 0 & 0 & 0 \\ 0 & 0 & 0 \end{bmatrix} \begin{bmatrix} x \\ y \\ z \end{bmatrix} = \begin{bmatrix} 0 \\ 0 \\ 0 \end{bmatrix}$$

Since the above equation holds for every vector (x, y, z), every nonzero vector is an eigenvector.

(d) Since the transformation leaves all vectors on the z-axis unchanged and alters (but does not reverse) the direction of all other vectors, its only eigenvalue is $\lambda = 1$ with corresponding eigenvectors $(0, 0, z)$ with $z \neq 0$. To verify this, observe that the characteristic equation is

$$\begin{vmatrix} \lambda - 1/\sqrt{2} & 1/\sqrt{2} & 0 \\ -1/\sqrt{2} & \lambda - 1/\sqrt{2} & 0 \\ 0 & 0 & \lambda - 1 \end{vmatrix} = 0$$

or

$$(\lambda - 1)\begin{vmatrix} \lambda - 1/\sqrt{2} & 1/\sqrt{2} \\ -1/\sqrt{2} & \lambda - 1/\sqrt{2} \end{vmatrix} = (\lambda - 1)\left[\left(\lambda - 1/\sqrt{2}\right)^2 + \left(1/\sqrt{2}\right)^2\right] = 0$$

Since the quadratic $(\lambda - 1/\sqrt{2})^2 + 1/2 = 0$ has no real roots, $\lambda = 1$ is the only real eigenvalue. If (x, y, z) is a corresponding eigenvector, then

$$\begin{bmatrix} 1 - 1/\sqrt{2} & 1/\sqrt{2} & 0 \\ -1/\sqrt{2} & 1 - 1/\sqrt{2} & 0 \\ 0 & 0 & 0 \end{bmatrix} \begin{bmatrix} x \\ y \\ z \end{bmatrix} = \begin{bmatrix} \left(1 - 1/\sqrt{2}\right)x + \left(1/\sqrt{2}\right)y \\ -\left(1/\sqrt{2}\right)x + \left(1 - 1/\sqrt{2}\right)y \\ 0 \end{bmatrix} = \begin{bmatrix} 0 \\ 0 \\ 0 \end{bmatrix}$$

You should verify that the above equation is valid if and only if $x = y = 0$. Therefore the corresponding eigenvectors are all of the form $(0, 0, z)$ with $z \neq 0$.

21. Since $T(x, y) = (0, 0)$ has the standard matrix $\begin{bmatrix} 0 & 0 \\ 0 & 0 \end{bmatrix}$, it is linear. If $T(x, y) = (1, 1)$ were linear, then we would have

$$(1, 1) = T(0, 0) = T(0 + 0, 0 + 0) = T(0, 0) + T(0, 0) = (1, 1) + (1, 1) = (2, 2)$$

Since this is a contradiction, T cannot be linear.

23. From Figure 1, we see that $T(\mathbf{e}_1) = (\cos 2\theta, \sin 2\theta)$ and from Figure 2, that $T(\mathbf{e}_2) =$

$$\left(\cos\left(\frac{3\pi}{2}+2\theta\right), \sin\left(\frac{3\pi}{2}+2\theta\right)\right) = (\sin 2\theta, -\cos 2\theta)$$

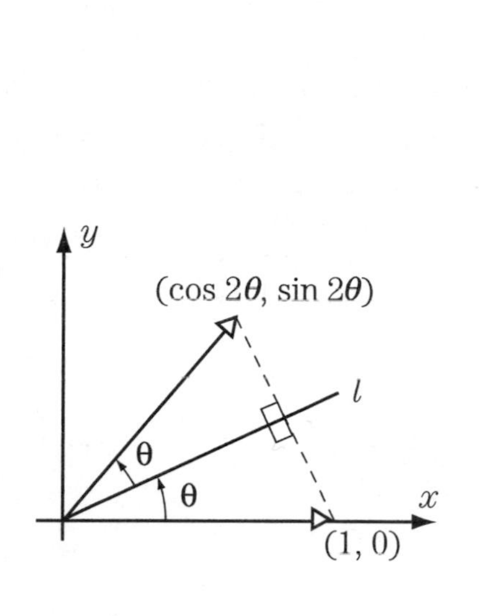

Figure 1

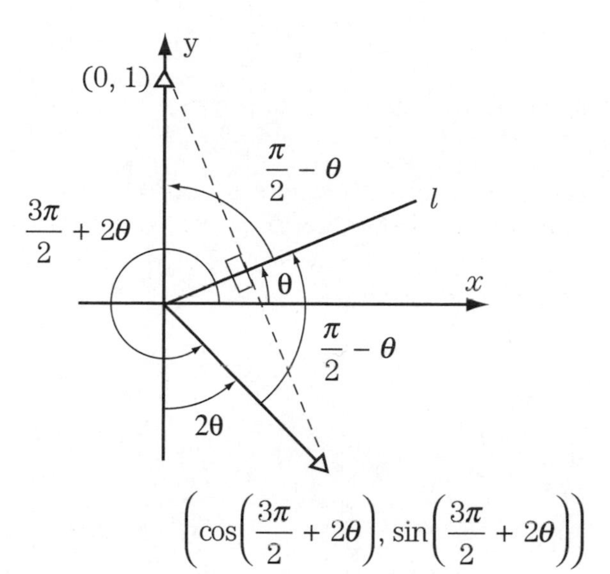

Figure 2

This, of course, should be checked for all possible diagrams, and in particular for the case $\frac{\pi}{2} < \theta < \pi$. The resulting standard matrix is

$$\begin{bmatrix} \cos 2\theta & \sin 2\theta \\ \sin 2\theta & -\cos 2\theta \end{bmatrix}$$

25. **(a)** False. The transformation $T(x_1, x_2) = x_1^2$ from R^2 to R^1 is not linear, but $T(\mathbf{0}) = \mathbf{0}$.

(b) True. If not, $T(\mathbf{u}) = T(\mathbf{v})$ where $\mathbf{u}$ and $\mathbf{v}$ are distinct. Why?

(c) False. One must also demand that $\mathbf{x} \neq \mathbf{0}$.

(d) True. If $c_1 = c_2 = 1$, we obtain equation (a) of Theorem 4.3.2 and if $c_2 = 0$, we obtain equation (b).

27. **(a)** The range of T cannot be all of R^n, since otherwise T would be invertible and $\det(A) \neq 0$. For instance, the matrix $\begin{bmatrix} 1 & 0 \\ 0 & 0 \end{bmatrix}$ sends the entire xy-plane to the x-axis.

(b) Since $\det(A) = 0$, the equation $T(\mathbf{x}) = A\mathbf{x} = \mathbf{0}$ will have a non-trivial solution and hence, T will map infinitely many vectors to $\mathbf{0}$.

EXERCISE SET 4.4

1. **(a)** $(x^2 + 2x - 1) - 2(3x^2 + 2) = -5x^2 + 2x - 5$

(b) $5/4x^2 + 3x) + 6(x^2 + 2x + 2) = 26x^2 + 27x + 6$

(c) $(x^4 + 2x^3 + x^2 - 2x + 1) - (2x^3 - 2x) = x^4 + x^2 + 1$

(d) $\pi(4x^3 - 3x^2 + 7x + 1) = 4\pi x^3 - 3\pi x^2 + 7\pi x + \pi$

3. **(a)** Note that the mapping $f\colon R_3 \to R$ given by $f(a, b, c) = |a|$ has $f(1, 0, 0) = 1, f(0, 1, 0) = 0$, and $f(0, 0, 1) = 0$. So If f were a linear mapping, the matrix would be $A = (1, 0, 0)$. Thus, $f(-1, 0, 0)$ would be found as $(1,0,0)\begin{pmatrix}-1\\0\\0\end{pmatrix} = -1$. Yet, $f(-1, 0, 0) = |-1| = 1 \neq -1$. Thus f is not linear.

(b) Yes, and here $A = (1, 0, 0)$ by reasoning as in (**a**).

5. **(a)**

$$A = \begin{pmatrix}3 & 0 & 0 & 0\\0 & 2 & 0 & 0\\0 & 0 & 1 & 0\end{pmatrix}$$

$$\begin{pmatrix}4 & 0 & 0 & 0 & 0\\0 & 3 & 0 & 0 & 0\\0 & 0 & 2 & 0 & 0\\0 & 0 & 0 & 1 & 0\end{pmatrix}$$

$$\begin{pmatrix}5 & 0 & 0 & 0 & 0 & 0\\0 & 4 & 0 & 0 & 0 & 0\\0 & 0 & 3 & 0 & 0 & 0\\0 & 0 & 0 & 2 & 0 & 0\\0 & 0 & 0 & 0 & 1 & 0\end{pmatrix}$$

7. **(a)** $T(ax + b) = (a + b)x + (a - b)$

$T: P_1 \to P_1$

(b) $T(ax + b) = ax^2 + (a + b)x + (2a - b)$

$T: P_1 \to P_2$

(c) $T(ax^3 + bx^2 + cx + d) = (a + 2c - d)x + (2a + b + c + 3d)$

$T: P_3 \to P_1$

(d) $T(ax^2 + bx + c) = bx$

$T: P_2 \to P_2$

(e) $T(ax^2 + bx + c) = b$

$T: P_2 \to P_0$

9. **(a)** $3e^t + 3e^{-t}$

(b) Yes, since $\cosh t = \frac{1}{2}e^t + \frac{1}{2}e^{-t}$, $\cosh t$ corresponds to the vector $(0, 0, \frac{1}{2}, \frac{1}{2})$.

(c) $A = \begin{pmatrix} 0 & 1 & 0 & 0 \\ 0 & 0 & 0 & 0 \\ 0 & 0 & 1 & 0 \\ 0 & 0 & 0 & -1 \end{pmatrix}$ $\quad (a, b, c, d) \to (b, 0, c, -d)$

11. If $S(u) = T(u) + f, f \neq 0$, then $S(0) = T(0) + f = f \neq 0$. Thus S is not linear.

13. **(a)** The Vandermonde system is

$$\begin{bmatrix} 1 & -2 & 4 \\ 1 & 0 & 0 \\ 1 & 1 & 1 \end{bmatrix} \begin{bmatrix} a_0 \\ a_1 \\ a_2 \end{bmatrix} = \begin{bmatrix} 1 \\ 1 \\ 4 \end{bmatrix}$$

Solving: $a_0 = 1, a_1 = 2, a_2 = 1$

$p(x) = x^2 + 2x + 1 = (x + 1)^2$

(b) The system is:

$$\begin{pmatrix} 1 & 0 & 0 \\ 1 & 2 & 0 \\ 1 & 3 & 3 \end{pmatrix}\begin{pmatrix} b_0 \\ b_1 \\ b_2 \end{pmatrix} = \begin{pmatrix} 1 \\ 1 \\ 4 \end{pmatrix}$$

Solving: $b_0 = 1, b_1 = 0, b_2 = 1.$

Thus, $p(x) = 1 \bullet (x + 2)(x) + 0 \bullet (x + 2) + 1 = (x + 2) \bullet x + 1$

$$= (x + 2) \bullet x + 1$$

$$= x^2 + 2x + 1$$

15. **(a)** The Vandermonde system is

$$\left(\begin{array}{cccc|c} 1 & -2 & 4 & -8 & -10 \\ 1 & -1 & 1 & -1 & 2 \\ 1 & 1 & 1 & 1 & 2 \\ 1 & 2 & 4 & 8 & 14 \end{array}\right) \quad \text{Solving,} \quad \begin{array}{l} a_0 = 2 \\ a_1 = -2 \\ a_2 = 0 \\ a_3 = 2 \end{array}$$

Thus, $p(x) = 2x^3 - 2x + 2$

(b) The system is

$$\left(\begin{array}{cccc|c} 1 & 0 & 0 & 0 & -10 \\ 1 & 1 & 0 & 0 & 2 \\ 1 & 3 & 6 & 0 & 2 \\ 1 & 4 & 12 & 12 & 14 \end{array}\right) \quad \text{Solving,} \quad \begin{array}{l} b_0 = -10 \\ b_1 = 12 \\ b_2 = -4 \\ b_3 = 2 \end{array}$$

Thus, $p(x) = 2(x + 2)(x + 1)(x - 1) - 4(x + 2)(x + 1) + 12(x + 2) - 10$

(c) We have $\begin{pmatrix} a_0 \\ a_1 \\ a_2 \\ a_3 \end{pmatrix} = \begin{pmatrix} 1 & 2 & 2 & -2 \\ 0 & 1 & 3 & -1 \\ 0 & 0 & 1 & 2 \\ 0 & 0 & 0 & 1 \end{pmatrix}\begin{pmatrix} -10 \\ 12 \\ -4 \\ 2 \end{pmatrix} = \begin{pmatrix} 2 \\ -2 \\ 0 \\ 2 \end{pmatrix}$

(d) We have $\begin{pmatrix} b_0 \\ b_1 \\ b_2 \\ b_3 \end{pmatrix} = \begin{pmatrix} 1 & -2 & 4 & -8 \\ 0 & 1 & -3 & 7 \\ 0 & 0 & 1 & -2 \\ 0 & 0 & 0 & 1 \end{pmatrix} \begin{pmatrix} 2 \\ -2 \\ 0 \\ 2 \end{pmatrix} = \begin{pmatrix} -10 \\ 12 \\ -4 \\ 2 \end{pmatrix}$

17. **(a)** $\begin{pmatrix} a_0 \\ a_1 \end{pmatrix} = \begin{pmatrix} 1 & -x_0 \\ 0 & 1 \end{pmatrix} \begin{pmatrix} b_0 \\ b_1 \end{pmatrix}$

(b) $\begin{pmatrix} a_0 \\ a_1 \\ a_2 \end{pmatrix} = \begin{pmatrix} 1 & -x_0 & x_0x_1 \\ 0 & 1 & -(x_0+x_1) \\ 0 & 0 & 1 \end{pmatrix} \begin{pmatrix} b_0 \\ b_1 \\ b_2 \end{pmatrix}$

(c)

$$\begin{pmatrix} a_0 \\ a_1 \\ a_2 \\ a_3 \\ a_4 \end{pmatrix} = \begin{bmatrix} 1 & -x_0 & x_0x_1 & -x_0x_1x_2 & x_0x_1x_2x_3 \\ 0 & 1 & -(x_0+x_1) & x_1x_2+x_0x_2+x_0x_1 & \blacklozenge_1 \\ 0 & 0 & 1 & -(x_0+x_1+x_2) & \blacklozenge_2 \\ 0 & 0 & 0 & 1 & -(x_2+x_3) \\ 0 & 0 & 0 & 0 & 1 \end{bmatrix} \begin{pmatrix} b_0 \\ b_1 \\ b_2 \\ b_3 \\ b_4 \end{pmatrix}$$

where

$$\blacklozenge_1 = -(x_1x_2x_3 + x_0x_2x_3 + x_0x_1x_3 + x_0x_1x_2)$$

$$\blacklozenge_2 = x_0x_1 + x_0x_2 + x_0x_3 + x_1x_2 + x_1x_3 + x_2x_3$$

19. **(a)** $D_2 = (2\ 0\ 0)$

(b) $D_2 = \begin{pmatrix} 6 & 0 & 0 & 0 \\ 0 & 2 & 0 & 0 \end{pmatrix}$

(c) No. For example, the matrix for first differentiation from $P_3 \to P_2$ is

$$D_2 = \begin{pmatrix} 3 & 0 & 0 & 0 \\ 0 & 2 & 0 & 0 \\ 0 & 0 & 1 & 0 \end{pmatrix}$$

D_1^2 cannot be formed.

21. **(a)** We first note $P = y_i$. Hence the Vandermonde system has a unique solution that are the coefficients of the polynomial of nth degree through the $n + 1$ data points. So, there is exactly one polynomial of the nth degree through $n + 1$ data points with the x_i unique. Thus, the Lagrange expression must be algebraically equivalent to the Vandermonde form.

(b) Since $c_i = y_i$, $i = 0, 1, \ldots, n$, then the linear systems for the Vandermonde and Newtons method remain the same.

(c) Newton's form allows for the easy addition of another point (x_{n+1}, y_{n+1}) that does not have to be in any order with respect to the other x_i values. This is done by adding a next term to $p(i)$, p_e.

$$\begin{aligned} p_{n+1}(x) &= b_{n+1}(x - x_0)(x - x_1)(x - x_2) \ldots (x - x_n) \\ &\quad + b_n(x - x_0)(x - x_1)(x - x_2) \ldots (x - x_{n-1}) \\ &\quad + \cdots + b_1(x - x_0) + b_0, \end{aligned}$$

where $P_n(x) = b_n(x - x_0)(x - x_1)(x - x_2) \ldots (x - x_{n-1}) + \ldots + b_1(x - x_0) + b_0$ is the interpolant to the points $(x_0, y_0) \ldots (x_n, y_n)$. The coefficients for $p_{n+1}(x)$ are found as in (4), giving an $n + 1$ degree polynomial. The extra point (x_{n+1}, y_{n+1}) may be the desired interpolating value.

23. We may assume in all cases that $\|x\| = 1$, since

$$\frac{\|T(2x)\|}{\|2x\|} = \frac{\|2T(x)\|}{\|2x\|} = \frac{\|T(x)\|}{\|x\|}$$

Let $(x_1, x_2) = (\cos\theta, \sin\theta) = x$ since $\|x\| = 1$.

(a) $\|T\| = \max\sqrt{4\cos^2\theta + \sin^2\theta} = \max\sqrt{1 + 3\cos^2\theta} = 2$

(b) $\|T\| = \max\sqrt{x_1^2 + x_2^2} = 1$

(c) $\|T\| = \max\sqrt{4x_1^2 + 9x_2^2} = \max\sqrt{4 + 5\sin^2\theta} = 3$

(d) $$\|T\| = \max\sqrt{\left(\frac{1}{\sqrt{2}}x_1 + \frac{1}{\sqrt{2}}x_2\right)^2 + \left(\frac{1}{\sqrt{2}}x_1 - \frac{1}{\sqrt{2}}x_2\right)^2}$$
$$= \max\sqrt{x_1^2 + x_2^2} = 1$$

EXERCISE SET 5.1

11. This is a vector space. We shall check only four of the axioms because the others follow easily from various properties of the real numbers.

(1) If f and g are real-valued functions defined everywhere, then so is $f + g$. We must also check that if $f(1) = g(1) = 0$, then $(f + g)(1) = 0$. But $(f + g)(1) = f(1) + g(1) = 0 + 0 = 0$.

(4) The zero vector is the function z which is zero everywhere on the real line. In particular, $z(1) = 0$.

(5) If f is a function in the set, then $-f$ is also in the set since it is defined for all real numbers and $-f(1) = -0 = 0$. Moreover, $f + (-f) = (-f) + f = z$.

(6) If f is in the set and k is any real number, then kf is a real valued function defined everywhere. Moreover, $kf(1) = k0 = 0$.

13. This is a vector space with $\mathbf{0} = (1, 0)$ and $-\mathbf{x} = (1, -x)$. The details are easily checked.

15. We must check all ten properties:

(1) If x and y are positive reals, so is $x + y = xy$.

(2) $x + y = xy = yx = y + x$

(3) $x + (y + z) = x(yz) = (xy)z = (x + y) + z$

(4) There is an object $\mathbf{0}$, the positive real number 1, which is such that

$$1 + x = 1 \cdot x = x = x \cdot 1 = x + 1$$

for all positive real numbers x.

(5) For each positive real x, the positive real $1/x$ acts as the negative:

$$x + (1/x) = x(1/x) = 1 = \mathbf{0} = 1 = (1/x)x = (1/x) + x$$

(6) If k is a real and x is a positive real, then $kx = x^k$ is again a positive real.

(7) $k(x + y) = (xy)^k = x^k y^k = kx + ky$

(8) $(k + \ell)x = x^{k+\ell} = x^k x^\ell = kx + \ell x$

(9) $k(\ell x) = (\ell x)^k = (\ell x)^k = x^{\ell k} = x^{k\ell} = (k\ell)x$

(10) $1x = x^1 = x$

17. **(a)** Only Axiom 8 fails to hold in this case. Let k and m be scalars. Then

$$(k+m)(x,y,z) = ((k+m)^2x, (k+m)^2y, (k+m)^2z) = (k^2x, k^2y, k^2z) + (2kmx, 2kmy, 2kmz) + (m^2x, m^2y, m^2z)$$
$$= k(x,y,z) + m(x,y,z) + (2kmx, 2kmy, 2kmz)$$
$$\neq k(x,y,z) + m(x,y,z),$$

and Axiom 8 fails to hold.

(b) Only Axioms 3 & 4 fail for this set.

Axiom 3: Using the obvious notation, we have

$$\begin{aligned}\mathbf{u} + (\mathbf{v} + \mathbf{w}) &= (u_1, u_2, u_3) + (v_3 + w_3, v_2 + w_2, v_1 + w_1)\\ &= (u_3 + v_1 + w_1, u_2 + v_2 + w_2, u_1 + v_3 + w_3)\end{aligned}$$

whereas

$$\begin{aligned}(\mathbf{u} + \mathbf{v}) + \mathbf{w} &= (u_3 + v_3, u_2 + v_2, u_1 + v_1) + (w_1, w_2, w_3)\\ &= (u_1 + v_1 + w_3, u_2 + v_2 + w_2, u_3 + v_3 + w_1)\end{aligned}$$

Thus, $\mathbf{u} + (\mathbf{v} + \mathbf{w}) \neq (\mathbf{u} + \mathbf{w}) + \mathbf{w}$.

Axiom 4: There is no zero vector in this set. If we assume that there is, and let $\mathbf{0} = (z_1, z_2, z_3)$, then for any vector (a, b, c), we have $(a, b, c) + (z_1, z_2, z_3) = (c + z_3, b + z_2, a + z_1) = (a, b, c)$. Solving for the z_i's, we have $z_3 = a - c$, $z_2 = 0$ and $z_1 = c - a$. Thus, there is no one zero vector that will work for every vector (a, b, c) in R^3.

(c) Let V be the set of all 2×2 invertible matrices and let A be a matrix in V. Since we are using standard matrix addition and scalar multiplication, the majority of axioms hold. However, the following axioms fail for this set V:

Axiom 1: Clearly if A is invertible, then so is $-A$. However, the matrix $A + (-A) = \mathbf{0}$ is not invertible, and thus $A + (-A)$ is not in V, meaning V is not closed under addition.

Axiom 4: We've shown that the zero matrix is not in V, so this axiom fails.

Axiom 6: For any 2×2 invertible matrix A, $\det(kA) = k^2 \det(A)$, so for $k \neq 0$, the matrix kA is also invertible. However, if $k = 0$, then kA is not invertible, so this axiom fails.

Thus, V is not a vector space.

19. **(a)** Let V be the set of all ordered pairs (x, y) that satisfy the equation $ax + by = c$, for fixed constants a, b and c. Since we are using the standard operations of addition and scalar multiplication, Axioms 2, 3, 5, 7, 8, 9, 10 will hold automatically. However, for Axiom 4 to hold, we need the zero vector $(0, 0)$ to be in V. Thus $a0 + b0 = c$, which forces $c = 0$. In this case, Axioms 1 and 6 are also satisfied. Thus, the set of all points in R^2 lying on a line is a vector space exactly in the case when the line passes through the origin.

(b) Let V be the set of all ordered triples (x, y, z) that satisfy the equation $ax + by + cz = d$, for fixed constants a, b, c and d. Since we are using the standard operations of addition and scalar multiplication, Axioms 2, 3, 5, 7, 8, 9, 10 will hold automatically. However, for Axiom 4 to hold, we need the zero vector (0, 0, 0) to be in V. Thus $a0 + b0 + c0 = d$, which forces $d = 0$. In this case, Axioms 1 and 6 are also satisfied. Thus, the set of all points in R^3 lying on a plane is a vector space exactly in the case when the plane passes through the origin.

25. No. Planes which do not pass through the origin do not contain the zero vector.

27. Since this space has only one element, it would have to be the zero vector. In fact, this is just the zero vector space.

33. Suppose that $\mathbf{u}$ has two negatives, $(-\mathbf{u})_1$ and $(-\mathbf{u})_2$. Then

$$(-\mathbf{u})_1 = (-\mathbf{u})_1 + \mathbf{0} = (-\mathbf{u})_1 + (\mathbf{u} + (-\mathbf{u})_2) = ((-\mathbf{u})_1 + \mathbf{u}) + (-\mathbf{u})_2 = \mathbf{0} + (-\mathbf{u})_2 = (-\mathbf{u})_2$$

Axiom 5 guarantees that $\mathbf{u}$ must have at least one negative. We have proved that it has at most one.

EXERCISE SET 5.2

1. **(a)** The set is closed under vector addition because

$$(a, 0, 0) + (b, 0, 0) = (a + b, 0, 0)$$

It is closed under scalar multiplication because

$$k(a, 0, 0) = (ka, 0, 0)$$

Therefore it is a subspace of R^3.

(b) This set is not closed under either vector addition or scalar multiplication. For example, $(a, 1, 1) + (b, 1, 1) = (a + b, 2, 2)$ and $(a + b, 2, 2)$ does not belong to the set. Thus it is not a subspace.

(c) This set is closed under vector addition because

$$(a_1, b_1, 0) + (a_2, b_2, 0) = (a_1 + a_2, b_1 + b_2, 0).$$

It is also closed under scalar multiplication because

$$k(a, b, 0) = (ka, kb, 0).$$

Therefore, it is a subspace of R^3.

3. **(a)** This is the set of all polynominals with degree ≤ 3 and with a constant term which is equal to zero. Certainly, the sum of any two such polynomials is a polynomial with degree ≤ 3 and with a constant term which is equal to zero. The same is true of a constant multiple of such a polynomial. Hence, this set is a subspace of P_3.

(c) The sum of two polynomials, each with degree ≤ 3 and each with integral coefficients, is again a polynomial with degree ≤ 3 and with integral coefficients. Hence, the subset is closed under vector addition. However, a constant multiple of such a polynomial will not necessarily have integral coefficients since the constant need not be an integer. Thus, the subset is not closed under scalar multiplication and is therefore not a subspace.

5. **(b)** If A and B are in the set, then $a_{ij} = -a_{ji}$ and $b_{ij} = -b_{ji}$ for all i and j. Thus $a_{ij} + b_{ij} = -(a_{ji} + b_{ji})$ so that $A + B$ is also in the set. Also $a_{ij} = -a_{ji}$ implies that $ka_{ij} = -(ka_{ji})$, so that kA is in the set for all real k. Thus the set is a subspace.

(c) For A and B to be in the set it is necessary and sufficient for both to be invertible, but the sum of 2 invertible matrices need not be invertible. (For instance, let $B = -A$.) Thus $A + B$ need not be in the set, so the set is not a subspace.

7. **(a)** We look for constants a and b such that $a\mathbf{u} + b\mathbf{v} = (2, 2, 2)$, or

$$a(0, -2, 2) + b(1, 3, -1) = (2, 2, 2)$$

Equating corresponding vector components gives the following system of equations:

$$b = 2$$

$$-2a + 3b = 2$$

$$2a - b = 2$$

From the first equation, we see that $b = 2$. Substituting this value into the remaining equations yields $a = 2$. Thus $(2, 2, 2)$ is a linear combination of $\mathbf{u}$ and $\mathbf{v}$.

(c) We look for constants a and b such that $a\mathbf{u} + b\mathbf{v} = (0, 4, 5)$, or

$$a(0, -2, 2) + b(1, 3, -1) = (0, 4, 5)$$

Equating corresponding components gives the following system of equations:

$$b = 0$$

$$-2a + 3b = 4$$

$$2a - b = 5$$

From the first equation, we see that $b = 0$. If we substitute this value into the remaining equations, we find that $a = -2$ and $a = 5/2$. Thus, the system of equations is inconsistent and therefore $(0, 4, 5)$ is not a linear combination of $\mathbf{u}$ and $\mathbf{v}$.

9. **(a)** We look for constants a, b, and c such that

$$a\mathbf{p}_1 + b\mathbf{p}_2 + c\mathbf{p}_3 = -9 - 7x - 15x^2$$

$$a(2+x+4x^2) + b(1-x+3x^2) + c(3+2x+5x^2) = -9-7x-15x^2$$

$$\left[\begin{array}{ccc|c} 2 & 1 & 3 & -9 \\ 1 & -1 & 2 & -7 \\ 4 & 3 & 5 & -15 \end{array}\right]$$

If we substitute the expressions for $\mathbf{p}_1$, $\mathbf{p}_2$, and $\mathbf{p}_3$ into the above equation and equate corresponding coefficients, we find that we have exactly the same system of equations that we had in Problem 8(a), above. Thus, we know that $a = -2$, $b = 1$, and $c = -2$ and thus $-2\mathbf{p}_1 + 1\mathbf{p}_2 - 2\mathbf{p}_3 = -9 - 7x - 15x^2$.

(c) Just as Problem 9(a) was Problem 8(a) in disguise, Problem 9(c) is Problem 8(c) in different dress. The constants are the same, so that $\mathbf{0} = 0\mathbf{p}_1 + 0\mathbf{p}_2 + 0\mathbf{p}_3$.

11. **(a)** Given any vector (x, y, z) in R^3, we must determine whether or not there are constants a, b, and c such that

$$\begin{aligned}(x, y, z) &= a\mathbf{v}_1 + b\mathbf{v}_2 + c\mathbf{v}_3\\ &= a(2, 2, 2) + b(0, 0, 3) + c(0, 1, 1)\\ &= (2a, 2a + c, 2a + 3b + c)\end{aligned}$$

or

$$\begin{aligned}x &= 2a\\ y &= 2a \qquad\;\; + c\\ z &= 2a + 3b + c\end{aligned}$$

This is a system of equations for a, b, and c. Since the determinant of the system is nonzero, the system of equations must have a solution for any values of x, y, and z, whatsoever. Therefore, $\mathbf{v}_1$, $\mathbf{v}_2$, and $\mathbf{v}_3$ do indeed span R^3.

Note that we can also show that the system of equations has a solution by solving for a, b, and c explicitly.

(c) We follow the same procedure that we used in Part (a). This time we obtain the system of equations

$$\begin{aligned}3a + 2b + 5c + d &= x\\ a - 3b - 2c + 4d &= y\\ 4a + 5b + 9c - d &= z\end{aligned}$$

The augmented matrix of this system is

$$\begin{bmatrix} 3 & 2 & 5 & 1 & x \\ 1 & -3 & -2 & 4 & y \\ 4 & 5 & 9 & -1 & z \end{bmatrix}$$

which reduces to

$$\begin{bmatrix} 1 & -3 & -2 & 4 & y \\ 0 & 1 & 1 & -1 & \dfrac{x-3y}{11} \\ 0 & 0 & 0 & 0 & \dfrac{z-4y}{17}-\dfrac{x-3y}{11} \end{bmatrix}$$

Thus the system is inconsistent unless the last entry in the last row of the above matrix is zero. Since this is not the case for all values of x, y, and z, the given vectors do not span R^3.

13. Given an arbitrary polynomial $a_0 + a_1x + a_2x^2$ in P_2, we ask whether there are numbers a, b, c and d such that

$$a_0 + a_1x + a_2x^2 = a\mathbf{p}_1 + b\mathbf{p}_2 + c\mathbf{p}_3 + d\mathbf{p}_4$$

If we equate coefficients, we obtain the system of equations:

$$\begin{aligned} a_0 &= a + 3b + 5c - 2d \\ a_1 &= -a + b - c - 2d \\ a_2 &= 2a + 4c + 2d \end{aligned}$$

A row-echelon form of the augmented matrix of this system is

$$\begin{bmatrix} 1 & 3 & 5 & -2 & a_0 \\ 0 & 1 & 1 & -1 & \dfrac{a_0+a_1}{4} \\ 0 & 0 & 0 & 0 & -a_0+3a_1+2a_2 \end{bmatrix}$$

Thus the system is inconsistent whenever $-a_0 + 3a_1 + 2a_2 \neq 0$ (for example, when $a_0 = 0$, $a_1 = 0$, and $a_2 = 1$). Hence the given polynomials do not span P_2.

15. The plane has the vector $\mathbf{u} \times \mathbf{v} = (0, 7, -7)$ as a normal and passes through the point $(0,0,0)$. Thus its equation is $y - z = 0$.

Alternatively, we look for conditions on a vector (x, y, z) which will insure that it lies in span $\{\mathbf{u}, \mathbf{v}\}$. That is, we look for numbers a and b such that

$$\begin{aligned}(x, y, z) &= a\mathbf{u} + b\mathbf{v} \\ &= a(-1, 1, 1) + b(3, 4, 4)\end{aligned}$$

If we expand and equate components, we obtain a system whose augmented matrix is

$$\begin{bmatrix} -1 & 3 & x \\ 1 & 4 & y \\ 1 & 4 & z \end{bmatrix}$$

This reduces to the matrix

$$\begin{bmatrix} 1 & -3 & -x \\ 0 & 1 & \dfrac{x+y}{7} \\ 0 & 0 & \dfrac{-y+z}{7} \end{bmatrix}$$

Thus the system is consistent if and only if $\dfrac{-y+z}{7} = 0$ or $y = z$.

17. The set of solution vectors of such a system does not contain the zero vector. Hence it cannot be a subspace of R^n.

19. Note that if we solve the system $\mathbf{v}_1 = a\mathbf{w}_1 + b\mathbf{w}_2$, we find that $\mathbf{v}_1 = \mathbf{w}_1 + \mathbf{w}_2$. Similarly, $\mathbf{v}_2 = 2\mathbf{w}_1 + \mathbf{w}_2$, $\mathbf{v}_3 = -\mathbf{w}_1 + 0\mathbf{w}_2$, $\mathbf{w}_1 = 0\mathbf{v}_1 + 0\mathbf{v}_2 - v_3$, and $\mathbf{w}_2 = \mathbf{v}_1 + 0\mathbf{v}_2 + \mathbf{v}_3$.

21. **(a)** We simply note that the sum of two continuous functions is a continuous function and that a constant times a continuous function is a continuous function.

(b) We recall that the sum of two differentiable functions is a differentiable function and that a constant times a differentiable function is a differentiable function.

23. **(a)** False. The system has the form $A\mathbf{x} = \mathbf{b}$ where $\mathbf{b}$ has at least one nonzero entry. Suppose that $\mathbf{x}_1$ and $\mathbf{x}_2$ are two solutions of this system; that is, $A\mathbf{x}_1 = \mathbf{b}$ and $A\mathbf{x}_2 = \mathbf{b}$. Then

$$A(\mathbf{x}_1 + \mathbf{x}_2) = A\mathbf{x}_1 + A\mathbf{x}_2 = \mathbf{b} + \mathbf{b} \neq \mathbf{b}$$

Thus the solution set is not closed under vector addition and so cannot form a subspace of R_n. Alternatively, we could show that it is not closed under scalar multiplication.

(b) True. Let $\mathbf{u}$ and $\mathbf{v}$ be vectors in W. Then we are given that $k\mathbf{u} + \mathbf{v}$ is in W for all scalars k. If $k = 1$, this shows that W is closed under addition. If $k = -1$ and $\mathbf{u} = \mathbf{v}$, then the zero vector of V must be in W. Thus, we can let $\mathbf{v} = \mathbf{0}$ to show that W is closed under scalar multiplication.

(d) True. Let W_1 and W_2 be subspaces of V. Then if $\mathbf{u}$ and $\mathbf{v}$ are in $W_1 \cap W_2$, we know that $\mathbf{u} + \mathbf{v}$ must be in both W_1 and W_2, as must $k\mathbf{u}$ for every scalar k. This follows from the closure of both W_1 and W_2 under vector addition and scalar multiplication.

(e) False. Span$\{\mathbf{v}\}$ = span$\{2\mathbf{v}\}$, but $\mathbf{v} \neq 2\,\mathbf{v}$ in general.

25. No. For instance, (1, 1) is in W_1 and (1, –1) is in W_2, but (1, 1) + (1, –1) = (2, 0) is in neither W_1 nor W_2.

27. They cannot all lie in the same plane.

EXERCISE SET 5.3

3. **(a)** Following the technique used in Example 4, we obtain the system of equations

$$\begin{aligned} 3k_1 + k_2 + 2k_3 + k_4 &= 0 \\ 8k_1 + 5k_2 - k_3 + 4k_4 &= 0 \\ 7k_1 + 3k_2 + 2k_3 \qquad &= 0 \\ -3k_1 - k_2 + 6k_3 + 3k_4 &= 0 \end{aligned}$$

Since the determinant of the coefficient matrix is nonzero, the system has *only* the trivial solution. Hence, the four vectors are linearly independent.

(b) Again following the technique of Example 4, we obtain the system of equations

$$\begin{aligned} 3k_2 + k_3 &= 0 \\ 3k_2 + k_3 &= 0 \\ 2k_1 \qquad\qquad &= 0 \\ 2k_1 \qquad - k_3 &= 0 \end{aligned}$$

The third equation, above, implies that $k_1 = 0$. This implies that k_3 and hence k_2 must also equal zero. Thus the three vectors are linearly independent.

5. **(a)** The vectors lie in the same plane through the origin if and only if they are linearly dependent. Since the determinant of the matrix

$$\begin{bmatrix} 2 & 6 & 2 \\ -2 & 1 & 0 \\ 0 & 4 & -4 \end{bmatrix}$$

is not zero, the matrix is invertible and the vectors are linearly independent. Thus they do not lie in the same plane.

7. **(a)** Note that $7\mathbf{v}_1 - 2\mathbf{v}_2 + 3\mathbf{v}_3 = \mathbf{0}$.

9. If there are constants a, b, and c such that

$$a(\lambda, -1/2, -1/2) + b(-1/2, \lambda, -1/2) + c(-1/2, -1/2, \lambda) = (0, 0, 0)$$

then

$$\begin{bmatrix} \lambda & -1/2 & -1/2 \\ -1/2 & \lambda & -1/2 \\ -1/2 & -1/2 & \lambda \end{bmatrix} \begin{bmatrix} a \\ b \\ c \end{bmatrix} = \begin{bmatrix} 0 \\ 0 \\ 0 \end{bmatrix}$$

The determinant of the coefficient matrix is

$$\lambda^3 - \frac{3}{4}\lambda - \frac{1}{4} = (\lambda - 1)\left(\lambda + \frac{1}{2}\right)^2$$

This equals zero if and only if $\lambda = 1$ or $\lambda = -1/2$. Thus the vectors are linearly dependent for these two values of λ and linearly independent for all other values.

11. Suppose that S has a linearly dependent subset T. Denote its vectors by $\mathbf{w}_1, \ldots, \mathbf{w}_m$. Then there exist constants k_i, not all zero, such that

$$k_1\mathbf{w}_1 + \cdots + k_m\mathbf{w}_m = \mathbf{0}$$

But if we let $\mathbf{u}_1, \ldots, \mathbf{u}_{n-m}$ denote the vectors which are in S but not in T, then

$$k_1\mathbf{w}_1 + \cdots + k_m\mathbf{w}_m + 0\mathbf{u}_1 + \cdots + 0\mathbf{u}_{n-m} = \mathbf{0}$$

Thus we have a linear combination of the vectors $\mathbf{v}_1, \ldots, \mathbf{v}_n$ which equals $\mathbf{0}$. Since not all of the constants are zero, it follows that S is not a linearly independent set of vectors, contrary to the hypothesis. That is, if S is a linearly independent set, then so is every non-empty subset T.

13. This is similar to Problem 10. Since $\{v_1, \mathbf{v}_2, \ldots, \mathbf{v}_r\}$ is a linearly dependent set of vectors, there exist constants $c_1, c_2, \ldots, c_r$ not all zero such that

$$c_1\mathbf{v}_1 + c_2\mathbf{v}_2 + \cdots + c_r\,\mathbf{v}_r = \mathbf{0}$$

But then

$$c_1\mathbf{v}_1 + c_2\mathbf{v}_2 + \cdots + c_r\mathbf{v}_r + 0\mathbf{v}_{r+1} + \cdots + 0\mathbf{v}_n = \mathbf{0}$$

The above equation implies that the vectors $\mathbf{v}_1, \ldots, \mathbf{v}_n$ are linearly dependent.

15. Suppose that $\{\mathbf{v}_1, \mathbf{v}_2, \mathbf{v}_3\}$ is linearly dependent. Then there exist constants a, b, and c not all zero such that

(*) $$a\mathbf{v}_1 + b\mathbf{v}_2 + c\mathbf{v}_3 = \mathbf{0}$$

Case 1: $c = 0$. Then (*) becomes

$$a\mathbf{v}_1 + b\mathbf{v}_2 = \mathbf{0}$$

where not both a and b are zero. But then $\{\mathbf{v}_1, \mathbf{v}_2\}$ is linearly dependent, contrary to hypothesis.

Case 2: $c \neq 0$. Then solving (*) for $\mathbf{v}_3$ yields

$$\mathbf{v}_3 = -\frac{a}{c}\mathbf{v}_1 - \frac{b}{c}\mathbf{v}_2$$

This equation implies that $\mathbf{v}_3$ is in span$\{\mathbf{v}_1, \mathbf{v}_2\}$, contrary to hypothesis. Thus, $\{\mathbf{v}_1, \mathbf{v}_2, \mathbf{v}_3\}$ is linearly independent.

21. **(a)** The Wronskian is

$$\begin{vmatrix} 1 & x & e^x \\ 0 & 1 & e^x \\ 0 & 0 & e^x \end{vmatrix} = e^x \not\equiv 0$$

Thus the vectors are linearly independent.

(b) The Wronskian is

$$\begin{vmatrix} \sin x & \cos x & x\sin x \\ \cos x & -\sin x & \sin x + x\cos x \\ -\sin x & -\cos x & 2\cos x - x\sin x \end{vmatrix} = \begin{vmatrix} \sin x & \cos x & x\sin x \\ \cos x & -\sin x & \sin x + x\cos x \\ 0 & 0 & 2\cos x \end{vmatrix}$$

$$= 2\cos x\,(-\sin^2 x - \cos^2 x) = -2\cos x \not\equiv 0$$

Thus the vectors are linearly independent.

23. Use Theorem 5.3.1, Part (a).

EXERCISE SET 5.4

3. **(a)** This set has the correct number of vectors and they are linearly independent because

$$\begin{vmatrix} 1 & 2 & 3 \\ 0 & 2 & 3 \\ 0 & 0 & 3 \end{vmatrix} = 6 \neq 0$$

Hence, the set is a basis.

(c) The vectors in this set are linearly dependent because

$$\begin{vmatrix} 2 & 4 & 0 \\ -3 & 1 & -7 \\ 1 & 1 & 1 \end{vmatrix} = 0$$

Hence, the set is not a basis.

5. The set has the correct number of vectors. To show that they are linearly independent, we consider the equation

$$a\begin{bmatrix} 3 & 6 \\ 3 & -6 \end{bmatrix} + b\begin{bmatrix} 0 & -1 \\ -1 & 0 \end{bmatrix} + c\begin{bmatrix} 0 & -8 \\ -12 & -4 \end{bmatrix} + d\begin{bmatrix} 1 & 0 \\ -1 & 2 \end{bmatrix} = \begin{bmatrix} 0 & 0 \\ 0 & 0 \end{bmatrix}$$

If we add matrices and equate corresponding entries, we obtain the following system of equations:

$$\begin{aligned} 3a \quad\quad\quad\quad\quad + d &= 0 \\ 6a - b - 8c \quad\quad\quad &= 0 \\ 3a - b - 12c - d &= 0 \\ -6a \quad\quad - 4c + 2d &= 0 \end{aligned}$$

Since the determinant of the coefficient matrix is nonzero, the system of equations has only the trivial solution; hence, the vectors are linearly independent.

7. **(a)** Clearly $\mathbf{w} = 3\mathbf{u}_1 - 7\mathbf{u}_2$, so the coordinate vector relative to $\{\mathbf{u}_1, \mathbf{u}_2\}$ is $(3, -7)$.

(b) If $\mathbf{w} = a\mathbf{u}_1 + b\mathbf{u}_2$, then equating coordinates yields the system of equations

$$\begin{aligned} 2a + 3b &= 1 \\ -4a + 8b &= 1 \end{aligned}$$

This system has the solution $a = 5/28$, $b = 3/14$. Thus the desired coordinate vector is $(5/28, 3/14)$.

9. **(a)** If $\mathbf{v} = a\mathbf{v}_1 + b\mathbf{v}_2 + c\mathbf{v}_3$, then

$$\begin{aligned} a + 2b + 3c &= 2 \\ 2b + 3c &= -1 \\ 3c &= 3 \end{aligned}$$

From the third equation, $c = 1$. Plugging this value into the second equation yields $b = -2$, and finally, the first equation yields $a = 3$. Thus the desired coordinate vector is $(3, -2, 1)$.

15. If we reduce the augmented matrix to row-echelon form, we obtain

$$\begin{bmatrix} 1 & -3 & 1 & 0 \\ 0 & 0 & 0 & 0 \\ 0 & 0 & 0 & 0 \end{bmatrix}$$

Thus $x_1 = 3r - s$, $x_2 = r$, and $x_3 = s$, and the solution vector is

$$\begin{bmatrix} x_1 \\ x_2 \\ x_3 \end{bmatrix} = \begin{bmatrix} 3r - s \\ r \\ s \end{bmatrix} = \begin{bmatrix} 3 \\ 1 \\ 0 \end{bmatrix} r + \begin{bmatrix} -1 \\ 0 \\ 1 \end{bmatrix} s$$

Since (3, 1, 0) and (–1, 0, 1) are linearly independent, they form a basis for the solution space and the dimension of the solution space is 2.

19. **(a)** Any two linearly independent vectors in the plane form a basis. For instance, (1, –1, –1) and (0, 5, 2) are a basis because they satisfy the plane equation and neither is a multiple of the other.

(c) Any nonzero vector which lies on the line forms a basis. For instance, (2, –1, 4) will work, as will any nonzero multiple of this vector.

(d) The vectors (1, 1, 0) and (0, 1, 1) form a basis because they are linearly independent and

$$a(1, 1, 0) + c(0, 1, 1) = (a, a + c, c)$$

21. **(a)** We consider the three linear systems

$$\begin{aligned} -k_1 + k_2 &= 1 \quad 0 \quad 0 \\ 2k_1 - 2k_2 &= 0 \quad 1 \quad 0 \\ 3k_1 - 2k_2 &= 0 \quad 0 \quad 1 \end{aligned}$$

which give rise to the matrix

$$\begin{bmatrix} -1 & 1 & 1 & 0 & 0 \\ 2 & -2 & 0 & 1 & 0 \\ 3 & -2 & 0 & 0 & 1 \end{bmatrix}$$

A row-echelon form of the matrix is

$$\begin{bmatrix} 1 & -1 & -1 & 0 & 0 \\ 0 & 1 & 3 & 0 & 1 \\ 0 & 0 & 1 & 1/2 & 0 \end{bmatrix}$$

from which we conclude that $\mathbf{e}_3$ is in the span of $\{\mathbf{v}_1, \mathbf{v}_2\}$, but $\mathbf{e}_1$ and $\mathbf{e}_2$ are not. Thus $\{\mathbf{v}_1, \mathbf{v}_2, \mathbf{e}_1\}$ and $\{\mathbf{v}_1, \mathbf{v}_2, \mathbf{e}_2\}$ are both bases for R^3.

23. Since $\{\mathbf{u}_1, \mathbf{u}_2, \mathbf{u}_3\}$ has the correct number of vectors, we need only show that they are linearly independent. Let

$$a\mathbf{u}_1 + b\mathbf{u}_2 + c\mathbf{u}_3 = \mathbf{0}$$

Thus

$$a\mathbf{v}_1 + b(\mathbf{v}_1 + \mathbf{v}_2) + c(\mathbf{v}_1 + \mathbf{v}_2 + \mathbf{v}_3) = \mathbf{0}$$

or

$$(a + b + c)\mathbf{v}_1 + (b + c)\mathbf{v}_2 + c\mathbf{v}_3 = \mathbf{0}$$

Since $\{\mathbf{v}_1, \mathbf{v}_2, \mathbf{v}_3\}$ is a linearly independent set, the above equation implies that $a + b + c = b + c = c = 0$. Thus, $a = b = c = 0$ and $\{\mathbf{u}_1, \mathbf{u}_2, \mathbf{u}_3\}$ is also linearly independent.

25. First notice that if $\mathbf{v}$ and $\mathbf{w}$ are vectors in V and a and b are scalars, then $(a\mathbf{v} + b\mathbf{w})_S = a(\mathbf{v})_S + b(\mathbf{w})_S$. This follows from the definition of coordinate vectors. Clearly, this result applies to any finite sum of vectors. Also notice that if $(\mathbf{v})_S = (\mathbf{0})_S$, then $\mathbf{v} = \mathbf{0}$. Why?

Now suppose that $k_1\mathbf{v}_1 + \cdots + k_r\mathbf{v}_r = \mathbf{0}$. Then

$$\begin{aligned}(k_1\mathbf{v}_1 + \cdots + k_r\mathbf{v}_r)_S &= k_1(\mathbf{v}_1)_S + \cdots + k_r(\mathbf{v}_r)_S \\ &= (\mathbf{0})_S\end{aligned}$$

Conversely, if $k_1(\mathbf{v}_1)_S + \cdots + k_r(\mathbf{v}_r)_S = (\mathbf{0})_S$, then

$$(k_1\mathbf{v}_1 + \cdots + k_r\mathbf{v}_r)_S = (\mathbf{0})_S, \quad \text{or} \quad k_1\mathbf{v}_1 + \cdots + k_r\mathbf{v}_r = \mathbf{0}$$

Thus the vectors $\mathbf{v}_1, \ldots, \mathbf{v}_r$ are linearly independent in V if and only if the coordinate vectors $(\mathbf{v}_1)_S, \ldots, (\mathbf{v}_r)_S$ are linearly independent in R^n.

27. **(a)** Let $\mathbf{v}_1$, $\mathbf{v}_2$, and $\mathbf{v}_3$ denote the vectors. Since $S = \{1, x, x^2\}$ is the standard basis for P_2, we have $(\mathbf{v}_1)_S = (-1, 1, -2)$, $(\mathbf{v}_2)_S = (3, 3, 6)$, and $(\mathbf{v}_3)_S = (9, 0, 0)$. Since $\{(-1, 1, -2), (3, 3, 6), (9, 0, 0)\}$ is a linearly independent set of three vectors in R^3, then it spans R^3. Thus, by Exercises 24 and 25, $\{\mathbf{v}_1, \mathbf{v}_2, \mathbf{v}_3\}$ is linearly independent and spans P_2. Hence it is a basis for P_2.

31. There is. Consider, for instance, the set of matrices

$$A = \begin{bmatrix} 0 & 1 \\ 1 & 1 \end{bmatrix} \quad B = \begin{bmatrix} 1 & 0 \\ 1 & 1 \end{bmatrix}$$

$$C = \begin{bmatrix} 1 & 1 \\ 0 & 1 \end{bmatrix} \quad \text{and} \quad D = \begin{bmatrix} 1 & 1 \\ 1 & 0 \end{bmatrix}$$

Each of these matrices is clearly invertible. To show that they are linearly independent, consider the equation

$$a\text{A} + b\text{B} + c\text{C} + d\text{D} = \begin{bmatrix} 0 & 0 \\ 0 & 0 \end{bmatrix}$$

This implies that

$$\begin{bmatrix} 0 & 1 & 1 & 1 \\ 1 & 0 & 1 & 1 \\ 1 & 1 & 0 & 1 \\ 1 & 1 & 1 & 0 \end{bmatrix} \begin{bmatrix} a \\ b \\ c \\ d \end{bmatrix} = \begin{bmatrix} 0 \\ 0 \\ 0 \\ 0 \end{bmatrix}$$

The above 4×4 matrix is invertible, and hence $a = b = c = d = 0$ is the only solution. And since the set $\{A, B, C, \text{ and } D\}$ consists of 4 linearly independent vectors, it forms a basis for M_{22}.

33. **(a)** The set has 10 elements in a 9 dimensional space.

35. **(b)** The equation $x_1 + x_2 + \cdots + x_n = 0$ can be written as $x_1 = -x_2 - x_3 - \cdots - x_n$ where x_2, $x_3, \ldots, x_n$ can all be assigned arbitrary values. Thus, its solution space should have dimension $n - 1$. To see this, we can write

$$\begin{bmatrix} x_1 \\ x_2 \\ x_3 \\ \vdots \\ x_n \end{bmatrix} = \begin{bmatrix} -x_2 - x_3 - \cdots - x_n \\ x_2 \\ x_3 \\ \vdots \\ x_n \end{bmatrix}$$

$$= x_2 \begin{bmatrix} -1 \\ 1 \\ 0 \\ 0 \\ \vdots \\ 0 \end{bmatrix} + x_3 \begin{bmatrix} -1 \\ 0 \\ 1 \\ 0 \\ \vdots \\ 0 \end{bmatrix} + \cdots + x_n \begin{bmatrix} -1 \\ 0 \\ 0 \\ 0 \\ \vdots \\ 1 \end{bmatrix}$$

The $n - 1$ vectors in the above equation are linearly independent, so the vectors do form a basis for the solution space.

EXERCISE SET 5.5

3. **(b)** Since the equation $A\mathbf{x} = \mathbf{b}$ has no solution, $\mathbf{b}$ is not in the column space of A.

(c) Since $A\begin{bmatrix} 1 \\ -3 \\ 1 \end{bmatrix} = \mathbf{b}$, we have $\mathbf{b} = \mathbf{c}_1 - 3\mathbf{c}_2 + \mathbf{c}_3$.

(d) Since $A\begin{bmatrix} 1 \\ t-1 \\ t \end{bmatrix} = \mathbf{b}$, we have $\mathbf{b} = \mathbf{c}_1 + (t-1)\mathbf{c}_2 + t\mathbf{c}_3$ for all real numbers t.

5. **(a)** The general solution is $x_1 = 1 + 3t$, $x_2 = t$. Its vector form is

$$\begin{bmatrix} 1 \\ 0 \end{bmatrix} + t\begin{bmatrix} 3 \\ 1 \end{bmatrix}$$

Thus the vector form of the general solution to $A\mathbf{x} = \mathbf{0}$ is

$$t\begin{bmatrix} 3 \\ 1 \end{bmatrix}$$

(c) The general solution is $x_1 = -1 + 2r - s - 2t$, $x_2 = r$, $x_3 = s$, $x_4 = t$. Its vector form is

$$\begin{bmatrix} -1 \\ 0 \\ 0 \\ 0 \end{bmatrix} + r\begin{bmatrix} 2 \\ 1 \\ 0 \\ 0 \end{bmatrix} + s\begin{bmatrix} -1 \\ 0 \\ 1 \\ 0 \end{bmatrix} + t\begin{bmatrix} -2 \\ 0 \\ 0 \\ 1 \end{bmatrix}$$

Thus the vector form of the general solution to $A\mathbf{x} = \mathbf{0}$ is

$$r\begin{bmatrix} 2 \\ 1 \\ 0 \\ 0 \end{bmatrix} + s\begin{bmatrix} -1 \\ 0 \\ 1 \\ 1 \end{bmatrix} + t\begin{bmatrix} -2 \\ 0 \\ 0 \\ 1 \end{bmatrix}$$

9. **(a)** One row-echelon form of A^T is

$$\begin{bmatrix} 1 & 5 & 7 \\ 0 & 1 & 1 \\ 0 & 0 & 0 \end{bmatrix}$$

Thus a basis for the column space of A is

$$\begin{bmatrix} 1 \\ 5 \\ 7 \end{bmatrix} \text{ and } \begin{bmatrix} 0 \\ 1 \\ 1 \end{bmatrix}$$

(c) One row-echelon form of A^T is

$$\begin{bmatrix} 1 & 2 & -1 \\ 0 & 1 & -1 \\ 0 & 0 & 0 \\ 0 & 0 & 0 \end{bmatrix}$$

Thus a basis for the column space of A is

$$\begin{bmatrix} 1 \\ 2 \\ -1 \end{bmatrix} \text{ and } \begin{bmatrix} 0 \\ 1 \\ -1 \end{bmatrix}$$

11. **(a)** The space spanned by these vectors is the row space of the matrix

$$\begin{bmatrix} 1 & 1 & -4 & -3 \\ 2 & 0 & 2 & -2 \\ 2 & -1 & 3 & 2 \end{bmatrix}$$

One row-echelon form of the above matrix is

$$\begin{bmatrix} 1 & 1 & -4 & -3 \\ 0 & 1 & -5 & -2 \\ 0 & 0 & 1 & -1/2 \end{bmatrix}$$

and the reduced row-echelon form is

$$\begin{bmatrix} 1 & 0 & 0 & -1/2 \\ 0 & 1 & 0 & -9/2 \\ 0 & 0 & 1 & -1/2 \end{bmatrix}$$

Thus {(1, 1, –4, –3), (0, 1, –5, –2), (0, 0, 1, –1/2)} is one basis. Another basis is {(1, 0, 0, –1/2), (0, 1, 0, –9/2), (0, 0, 1, –1/2)}.

13. Let A be an $n \times n$ invertible matrix. Since A^T is also invertible, it is row equivalent to I_n. It is clear that the column vectors of I_n are linearly independent. Hence, by virtue of Theorem 5.5.5, the column vectors of A^T, which are just the row vectors of A, are also linearly independent. Therefore the rows of A form a set of n linearly independent vectors in R^n, and consequently form a basis for R^n.

15. **(a)** We are looking for a matrix so that the only solution to the equation $A\mathbf{x} = \mathbf{0}$ is $\mathbf{x} = \mathbf{0}$.

Any invertible matrix will satisfy this condition. For example, the nullspace of the matrix $A = \begin{bmatrix} 1 & 0 & 0 \\ 0 & 1 & 0 \\ 0 & 0 & 1 \end{bmatrix}$ is the single point (0, 0, 0).

(b) In this case, we are looking for a matrix so that the solution of $A\mathbf{x} = \mathbf{0}$ is one-dimensional. Thus, the reduced row-echelon form of A has one column without a leading one. As an example, the nullspace of the matrix $A = \begin{bmatrix} 1 & 0 & -1 \\ 0 & 1 & -1 \\ 0 & 0 & 0 \end{bmatrix}$ is span $\left\{ \begin{bmatrix} 1 \\ 1 \\ 1 \end{bmatrix} \right\}$, a line in R^3.

(c) In this case, we are looking for a matrix so that the solution space of $A\mathbf{x} = \mathbf{0}$ is two-dimensional. Thus, the reduced row-echelon form of A has two columns without leading ones. As an example, the nullspace of the matrix $A = \begin{bmatrix} 1 & 1 & -1 \\ 0 & 0 & 0 \\ 0 & 0 & 0 \end{bmatrix}$ is $\text{span}\left\{\begin{bmatrix} -1 \\ 1 \\ 0 \end{bmatrix}, \begin{bmatrix} 1 \\ 0 \\ 1 \end{bmatrix}\right\}$, a plane in R^3.

17. **(a)** The matrices will all have the form $\begin{bmatrix} 3s-5s \\ 3t-5t \end{bmatrix} = s\begin{bmatrix} 3 & -5 \\ 0 & 0 \end{bmatrix} + t\begin{bmatrix} 0 & 0 \\ 3 & -5 \end{bmatrix}$ where s and t are any real numbers.

(b) Since A and B are invertible, their nullspaces are the origin. The nullspace of C is the line $3x + y = 0$. The nullspace of D is the entire xy-plane.

19. Theorem: If A and B are $n \times n$ matrices and A is invertible, then the row space of AB is the row space of B.

Proof: If A is invertible, then there exist elementary matrices $E_1, E_2, \ldots, E_k$ such that

$$A = E_1E_2 \ldots E_kI_n$$

or

$$AB = E_1E_2 \ldots E_kB$$

Thus, Theorem 5.5.4 guarantees that AB and B will have the same row spaces.

EXERCISE SET 5.6

7. Use Theorems 5.6.5 and 5.6.7.

(a) The system is consistent because the two ranks are equal. Since $n = r = 3$, $n - r = 0$ and therefore the number of parameters is 0.

(b) The system is inconsistent because the two ranks are not equal.

(d) The system is consistent because the two ranks are equal. Here $n = 9$ and $r = 2$, so that $n - r = 7$ parameters will appear in the solution.

(f) Since the ranks are equal, the system is consistent. However A must be the zero matrix, so the system gives no information at all about its solution. This is reflected in the fact that $n - r = 4 - 0 = 4$, so that there will be 4 parameters in the solution for the 4 variables.

9. The system is of the form $A\mathbf{x} = \mathbf{b}$ where rank$(A) = 2$. Therefore it will be consistent if and only if rank$([A|\mathbf{b}]) = 2$. Since $[A|\mathbf{b}]$ reduces to

$$\begin{bmatrix} 1 & -3 & b_1 \\ 0 & 1 & b_2 - b_1 \\ 0 & 0 & b_3 - 4b_2 + 3b_1 \\ 0 & 0 & b_4 + b_2 - 2b_1 \\ 0 & 0 & b_5 - 8b_2 + 7b_1 \end{bmatrix}$$

the system will be consistent if and only if $b_3 = 4b_2 - 3b_1$, $b_4 = -b_2 + 2b_1$, and $b_5 = 8b_2 - 7b_1$, where b_1 and b_2 can assume any values.

11. If the nullspace of A is a line through the origin, then it has the form $x = at$, $y = bt$, $z = ct$ where t is the only parameter. Thus nullity$(A) = 3 - \text{rank}(A) = 1$. That is, the row and column spaces of A have dimension 2, so neither space can be a line. Why?

13. Call the matrix A. If $r = 2$ and $s = 1$, then clearly rank$(A) = 2$. Otherwise, either $r - 2$ or $s - 1 \neq 0$ and rank$(A) = 3$. Rank(A) can never be 1.

17. **(a)** False. Let $A = \begin{bmatrix} 1 & 0 & 0 \\ 0 & 1 & 0 \end{bmatrix}$

(c) True. If A were an $m \times n$ matrix where, say, $m > n$, then it would have m rows, each of which would be a vector in R_n. Thus, by Theorem 5.4.2, they would form a linearly dependent set.

SUPPLEMENTARY EXERCISES 5

1. **(b)** The augmented matrix of this system reduces to

$$\begin{bmatrix} 2 & -3 & 1 & 0 \\ 0 & 0 & 0 & 0 \\ 0 & 0 & 0 & 0 \end{bmatrix}$$

Therefore, the solution space is a plane with equation $2x - 3y + z = 0$

(c) The solution is $x = 2t$, $y = t$, $z = 0$, which is a line.

5. **(a)** We look for constants a, b, and c such that $\mathbf{v} = a\mathbf{v}_1 + b\mathbf{v}_2 + c\mathbf{v}_3$, or

$$\begin{aligned} a + 3b + 2c &= 1 \\ -a \qquad\quad + c &= 1 \end{aligned}$$

This system has the solution

$$a = t - 1 \qquad b = \frac{2}{3} - t \qquad c = t$$

where t is arbitrary. If we set $t = 0$ and $t = 1$, we obtain $\mathbf{v} = (-1)\mathbf{v}_1 + (2/3)\mathbf{v}_2$ and $\mathbf{v} = (-1/3)\mathbf{v}_2 + \mathbf{v}_3$, respectively. There are infinitely many other possibilities.

(b) Since $\mathbf{v}_1$, $\mathbf{v}_2$, and $\mathbf{v}_3$ all belong to R^2 and $\dim(R^2) = 2$, it follows from Theorem 5.4.2 that these three vectors do not form a basis for R^2. Hence, Theorem 5.4.1 does not apply.

7. Consider the polynomials x and $x + 1$ in P_1. Verify that these polynomials form a basis for P_1.

13. **(a)** Since $\begin{vmatrix} 1 & 0 \\ 2 & -1 \end{vmatrix} = -1 \neq 0$, the rank is 2.

(b) Since all three 2×2 subdeterminants are zero, the rank is 1.

(c) Since the determinant of the matrix is zero, its rank is less than 3. Since $\begin{vmatrix} 1 & 0 \\ 2 & -1 \end{vmatrix} = -1 \neq 0$, the rank is 2.

(d) Since the determinant of the 3×3 submatrix obtained by deleting the last column is $30 \neq 0$, the rank of the matrix is 3.

15. **(b)** Let $S = \{\mathbf{v}_1, \ldots, \mathbf{v}_n\}$ and let $\mathbf{u} = u_1\mathbf{v}_1 + \cdots + \mathbf{u}_n\mathbf{v}_n$. Thus $(\mathbf{u})_S = (u_1, \ldots, u_n)$. We have

$$k\mathbf{u} = ku_1\mathbf{v}_1 + \cdots + ku_n\mathbf{v}_n$$

so that $(k\mathbf{u})_S = (ku_1, \ldots, ku_n) = k(\mathbf{u}_1, \ldots, \mathbf{u}_n)$. Therefore $(\mathbf{ku})_S = \mathrm{k}(\mathbf{u})_S$.

EXERCISE SET 6.1

1. **(c)** Since $\mathbf{v} + \mathbf{w} = (3, 11)$, we have

$$\langle \mathbf{u}, \mathbf{v} + \mathbf{w} \rangle = 3(3) + (-2)(11) = -13$$

On the other hand,

$$\langle \mathbf{u}, \mathbf{v} \rangle = 3(4) + (-2)(5) = 2$$

and

$$\langle \mathbf{u}, \mathbf{w} \rangle = 3(-1) + (-2)(6) = -15$$

(d) Since $k\mathbf{u} = (-12, 8)$ and $k\mathbf{v} = (-16, -20)$, we have

$$\langle k\mathbf{u}, \mathbf{v} \rangle = (-12)(4) + (8)(5) = -8$$

and

$$\langle \mathbf{u}, k\mathbf{v} \rangle = 3(-16) + (-2)(-20) = -8$$

Since $\langle \mathbf{u},\mathbf{v} \rangle = 2$, $k\langle \mathbf{u}, \mathbf{v} \rangle = -8$.

3. **(a)** $\langle \mathbf{u}, \mathbf{v} \rangle = 3(-1) - 2(3) + 4(1) + 8(1) = 3$

5. **(a)** By Formula (4),

$$\langle \mathbf{u}, \mathbf{v} \rangle = \begin{bmatrix} v_1 & v_2 \end{bmatrix} \begin{bmatrix} 3 & 0 \\ 0 & 2 \end{bmatrix} \begin{bmatrix} 3 & 0 \\ 0 & 2 \end{bmatrix} \begin{bmatrix} u_1 \\ u_2 \end{bmatrix}$$

$$= \begin{bmatrix} v_1 & v_2 \end{bmatrix} \begin{bmatrix} 9 & 0 \\ 0 & 4 \end{bmatrix} \begin{bmatrix} u_1 \\ u_2 \end{bmatrix}$$

$$= \begin{bmatrix} 9v_1 & 4v_2 \end{bmatrix} \begin{bmatrix} u_1 \\ u_2 \end{bmatrix}$$

$$= 9u_1v_1 + 4u_2v_2$$

(b) We have $\langle \mathbf{u}, \mathbf{v} \rangle = 9(-3)(1) + 4(2)(7) = 29$.

7. **(a)** By Formula (4), we have $\langle \mathbf{u}, \mathbf{v} \rangle = \mathbf{v}^T A^T A\mathbf{u}$ where

$$A = \begin{bmatrix} \sqrt{3} & 0 \\ 0 & \sqrt{5} \end{bmatrix}$$

9. **(b)** Axioms 1 and 4 are easily checked. However, if $\mathbf{w} = (w_1, w_2, w_3)$, then

$$\langle \mathbf{u} + \mathbf{v}, \mathbf{w} \rangle = (u_1 + v_1)^2 w_1^2 + (u_2 + v_2)^2 w_2^2 + (u_3 + v_3)^2 w_3^2$$

$$= \langle \mathbf{u}, \mathbf{w} \rangle + \langle \mathbf{v}, \mathbf{w} \rangle + 2u_1v_1w_1^2 + 2u_2v_2w_2^2 + 2u_3v_3w_3^2$$

If, for instance, $\mathbf{u} = \mathbf{v} = \mathbf{w} = (1, 0, 0)$, then Axiom 2 fails.

To check Axiom 3, we note that $\langle k\mathbf{u}, \mathbf{v} \rangle = k^2 \langle \mathbf{u}, \mathbf{v} \rangle$. Thus $\langle k\mathbf{u}, \mathbf{v} \rangle \neq k\langle \mathbf{u}, \mathbf{v} \rangle$ unless $k = 0$ or $k = 1$, so Axiom 3 fails.

(c) (1) Axiom 1 follows from the commutativity of multiplication in R.

(2) If $\mathbf{w} = (w_1, w_2, w_3)$, then

$$\langle \mathbf{u} + \mathbf{v}, \mathbf{w} \rangle = 2(u_1 + v_1)w_1 + (u_2 + v_2)w_2 + 4(u_3 + v_3)w_3$$

$$= 2u_1w_1 + u_2w_2 + 4u_3w_3 + 2v_1w_1 + v_2w_2 + 4v_3w_3$$

$$= \langle \mathbf{u}, \mathbf{w} \rangle + \langle \mathbf{v}, \mathbf{w} \rangle$$

(3) $\langle k\mathbf{u}, \mathbf{v} \rangle = 2(ku_1)v_1 + (ku_2)v_2 + 4(ku_3)v_3 = k\langle \mathbf{u}, \mathbf{v} \rangle$

(4) $\langle \mathbf{v}, \mathbf{v} \rangle = 2v_1^2 + v_2^2 + 4v_3^2 \geq 0$

$= 0$ if and only if $v_1 = v_2 = v_3 = 0$, or $\mathbf{v} = \mathbf{0}$

Thus this is an inner product for R^3.

11. We have $\mathbf{u} - \mathbf{v} = (-3, -3)$.

(b) $d(\mathbf{u}, \mathbf{v}) = \|(-3, -3)\| = [3(9) + 2(9)]^{1/2} = \sqrt{45} = 3\sqrt{5}$

(c) From Problem 10(c), we have

$$[d(\mathrm{u}, \mathrm{v})]^2 = \begin{bmatrix} -3 & -3 \end{bmatrix} \begin{bmatrix} 2 & -1 \\ -1 & 13 \end{bmatrix} \begin{bmatrix} -3 \\ -3 \end{bmatrix} = 117$$

Thus

$$d(\mathbf{u}, \mathbf{v}) = \sqrt{117} = 3\sqrt{13}$$

13. **(a)** $\|A\| = [(-2)^2 + (5)^2 + (3)^2 + (6)^2]^{1/2} = \sqrt{74}$

15. **(a)** Since $A - B = \begin{bmatrix} 6 & -1 \\ 8 & -2 \end{bmatrix}$, we have

$$d(A, B) = \langle A - B, A - B \rangle^{1/2} = [6^2 + (-1)^2 + 8^2 + (-2)^2]^{1/2} = \sqrt{105}$$

17. **(a)** For instance, $\|x\| = \left(\int_{-1}^{1} x^2 dx \right)^{1/2} = \left(\left. \frac{x^3}{3} \right]_{-1}^{1} \right)^{1/2} = \left(\frac{2}{3} \right)^{1/2}$

(b) We have

$$\begin{aligned} d(\mathbf{p}, \mathbf{q}) &= \|\mathbf{p} - \mathbf{q}\| \\ &= \|1 - x\| \\ &= \left(\int_{-1}^{1} (1-x)^2 dx \right)^{1/2} \\ &= \left(\int_{-1}^{1} (1 - 2x + x^2) dx \right)^{1/2} \\ &= \left(\left. \left(x - x^2 + \frac{x^3}{3} \right) \right]_{-1}^{1} \right)^{1/2} \\ &= 2\left(\frac{2}{3} \right)^{1/2} = \frac{2}{3}\sqrt{6} \end{aligned}$$

21. If, in the solution to Exercise 20, we subtract (**) from (*) and divide by 4, we obtain the desired result.

23. Axioms 1 and 3 are easily verified. So is Axiom 2, as shown: Let $\mathbf{r} = r(x)$ be a polynomial in P_2. Then

$$\begin{aligned}\langle \mathbf{p} + \mathbf{q}, \mathbf{r}\rangle &= [(p + q)(0)]r(0) + [(p + q)(1/2)]r(1/2) + [(p + q)(1)]r(1) \\ &= p(0)r(0) + p(1/2)r(1/2) + p(1)r(1) + q(0)r(0) + q(1/2)r(1/2) + q(1)r(1) \\ &= \langle \mathbf{p}, \mathbf{r}\rangle + \langle \mathbf{q}, \mathbf{r}\rangle\end{aligned}$$

It remains to verify Axiom 4:

$$\langle \mathbf{p}, \mathbf{p}\rangle = [p(0)]^2 + [p(1/2)]^2 + [p(1)]^2 \geq 0$$

and

$$\langle \mathbf{p}, \mathbf{p}\rangle = 0 \quad \text{if and only if} \quad p(0) = p(1/2) = p(1) = 0$$

But a quadratic polynomial can have at most two zeros unless it is identically zero. Thus $\langle \mathbf{p}, \mathbf{p}\rangle = 0$ if and only if $\mathbf{p}$ is identically zero, or $\mathbf{p} = \mathbf{0}$.

27. **(b)** $\langle \mathbf{p}, \mathbf{q}\rangle = \int_{-1}^{1} (x - 5x^3)(2 + 8x^2)\, dx = \int_{-1}^{1} (2x - 2x^3 - 40x^5)\, dx$

$$= x^2 - x^4/2 - 20x^6/3\Big]_{-1}^{1} = 0$$

29. We have $\langle U, V\rangle = u_1v_1 + u_2v_2 + u_3v_3 + u_4v_4$ and

$$\begin{aligned}\text{tr}(U^TV) &= \text{tr}\left(\begin{bmatrix} u_1 & u_3 \\ u_2 & u_4 \end{bmatrix}\begin{bmatrix} v_1 & v_2 \\ v_3 & v_4 \end{bmatrix}\right) \\ &= \text{tr}\left(\begin{bmatrix} u_1v_1 + u_3v_3 & u_1v_2 + u_3v_4 \\ u_2v_1 + u_4v_3 & u_2v_2 + u_4v_4 \end{bmatrix}\right) \\ &= u_1v_1 + u_3v_3 + u_2v_2 + u_4v_4\end{aligned}$$

which does, indeed, equal $\langle U, V\rangle$.

31. Calling the matrix A, we have

$$\langle \mathbf{u}, \mathbf{v} \rangle = \mathbf{v}^T A^T A\mathbf{u} = \mathbf{v}^T A^2\mathbf{u} = w_1 u_1 v_1 + \cdots + w_n u_n v_n$$

33. To prove Part (a) of Theorem 6.1.1 first observe that $\langle \mathbf{0}, \mathbf{v} \rangle = \langle \mathbf{v}, \mathbf{0} \rangle$ by the symmetry axiom. Moreover,

$$\begin{aligned} \langle \mathbf{0}, \mathbf{v} \rangle &= \langle 0\mathbf{0}, \mathbf{v} \rangle && \text{by Theorem 5.1.1} \\ &= 0\langle \mathbf{0}, \mathbf{v} \rangle && \text{by the homogeneity axiom} \\ &= 0 \end{aligned}$$

Alternatively,

$$\begin{aligned} \langle \mathbf{0}, \mathbf{v} \rangle + \langle \mathbf{0}, \mathbf{v} \rangle &= \langle \mathbf{0} + \mathbf{0}, \mathbf{v} \rangle && \text{by additivity} \\ &= \langle \mathbf{0}, \mathbf{v} \rangle && \text{by definition of the zero vector} \end{aligned}$$

But $\langle \mathbf{0}, \mathbf{v} \rangle = 2\langle \mathbf{0}, \mathbf{v} \rangle$ only if $\langle \mathbf{0}, \mathbf{v} \rangle = 0$.

To prove Part (d), observe that, by Theorem 5.1.1, $-\mathbf{v}$ (the inverse of $\mathbf{v}$) and $(-1)\mathbf{v}$ are the same vector. Thus,

$$\begin{aligned} \langle \mathbf{u} - \mathbf{v}, \mathbf{w} \rangle &= \langle \mathbf{u} + (-\mathbf{v}), \mathbf{w} \rangle \\ &= \langle \mathbf{u}, \mathbf{w} \rangle + \langle -\mathbf{v}, \mathbf{w} \rangle && \text{by additivity} \\ &= \langle \mathbf{u}, \mathbf{w} \rangle - \langle \mathbf{v}, \mathbf{w} \rangle && \text{by homogeneity} \end{aligned}$$

EXERCISE SET 6.2

1. **(e)** Since $\mathbf{u} \cdot \mathbf{v} = 0 + 6 + 2 + 0 = 8$, the vectors are not orthogonal.

3. We have $k\mathbf{u} + \mathbf{v} = (k + 6, k + 7, -k - 15)$, so

$$\begin{aligned}\|k\mathbf{u} + \mathbf{v}\| &= \langle (k\mathbf{u} + \mathbf{v}), (k\mathbf{u} + \mathbf{v})\rangle^{1/2} \\ &= \left[(k+6)^2 + (k+7)^2 + (-k-15)^2\right]^{1/2} \\ &= (3k^2 + 56k + 310)^{1/2}\end{aligned}$$

Since $\|k\mathbf{u} + \mathbf{v}\| = 13$ exactly when $\|k\mathbf{u} + \mathbf{v}\|^2 = 169$, we need to solve the quadratic equation $3k^2 + 56k + 310 = 169$ to find k. Thus, values of k that give $\|k\mathbf{u} + \mathbf{v}\| = 13$ are $k = -3$ or $k = -47/3$.

5. **(a)** $$\cos\theta = \frac{\langle (1,-3),(2,4)\rangle}{\|(1,-3)\|\,\|(2,4)\|} = \frac{2-12}{\sqrt{10}\sqrt{20}} = \frac{-1}{\sqrt{2}}$$

(c) $$\cos\theta = \frac{\langle (-1,5,2),\ (2,4,-9)\rangle}{\|(-1,5,2)\|\,\|(2,4,-9)\|} = \frac{-2+20-18}{\sqrt{30}\,\sqrt{101}} = 0$$

(e) $$\cos\theta = \frac{\langle (1,0,1,0),(-3,-3,-3,-3)\rangle}{\|(1,0,1,0)\|\,\|(-3,-3,-3,-3)\|} = \frac{-3-3}{\sqrt{2}\sqrt{36}} = \frac{-1}{\sqrt{2}}$$

7. $\langle \mathbf{p}, \mathbf{q}\rangle = (1)(0) + (-1)(2) + (2)(1) = 0$

9. **(b)** $$\left\langle \begin{bmatrix} 2 & 1 \\ -1 & 3 \end{bmatrix}, \begin{bmatrix} 1 & 1 \\ 0 & -1 \end{bmatrix} \right\rangle = (2)(1) + (1)(1) + (-1)(0) + (3)(-1) = 0$$

Thus the matrices are orthogonal.

(d) $\left\langle \begin{bmatrix} 2 & 1 \\ -1 & 3 \end{bmatrix}, \begin{bmatrix} 2 & 1 \\ 5 & 2 \end{bmatrix} \right\rangle = 4 + 1 - 5 + 6 = 6 \neq 0$

Thus the matrices are not orthogonal.

11. We must find two vectors $\mathbf{x} = (x_1, x_2, x_3, x_4)$ such that $\langle \mathbf{x}, \mathbf{x} \rangle = 1$ and $\langle \mathbf{x}, \mathbf{u} \rangle = \langle \mathbf{x}, \mathbf{v} \rangle = \langle \mathbf{x}, \mathbf{w} \rangle = 0$. Thus x_1, x_2, x_3, and x_4 must satisfy the equations

$$\begin{aligned} x_1^2 + x_2^2 + x_3^2 + x_4^2 &= 1 \\ 2x_1 + x_2 - 4x_3 &= 0 \\ -x_1 - x_2 + 2x_3 + 2x_4 &= 0 \\ 3x_1 + 2x_2 + 5x_3 + 4x_4 &= 0 \end{aligned}$$

The solution to the three linear equations is $x_1 = -34t$, $x_2 = 44t$, $x_3 = -6t$, and $x_4 = 11t$. If we substitute these values into the quadratic equation, we get

$$[(-34)^2 + (44)^2 + (-6)^2 + (11)^2]\, t^2 = 1$$

or

$$t = \pm\frac{1}{\sqrt{57}}$$

Therefore, the two vectors are

$$\pm\frac{1}{\sqrt{57}}(-34, 44, -6, 11)$$

13. **(a)** Here $\langle \mathbf{u}, \mathbf{v} \rangle^2 = (3(-2)(1) + 2(1)(0))^2 = 36$, while, on the other hand, $\langle \mathbf{u}, \mathbf{u} \rangle \langle \mathbf{v}, \mathbf{v} \rangle = (3(-2)^2 + 2(1)^2)\,(3(1)^2 + 2(0)^2) = 42$.

15. **(a)** Here $W^{\perp}$ is the line which is normal to the plane and which passes through the origin. By inspection, a normal vector to the plane is $(1, -2, -3)$. Hence this line has parametric equations $x = t$, $y = -2t$, $z = -3t$.

17. **(a)** The subspace of R^3 spanned by the given vectors is the row space of the matrix

$$\begin{bmatrix} 1 & -1 & 3 \\ 5 & -4 & -4 \\ 7 & -6 & 2 \end{bmatrix} \quad \text{which reduces to} \quad \begin{bmatrix} 1 & -1 & 3 \\ 0 & 1 & -19 \\ 0 & 0 & 0 \end{bmatrix}$$

The space we are looking for is the nullspace of this matrix. From the reduced form, we see that the nullspace consists of all vectors of the form $(16, 19, 1)t$, so that the vector $(16, 19, 1)$ is a basis for this space.

Alternatively the vectors $\mathbf{w}_1 = (1, -1, 3)$ and $\mathbf{w}_2 = (0, 1, -19)$ form a basis for the row space of the matrix. They also span a plane, and the orthogonal complement of this plane is the line spanned by the normal vector $\mathbf{w}_1 \times \mathbf{w}_2 = (16, 19, 1)$.

19. If $\mathbf{u}$ and $\mathbf{v}$ are orthogonal vectors with norm 1, then

$$\begin{aligned} \|\mathbf{u} - \mathbf{v}\| &= \langle \mathbf{u} - \mathbf{v}, \mathbf{u} - \mathbf{v} \rangle^{1/2} \\ &= [\langle \mathbf{u}, \mathbf{u} \rangle - 2\langle \mathbf{u}, \mathbf{v} \rangle + \langle \mathbf{v}, \mathbf{v} \rangle]^{1/2} \\ &= [1 - 2(0) + 1]^{1/2} \\ &= \sqrt{2} \end{aligned}$$

21. By definition, $\mathbf{u}$ is in span $\{\mathbf{u}_1, \mathbf{u}_2, \ldots, \mathbf{u}_r\}$ if and only if there exist constants $c_1, c_2, \ldots, c_r$ such that

$$\mathbf{u} = c_1\mathbf{u}_1 + c_2\mathbf{u}_2 + \cdots + c_r\mathbf{u}_r$$

But if $\langle \mathbf{w}, \mathbf{u}_1 \rangle = \langle \mathbf{w}, \mathbf{u}_2 \rangle = \cdots = \langle \mathbf{w}, \mathbf{u}_r \rangle = 0$, then $\langle \mathbf{w}, \mathbf{u} \rangle = 0$.

23. We have that $W = \text{span}\{\mathbf{w}_1, \mathbf{w}_2, \ldots, \mathbf{w}_k\}$

Suppose that $\mathbf{w}$ is in $W^\perp$. Then, by definition, $\langle \mathbf{w}, \mathbf{w}_i \rangle = 0$ for each basis vector $\mathbf{w}_i$ of W.

Conversely, if a vector $\mathbf{w}$ of V is orthogonal to each basis vector of W, then, by Problem 20, it is orthogonal to every vector in W.

25. **(c)** By Property (3) in the definition of inner product, we have

$$\|k\mathbf{u}\|^2 = \langle k\mathbf{u}, k\mathbf{u} \rangle = k^2\langle \mathbf{u}, \mathbf{u} \rangle = k^2\|\mathbf{u}\|^2$$

Therefore $\|k\mathbf{u}\| = |k|\,\|\mathbf{u}\|$.

27. This is just the Cauchy-Schwarz inequality using the inner product on R^n generated by A (see Formula (4) of Section 6.1).

31. We wish to show that $\angle ABC$ is a right angle, or that $\overrightarrow{AB}$ and $\overrightarrow{BC}$ are orthogonal. Observe that $\overrightarrow{AB} = \mathbf{u} - (-\mathbf{v})$ and $\overrightarrow{BC} = \mathbf{v} - \mathbf{u}$ where $\mathbf{u}$ and $\mathbf{v}$ are radii of the circle, as shown in the figure. Thus $\|\mathbf{u}\| = \|\mathbf{v}\|$. Hence

$$\begin{aligned}\langle \overrightarrow{AB}, \overrightarrow{BC} \rangle &= \langle \mathbf{u} + \mathbf{v}, \mathbf{v} - \mathbf{u} \rangle \\ &= \langle \mathbf{u}, \mathbf{v} \rangle + \langle \mathbf{v}, \mathbf{v} \rangle + \langle \mathbf{u}, -\mathbf{u} \rangle + \langle \mathbf{v}, -\mathbf{u} \rangle \\ &= \langle \mathbf{v}, \mathbf{u} \rangle + \langle \mathbf{v}, \mathbf{v} \rangle - \langle \mathbf{u}, \mathbf{u} \rangle - \langle \mathbf{v}, \mathbf{u} \rangle \\ &= \|\mathbf{v}\|^2 - \|\mathbf{u}\|^2 \\ &= 0\end{aligned}$$

33. **(a)** As noted in Example 9 of Section 6.1, $\int_0^1 f(x)g(x)dx$ is an inner product on $C[0, 1]$. Thus the Cauchy-Schwarz Inequality must hold, and that is exactly what we're asked to prove.

(b) In the inner product notation, we must show that

$$\langle \mathbf{f} + \mathbf{g}, \mathbf{f} + \mathbf{g} \rangle^{1/2} \le \langle \mathbf{f}, \mathbf{f} \rangle^{1/2} + \langle \mathbf{g}, \mathbf{g} \rangle^{1/2}$$

or, squaring both sides, that

$$\langle \mathbf{f} + \mathbf{g}, \mathbf{f} + \mathbf{g} \rangle \le \langle \mathbf{f}, \mathbf{f} \rangle + 2\langle \mathbf{f}, \mathbf{f} \rangle^{1/2} \langle \mathbf{g}, \mathbf{g} \rangle^{1/2} + \langle \mathbf{g}, \mathbf{g} \rangle$$

For any inner product, we know that

$$\langle \mathbf{f} + \mathbf{g}, \mathbf{f} + \mathbf{g} \rangle = \langle \mathbf{f}, \mathbf{f} \rangle + 2\langle \mathbf{f}, \mathbf{g} \rangle + \langle \mathbf{g}, \mathbf{g} \rangle$$

By the Cauchy-Schwarz Inequality

$$\langle \mathbf{f}, \mathbf{g} \rangle^2 \le \langle \mathbf{f}, \mathbf{f} \rangle \langle \mathbf{g}, \mathbf{g} \rangle$$

or

$$\langle \mathbf{f}, \mathbf{g} \rangle \le \langle \mathbf{f}, \mathbf{f} \rangle^{1/2} \langle \mathbf{g}, \mathbf{g} \rangle^{1/2}$$

If we substitute the above inequality into the equation for $\langle \mathbf{f} + \mathbf{g}, \mathbf{f} + \mathbf{g} \rangle$, we obtain

$$\langle \mathbf{f} + \mathbf{g}, \mathbf{f} + \mathbf{g} \rangle \leq \langle \mathbf{f}, \mathbf{f} \rangle + 2\langle \mathbf{f}, \mathbf{f} \rangle^{1/2} \langle \mathbf{g}, \mathbf{g} \rangle^{1/2} + \langle \mathbf{g}, \mathbf{g} \rangle$$

as required.

35. **(a)** $W^{\perp}$ is the line $y = -x$.

(b) $W^{\perp}$ is the xz-plane.

(c) $W^{\perp}$ is the x-axis.

37. **(b)** False. Let $n = 3$, let V be the xy-plane, and let W be the x-axis. Then $V^{\perp}$ is the z-axis and $W^{\perp}$ is the yz-plane. In fact $V^{\perp}$ is a subspace of $W^{\perp}$

(c) True. The two spaces are orthogonal complements and the only vector orthogonal to itself is the zero vector.

(d) False. For instance, if A is invertible, then both its row space and its column space are all of R^n.

EXERCISE SET 6.3

5. See Exercise 3, Parts (b) and (c).

7. **(b)** Call the vectors $\mathbf{u}_1$, $\mathbf{u}_2$ and $\mathbf{u}_3$. Then $\langle \mathbf{u}_1, \mathbf{u}_2 \rangle = 2 - 2 = 0$ and $\langle \mathbf{u}_1, \mathbf{u}_3 \rangle = \langle \mathbf{u}_2, \mathbf{u}_3 \rangle = 0$. The set is therefore orthogonal. Moreover, $\|\mathbf{u}_1\| = \sqrt{2}$, $\|\mathbf{u}_2\| = \sqrt{8} = 2\sqrt{2}$, and $\|\mathbf{u}_3\| = \sqrt{25} = 5$. Thus $\left\{ \frac{1}{\sqrt{2}}\mathbf{u}_1, \frac{1}{2\sqrt{2}}\mathbf{u}_2, \frac{1}{5}\mathbf{u}_3 \right\}$ is an orthonormal set.

9. It is easy to verify that $\mathbf{v}_1 \cdot \mathbf{v}_2 = \mathbf{v}_1 \cdot \mathbf{v}_3 = \mathbf{v}_2 \cdot \mathbf{v}_3 = 0$ and that $\|\mathbf{v}_3\| = 1$. Moreover, $\|\mathbf{v}_1\|^2 = (-3/5)^2 + (4/5)^2 = 1$ and $\|\mathbf{v}_2\| = (4/5)^2 + (3/5)^2 = 1$. Thus $\{\mathbf{v}_1, \mathbf{v}_2, \mathbf{v}_3\}$ is an orthonormal set in R^3. It will be an orthonormal basis provided that the three vectors are linearly independent, which is guaranteed by Theorem 6.3.3.

(b) By Theorem 6.3.1, we have

$$\begin{aligned}(3,-7,4) &= \left(-\frac{9}{5} - \frac{28}{5} + 0\right)\mathbf{v}_1 + \left(\frac{12}{5} - \frac{21}{5} + 0\right)\mathbf{v}_2 + 4\mathbf{v}_3 \\ &= (-37/5)\mathbf{v}_1 + (-9/5)\mathbf{v}_2 + 4\mathbf{v}_3\end{aligned}$$

11. **(a)** We have $(\mathbf{w})_S = (\langle \mathbf{w}, \mathbf{u}_1 \rangle, \langle \mathbf{w}, \mathbf{u}_2 \rangle) = \left(\frac{-4}{\sqrt{2}}, \frac{10}{\sqrt{2}} \right) = \left(-2\sqrt{2}, 5\sqrt{2} \right)$.

17. **(a)** Let

$$\mathbf{v}_1 = \frac{\mathbf{u}_1}{\|\mathbf{u}_1\|} = \left(\frac{1}{\sqrt{3}}, \frac{1}{\sqrt{3}}, \frac{1}{\sqrt{3}} \right)$$

Since $\langle \mathbf{u}_2, \mathbf{v}_1 \rangle = 0$, we have

$$\mathbf{v}_2 = \frac{\mathbf{u}_2}{\|\mathbf{u}_2\|} = \left(-\frac{1}{\sqrt{2}}, \frac{1}{\sqrt{2}}, 0\right)$$

Since $\langle \mathbf{u}_3, \mathbf{v}_1 \rangle = \frac{4}{\sqrt{3}}$ and $\langle \mathbf{u}_3, \mathbf{v}_2 \rangle = \frac{1}{\sqrt{2}}$, we have

$$\mathbf{u}_3 - \langle \mathbf{u}_3, \mathbf{v}_1 \rangle \mathbf{v}_1 - \langle \mathbf{u}_3, \mathbf{v}_2 \rangle \mathbf{v}_2$$

$$= (1, 2, 1) - \frac{4}{\sqrt{3}}\left(\frac{1}{\sqrt{3}}, \frac{1}{\sqrt{3}}, \frac{1}{\sqrt{3}}\right) - \frac{1}{\sqrt{2}}\left(-\frac{1}{\sqrt{2}}, \frac{1}{\sqrt{2}}, 0\right)$$

$$= \left(\frac{1}{6}, \frac{1}{6}, -\frac{1}{3}\right)$$

This vector has norm $\left\|\left(\frac{1}{6}, \frac{1}{6}, -\frac{1}{3}\right)\right\| = \frac{1}{\sqrt{6}}$. Thus

$$\mathbf{v}_3 = \left(\frac{1}{\sqrt{6}}, \frac{1}{\sqrt{6}}, -\frac{2}{\sqrt{6}}\right)$$

and $\{\mathbf{v}_1, \mathbf{v}_2, \mathbf{v}_3\}$ is the desired orthonormal basis.

19. Since the third vector is the sum of the first two, we ignore it. Let $\mathbf{u}_1 = (0, 1, 2)$ and $\mathbf{u}_2 = (-1, 0, 1)$. Then

$$\mathbf{v}_1 = \frac{\mathbf{u}_1}{\|\mathbf{u}_1\|} = \left(0, \frac{1}{\sqrt{5}}, \frac{2}{\sqrt{5}}\right)$$

Since $\langle \mathbf{u}_2, \mathbf{v}_1 \rangle = \frac{2}{\sqrt{5}}$, then

$$\mathbf{u}_2 - \langle \mathbf{u}_2, \mathbf{v}_1 \rangle \mathbf{v}_1 = \left(-1, -\frac{2}{5}, \frac{1}{5}\right)$$

where $\left\|\left(-1, -\frac{2}{5}, \frac{1}{5}\right)\right\| = \frac{\sqrt{30}}{5}$. Hence

$$\mathbf{v}_2 = \left(-\frac{5}{\sqrt{30}}, \frac{-2}{\sqrt{30}}, \frac{1}{\sqrt{30}}\right)$$

Thus $\{\mathbf{v}_1, \mathbf{v}_2\}$ is an orthonormal basis.

21. Note that $\mathbf{u}_1$ and $\mathbf{u}_2$ are orthonormal. Thus we apply Theorem 6.3.5 to obtain

$$\begin{aligned}\mathbf{w}_1 &= \langle \mathbf{w}, \mathbf{u}_1 \rangle \mathbf{u}_1 + \langle \mathbf{w}, \mathbf{u}_2 \rangle \mathbf{u}_2 \\ &= -\left(\frac{4}{5}, 0, -\frac{3}{5}\right) + 2(0, 1, 0) \\ &= \left(-\frac{4}{5}, 2, \frac{3}{5}\right)\end{aligned}$$

and

$$\begin{aligned}\mathbf{w}_2 &= \mathbf{w} - \mathbf{w}_1 \\ &= \left(\frac{9}{5}, 0, \frac{12}{5}\right)\end{aligned}$$

25. By Theorem 6.3.1, we know that

$$\mathbf{w} = a_1\mathbf{v}_1 + a_2\mathbf{v}_2 + a_3\mathbf{v}_3$$

where $a_i = \langle \mathbf{w}, \mathbf{v}_i \rangle$. Thus

$$\begin{aligned}\|\mathbf{w}\|^2 &= \langle \mathbf{w}, \mathbf{w} \rangle \\ &= \sum_{i=1}^{3} a_i^2 \langle \mathbf{v}_i, \mathbf{v}_i \rangle + \sum_{i \neq 1} a_i a_j \langle \mathbf{v}_i, \mathbf{v}_j \rangle\end{aligned}$$

But $\langle \mathbf{v}_i, \mathbf{v}_j \rangle = 0$ if $i \neq j$ and $\langle \mathbf{v}_i, \mathbf{v}_i \rangle = 1$ because the set $\{\mathbf{v}_1, \mathbf{v}_2, \mathbf{v}_3\}$ is orthonormal. Hence

$$\begin{aligned}\|\mathbf{w}\|^2 &= a_1^2 + a_2^2 + a_3^2 \\ &= \langle \mathbf{w}, \mathbf{v}_1 \rangle^2 + \langle \mathbf{w}, \mathbf{v}_2 \rangle^2 + \langle \mathbf{w}, \mathbf{v}_3 \rangle^2\end{aligned}$$

27. Suppose the contrary; that is, suppose that

$$(*) \qquad \mathbf{u}_3 - \langle \mathbf{u}_3, \mathbf{v}_1 \rangle \mathbf{v}_1 - \langle \mathbf{u}_3, \mathbf{v}_2 \rangle \mathbf{v}_2 = \mathbf{0}$$

Then (*) implies that $\mathbf{u}_3$ is a linear combination of $\mathbf{v}_1$ and $\mathbf{v}_2$. But $\mathbf{v}_1$ is a multiple of $\mathbf{u}_1$ while $\mathbf{v}_2$ is a linear combination of $\mathbf{u}_1$ and $\mathbf{u}_2$. Hence, (*) implies that $\mathbf{u}_3$ is a linear combination of $\mathbf{u}_1$ and $\mathbf{u}_2$ and therefore that $\{\mathbf{u}_1, \mathbf{u}_2, \mathbf{u}_3\}$ is linearly dependent, contrary to the hypothesis that $\{\mathbf{u}_1, \ldots, \mathbf{u}_n\}$ is linearly independent. Thus, the assumption that (*) holds leads to a contradiction.

29. We have $\mathbf{u}_1 = 1$, $\mathbf{u}_2 = x$, and $\mathbf{u}_3 = x^2$. Since

$$\|\mathbf{u}_1\|^2 = \langle \mathbf{u}_1, \mathbf{u}_1 \rangle = \int_{-1}^{1} 1\, dx = 2$$

we let

$$\mathbf{v}_1 = \frac{1}{\sqrt{2}}$$

Then

$$\langle \mathbf{u}_2, \mathbf{v}_1 \rangle = \frac{1}{\sqrt{2}} \int_{-1}^{1} x\, dx = 0$$

and thus $\mathbf{v}_2 = \mathbf{u}_2/\|\mathbf{u}_2\|$ where

$$\|\mathbf{u}_2\|^2 = \int_{-1}^{1} x\, dx = \frac{2}{3}$$

Hence

$$\mathbf{v}_2 = \sqrt{\frac{3}{2}}\, x$$

In order to compute $\mathbf{v}_3$, we note that

$$\langle \mathbf{u}_3, \mathbf{v}_1 \rangle = \frac{1}{\sqrt{2}} \int_{-1}^{1} x^2\, dx = \frac{\sqrt{2}}{3}$$

and

$$\langle \mathbf{u}_3, \mathbf{v}_2 \rangle = \sqrt{\frac{3}{2}} \int_{-1}^{1} x^3 \, dx = 0$$

Thus

$$\mathbf{u}_3 - \langle \mathbf{u}_3, \mathbf{v}_1 \rangle \mathbf{v}_1 - \langle \mathbf{u}_3, \mathbf{v}_2 \rangle \mathbf{v}_2 = x^2 - \frac{1}{3}$$

and

$$\left\| x^2 - \frac{1}{3} \right\|^2 = \int_{-1}^{1} \left[x^2 - \frac{1}{3} \right]^2 dx = \frac{8}{45}$$

Hence,

$$\mathbf{v}_3 = \sqrt{\frac{45}{8}} \left(x^2 - \frac{1}{3} \right) \quad \text{or} \quad \mathbf{v}_3 = \frac{\sqrt{5}}{2\sqrt{2}} \left(3x^2 - 1 \right)$$

31. This is similar to Exercise 29 except that the lower limit of integration is changed from –1 to 0. If we again set $\mathbf{u}_1 = 1$, $\mathbf{u}_2 = x$, and $\mathbf{u}_3 = x^2$, then $\|\mathbf{u}_1\| = 1$ and thus

$$\mathbf{v}_1 = 1$$

Then $\langle \mathbf{u}_2, \mathbf{v}_1 \rangle = \int_0^1 x \, dx = \frac{1}{2}$ and thus

$$\mathbf{v}_2 = \frac{x - 1/2}{\|x - 1/2\|} = \sqrt{12}(x - 1/2)$$

or

$$\mathbf{v}_2 = \sqrt{3}(2x - 1)$$

Finally,

$$\langle \mathbf{u}_3, \mathbf{v}_1 \rangle = \int_0^1 x^2 dx = \frac{1}{3}$$

and

$$\langle \mathbf{u}_3, \mathbf{v}_2 \rangle = \sqrt{3}\int_0^1 (2x^3 - x^2)\,dx = \frac{\sqrt{3}}{6}$$

Thus

$$\mathbf{v}_3 = \frac{x^2 - \frac{1}{3} - \frac{1}{2}(2x-1)}{\left\| x^2 - \frac{1}{3} - \frac{1}{2}(2x-1) \right\|} = 6\sqrt{5}\left(x^2 - x + \frac{1}{6}\right)$$

or

$$\mathbf{v}_3 = \sqrt{5}\,(6x^2 - 6x + 1)$$

33. Let W be a finite dimensional subspace of the inner product space V and let $\{\mathbf{v}_1, \mathbf{v}_2, \ldots, \mathbf{v}_r\}$ be an orthonormal basis for W. Then if $\mathbf{u}$ is any vector in V, we know from Theorem 6.3.4 that $\mathbf{u} = \mathbf{w}_1 + \mathbf{w}_2$ where $\mathbf{w}_1$ is in W and $\mathbf{w}_2$ is in $W^\perp$. Moreover, this decomposition of $\mathbf{u}$ is unique. Theorem 6.3.5 gives us a candidate for $\mathbf{w}_1$. To prove the theorem, we must show that if $\mathbf{w}_1 = \langle \mathbf{u}, \mathbf{v}_1 \rangle \mathbf{v}_1 + \cdots + \langle \mathbf{u}, \mathbf{v}_r \rangle \mathbf{v}_r$ and, therefore, that $\mathbf{w}_2 = \mathbf{u} - \mathbf{w}_1$ then

(i) $\quad \mathbf{w}_1$ is in W

and

(ii) $\quad \mathbf{w}_2$ is orthogonal to W.

That is, we must show that this candidate "works." Then, since $\mathbf{w}_1$ is unique, it will be $\text{proj}_W \mathbf{u}$.

Part (i) follows immediately because $\mathbf{w}_1$ is, by definition, a linear combination of the vectors $\mathbf{v}_1, \mathbf{v}_2, \ldots, \mathbf{v}_r$.

$$\begin{aligned}
\langle \mathbf{w}_2, \mathbf{v}_i \rangle &= \langle \mathbf{u} - \mathbf{w}_1, \mathbf{v}_i \rangle \\
&= \langle \mathbf{u}, \mathbf{v}_i \rangle - \langle \mathbf{w}_1, \mathbf{v}_i \rangle \\
&= \langle \mathbf{u}, \mathbf{v}_i \rangle - \langle \mathbf{u}, \mathbf{v}_i \rangle \langle \mathbf{v}_i, \mathbf{v}_i \rangle \\
&= \langle \mathbf{u}, \mathbf{v}_i \rangle - \langle \mathbf{u}, \mathbf{v}_i \rangle \\
&= 0
\end{aligned}$$

Thus, $\mathbf{w}_2$ is orthogonal to each of the vectors $\mathbf{v}_1, \mathbf{v}_2, \ldots, \mathbf{v}_r$ and hence $\mathbf{w}_2$ is in $W^{\perp}$.

If the vectors $\mathbf{v}_i$ form an orthogonal set, not necessarily orthonormal, then we must normalize them to obtain Part (b) of the theorem.

35. The vectors $\mathbf{x} = (1/\sqrt{3}, 0)$ and $\mathbf{y} = (0, 1/\sqrt{2})$ are orthonormal with respect to the given inner product. However, although they are orthogonal with respect to the Euclidean inner product, they are not orthonormal.

The vectors $\mathbf{x} = (2/\sqrt{30}, 3/\sqrt{30})$ and $\mathbf{y} = (1/\sqrt{5}, -1/\sqrt{5})$ are orthonormal with respect to the given inner product. However, they are neither orthogonal nor of unit length with respect to the Euclidean inner product.

37. **(a)** True. Suppose that $\mathbf{v}_1, \mathbf{v}_2, \ldots, \mathbf{v}_n$ is an orthonormal set of vectors. If they were linearly dependent, then there would be a linear combination

$$c_1\mathbf{v}_1 + c_2\mathbf{v}_2 + \cdots + c_n\mathbf{v}_n = 0$$

where at least one of the numbers $c_i \neq 0$. But

$$c_i = \langle \mathbf{v}_i, c_1\mathbf{v}_1 + c_2\mathbf{v}_2 + \cdots + c_n\mathbf{v}_n \rangle = \langle \mathbf{v}_i, 0 \rangle = 0$$

for $i = 1, \ldots, n$. Thus, the orthonormal set of vectors cannot be linearly dependent.

(b) False. The zero vector space has no basis $\mathbf{0}$. This vector cannot be linearly independent.

(c) True, since $\text{proj}_W \mathbf{u}$ is in W and $\text{proj}_{W^{\perp}} \mathbf{u}$ is in $W^{\perp}$.

(d) True. If A is a (necessarily square) matrix with a nonzero determinant, then A has linearly independent column vectors. Thus, by Theorem 6.3.7, A has a QR decomposition.

EXERCISE SET 6.4

1. **(a)** If we call the system $A\mathbf{x} = \mathbf{b}$, then the associated normal system is $A^T A\mathbf{x} = A^T\mathbf{b}$, or

$$\begin{bmatrix} 1 & 2 & 4 \\ -1 & 3 & 5 \end{bmatrix}\begin{bmatrix} 1 & -1 \\ 2 & 3 \\ 4 & 5 \end{bmatrix}\begin{bmatrix} x_1 \\ x_2 \end{bmatrix} = \begin{bmatrix} 1 & 2 & 4 \\ -1 & 3 & 5 \end{bmatrix}\begin{bmatrix} 2 \\ -1 \\ 5 \end{bmatrix}$$

which simplifies to

$$\begin{bmatrix} 21 & 25 \\ 25 & 35 \end{bmatrix}\begin{bmatrix} x_1 \\ x_2 \end{bmatrix} = \begin{bmatrix} 20 \\ 20 \end{bmatrix}$$

3. **(a)** The associated normal system is $A^T A\mathbf{x} = A^T\mathbf{b}$, or

$$\begin{bmatrix} 1 & -1 & -1 \\ 1 & 1 & 2 \end{bmatrix}\begin{bmatrix} 1 & 1 \\ -1 & 1 \\ -1 & 2 \end{bmatrix}\begin{bmatrix} x_1 \\ x_2 \end{bmatrix} = \begin{bmatrix} 1 & -1 & -1 \\ 1 & 1 & 2 \end{bmatrix}\begin{bmatrix} 7 \\ 0 \\ -7 \end{bmatrix}$$

or

$$\begin{bmatrix} 3 & -2 \\ -2 & 6 \end{bmatrix}\begin{bmatrix} x_1 \\ x_2 \end{bmatrix} = \begin{bmatrix} 14 \\ -7 \end{bmatrix}$$

This system has solution $x_1 = 5$, $x_2 = 1/2$, which is the least squares solution of $A\mathbf{x} = \mathbf{b}$.

The orthogonal projection of $\mathbf{b}$ on the column space of A is $A\mathbf{x}$, or

$$\begin{bmatrix} 1 & 1 \\ -1 & 1 \\ -1 & 2 \end{bmatrix}\begin{bmatrix} 5 \\ 1/2 \end{bmatrix} = \begin{bmatrix} 11/2 \\ -9/2 \\ -4 \end{bmatrix}$$

3. **(c)** The associated normal system is

$$\begin{bmatrix} 1 & 2 & 1 & 1 \\ 0 & 1 & 1 & 1 \\ -1 & -2 & 0 & -1 \end{bmatrix}\begin{bmatrix} 1 & 0 & -1 \\ 2 & 1 & -2 \\ 1 & 1 & 0 \\ 1 & 1 & -1 \end{bmatrix}\begin{bmatrix} x_1 \\ x_2 \\ x_3 \end{bmatrix}$$

$$= \begin{bmatrix} 1 & 2 & 1 & 1 \\ 0 & 1 & 1 & 1 \\ -1 & -2 & 0 & -1 \end{bmatrix}\begin{bmatrix} 6 \\ 0 \\ 9 \\ 3 \end{bmatrix}$$

or

$$\begin{bmatrix} 7 & 4 & -6 \\ 4 & 3 & -3 \\ -6 & -3 & 6 \end{bmatrix}\begin{bmatrix} x_1 \\ x_2 \\ x_3 \end{bmatrix} = \begin{bmatrix} 18 \\ 12 \\ -9 \end{bmatrix}$$

This system has solution $x_1 = 12$, $x_2 = -3$, $x_3 = 9$, which is the least squares solution of $A\mathbf{x} = \mathbf{b}$.

The orthogonal projection of $\mathbf{b}$ on the column space of A is $A\mathbf{x}$, or

$$\begin{bmatrix} 1 & 0 & -1 \\ 2 & 1 & -2 \\ 1 & 1 & 0 \\ 1 & 1 & -1 \end{bmatrix}\begin{bmatrix} 12 \\ -3 \\ 9 \end{bmatrix} = \begin{bmatrix} 3 \\ 3 \\ 9 \\ 0 \end{bmatrix}$$

which can be written as (3, 3, 9, 0).

5. (a) First we find a least squares solution of $A\mathbf{x} = \mathbf{u}$ where $A = [\mathbf{v}_1^T|\mathbf{v}_2^T|\mathbf{v}_3^T]$. The associated normal system is

$$\begin{bmatrix} 2 & 1 & 1 & 1 \\ 1 & 0 & 1 & 1 \\ -2 & -1 & 0 & -1 \end{bmatrix}\begin{bmatrix} 2 & 1 & -2 \\ 1 & 0 & -1 \\ 1 & 1 & 0 \\ 1 & 1 & -1 \end{bmatrix}\begin{bmatrix} x_1 \\ x_2 \\ x_3 \end{bmatrix}$$

$$= \begin{bmatrix} 2 & 1 & 1 & 1 \\ 1 & 0 & 1 & 1 \\ -2 & -1 & 0 & -1 \end{bmatrix} \begin{bmatrix} 6 \\ 3 \\ 9 \\ 6 \end{bmatrix}$$

or

$$\begin{bmatrix} 7 & 4 & -6 \\ 4 & 3 & -3 \\ -6 & -3 & 6 \end{bmatrix} \begin{bmatrix} x_1 \\ x_2 \\ x_3 \end{bmatrix} = \begin{bmatrix} 30 \\ 21 \\ -21 \end{bmatrix}$$

This system has solution $x_1 = 6$, $x_2 = 3$, $x_3 = 4$, which is the least squares solution. The desired orthogonal projection is $A\mathbf{x}$, or

$$\begin{bmatrix} 2 & 1 & -2 \\ 1 & 0 & -1 \\ 1 & 1 & 0 \\ 1 & 1 & -1 \end{bmatrix} \begin{bmatrix} 6 \\ 3 \\ 4 \end{bmatrix} = \begin{bmatrix} 7 \\ 2 \\ 9 \\ 5 \end{bmatrix}$$

or (7, 2, 9, 5).

7. **(a)** If we use the vector (1, 0) as a basis for the x-axis and let $A = \begin{bmatrix} 1 \\ 0 \end{bmatrix}$, then we have

$$[P] = A(A^T A)^{-1} A^T = \begin{bmatrix} 1 \\ 0 \end{bmatrix} [1]\,[1\ 0] = \begin{bmatrix} 1 & 0 \\ 0 & 0 \end{bmatrix}$$

11. **(a)** The vector $\mathbf{v} = (2, -1, 4)$ forms a basis for the line W.

(b) If we let $A = [\mathbf{v}^T]$, then the standard matrix for the orthogonal projection on W is

$$[P] = A(A^TA)^{-1}A^T = \begin{bmatrix} 2 \\ -1 \\ 4 \end{bmatrix}\left(\begin{bmatrix} 2 & -1 & 4 \end{bmatrix}\begin{bmatrix} 2 \\ -1 \\ 4 \end{bmatrix}\right)^{-1}\begin{bmatrix} 2 & -1 & 4 \end{bmatrix}$$

$$= \begin{bmatrix} 2 \\ -1 \\ 4 \end{bmatrix}\begin{bmatrix} \frac{1}{21} \end{bmatrix}\begin{bmatrix} 2 & -1 & 4 \end{bmatrix}$$

$$= \frac{1}{21}\begin{bmatrix} 4 & -2 & 8 \\ -2 & 1 & -4 \\ 8 & -4 & 16 \end{bmatrix}$$

(c) By Part (b), the point (x_0, y_0, z_0) projects to the point on the line W given by

$$\frac{1}{21}\begin{bmatrix} 4 & -2 & 8 \\ -2 & 1 & -4 \\ 8 & -4 & 16 \end{bmatrix}\begin{bmatrix} x_0 \\ y_0 \\ z_0 \end{bmatrix} = \begin{bmatrix} (4x_0 - 2y_0 + 8z_0)/21 \\ (-2x_0 + y_0 - 4z_0)/21 \\ (8x_0 - 4y_0 + 16z_0)/21 \end{bmatrix}$$

(d) By the result in Part (c), the point $(2, 1, -3)$ projects to the point $(-6/7, 3/7, -12/7)$. The distance between these two points is $\sqrt{497}/7$.

13. **(a)** Using horizontal vector notation, we have $\mathbf{b} = (7, 0, -7)$ and $A\mathbf{x} = (11/2, -9/2, -4)$. Therefore $A\mathbf{x} - \mathbf{b} = (-3/2, -9/2, 3)$, which is orthogonal to both of the vectors $(1, -1, -1)$ and $(1, 1, 2)$ which span the column space of A. Hence the error vector is orthogonal to the column space of A.

(c) In horizontal vector notation, $\mathbf{b} = (6, 0, 9, 3)$ and $A\mathbf{x} = (3, 3, 9, 0)$. Hence $A\mathbf{x} - \mathbf{b} = (-3, 3, 0, -3)$, which is orthogonal to the three vectors $(1, 2, 1, 1)$, $(0, 1, 1, 1)$, and $(-1, -2, 0, -1)$ which span the column space of A. Therefore $A\mathbf{x} - \mathbf{b}$ is orthogonal to the column space of A.

15. Recall that if $\mathbf{b}$ is orthogonal to the column space of A, then $\text{proj}_W\mathbf{b} = 0$.

17. If A is an $m \times n$ matrix with linearly independent row vectors, then A^T is an $n \times m$ matrix with linearly independent column vectors which span the row space of A. Therefore, by Formula (6) and the fact that $(A^T)^T = A$, the standard matrix for the orthogonal projection, S, of R^n on the row space of A is $[S] = A^T(AA^T)^{-1}A$.

19. If we assume a relationship $V = IR + c$, we have the linear system

$$1 = 0.1\,R + c$$

$$2.1 = 0.2\,R + c$$

$$2.9 = 0.3\,R + c$$

$$4.2 = 0.4\,R + c$$

$$5.1 = 0.5\,R + c$$

This system can be written as $A\mathbf{x} = \mathbf{b}$, where

$$A = \begin{bmatrix} 0.1 & 1 \\ 0.2 & 1 \\ 0.3 & 1 \\ 0.4 & 1 \\ 0.5 & 1 \end{bmatrix} \quad \textit{and} \quad \mathbf{b} = \begin{bmatrix} 1 \\ 2.1 \\ 2.9 \\ 4.2 \\ 5.1 \end{bmatrix}.$$

Then, we have the least squares solution

$$\mathbf{x} = (A^TA)^{-1}A^T\mathbf{b} = \begin{bmatrix} 0.55 & 1.55 \\ 1.5 & 5 \end{bmatrix}^{-1} \begin{bmatrix} 5.62 \\ 15.3 \end{bmatrix} = \begin{bmatrix} 10.3 \\ -0.03 \end{bmatrix}.$$

Thus, we have the relationship $V = 10.3\,R - 0.03$.

EXERCISE SET 6.5

1. **(b)** We have $(\mathbf{w})_S = (a, b)$ where $\mathbf{w} = a\mathbf{u}_1 + b\mathbf{u}_2$. Thus

$$2a + 3b = 1$$
$$-4a + 8b = 1$$

or $a = \dfrac{5}{28}$ and $b = \dfrac{3}{14}$. Hence $(\mathrm{w})_S = \left(\dfrac{5}{28}, \dfrac{3}{14}\right)$ and

$$[\mathrm{w}]_S = \begin{bmatrix} \dfrac{5}{28} \\ \dfrac{3}{14} \end{bmatrix}$$

3. **(b)** Let $\mathbf{p} = a\mathbf{p}_1 + b\mathbf{p}_2 + c\mathbf{p}_3$. Then

$$\begin{aligned} a + b \phantom{{}+c} &= 2 \\ a \phantom{{}+b} + c &= -1 \\ b + c &= 1 \end{aligned}$$

or $a = 0$, $b = 2$, and $c = -1$. Thus $(\mathbf{v})_S = (0, 2, -1)$ and

$$[\mathbf{v}]_S = \begin{bmatrix} 0 \\ 2 \\ -1 \end{bmatrix}$$

5. **(a)** We have $\mathbf{w} = 6\mathbf{v}_1 - \mathbf{v}_2 + 4\mathbf{v}_3 = (16, 10, 12)$.

(c) We have $B = -8A_1 + 7A_2 + 6A_3 + 3A_4 = \begin{bmatrix} 15 & -1 \\ 6 & 3 \end{bmatrix}$.

7. **(a)** Since $\mathbf{v}_1 = \frac{13}{10}\mathbf{u}_1 - \frac{2}{5}\mathbf{u}_2$ and $\mathbf{v}_2 = -\frac{1}{2}\mathbf{u}_1 + 0\mathbf{u}_2$, the transition matrix is

$$Q = \begin{bmatrix} \frac{13}{10} & -\frac{1}{2} \\ -\frac{2}{5} & 0 \end{bmatrix}$$

(b) Since $\mathbf{u}_1 = 0\mathbf{v}_1 - 2\mathbf{v}_2$ and $\mathbf{u}_2 = -\frac{5}{2}\mathbf{v}_1 - \frac{13}{2}\mathbf{v}_2$, the transition matrix is

$$P = \begin{bmatrix} 0 & -\frac{5}{2} \\ -2 & -\frac{13}{2} \end{bmatrix}$$

Note that $P = Q^{-1}$.

(c) We find that $\mathbf{w} = -\frac{17}{10}\mathbf{u}_1 + \frac{8}{5}\mathbf{u}_2$; that is

$$[\mathbf{w}]_B = \begin{bmatrix} -\frac{17}{10} \\ \frac{8}{5} \end{bmatrix}$$

and hence

$$[\mathbf{w}]_{B'} = \begin{bmatrix} 0 & -\frac{5}{2} \\ -2 & -\frac{13}{2} \end{bmatrix} \begin{bmatrix} -\frac{17}{10} \\ \frac{8}{5} \end{bmatrix} = \begin{bmatrix} -4 \\ -7 \end{bmatrix}$$

(d) Verify that $\mathbf{w} = (-4)\mathbf{v}_1 + (-7)\mathbf{v}_2$.

11. (a) By hypothesis, $\mathbf{f}_1$ and $\mathbf{f}_2$ span V. Since neither is a multiple of the other, then $\{\mathbf{f}_1, \mathbf{f}_2\}$ is a linearly independent set and hence is a basis for V. Now by inspection, $\mathbf{f}_1 = \frac{1}{2}\mathbf{g}_1 + \left(-\frac{1}{6}\right)\mathbf{g}_2$ and $\mathbf{f}_2 = \frac{1}{3}\mathbf{g}_2$. Therefore, $\{\mathbf{g}_1, \mathbf{g}_2\}$ must also be a basis for V because it is a spanning set which contains the correct number of vectors.

(b) The transition matrix is

$$\begin{bmatrix} \frac{1}{2} & 0 \\ -\frac{1}{6} & \frac{1}{3} \end{bmatrix}^{-1} = \begin{bmatrix} 2 & 0 \\ 1 & 3 \end{bmatrix}$$

(c) From the observations in Part (a), we have

$$P = \begin{bmatrix} \frac{1}{2} & 0 \\ -\frac{1}{6} & \frac{1}{3} \end{bmatrix}$$

(d) Since $\mathbf{h} = 2\mathbf{f}_1 + (-5)\mathbf{f}_2$, we have $[\mathbf{h}]_B = \begin{bmatrix} 2 \\ -5 \end{bmatrix}$; thus

$$[\mathbf{h}]_{B'} = \begin{bmatrix} \frac{1}{2} & 0 \\ -\frac{1}{6} & \frac{1}{3} \end{bmatrix} \begin{bmatrix} 2 \\ -5 \end{bmatrix} = \begin{bmatrix} 1 \\ -2 \end{bmatrix}$$

EXERCISE SET 6.6

3. **(b)** Since the row vectors form an orthonormal set, the matrix is orthogonal. Therefore its inverse is its transpose,

$$\begin{bmatrix} 1/\sqrt{2} & 1/\sqrt{2} \\ -1/\sqrt{2} & 1/\sqrt{2} \end{bmatrix}$$

(c) Since the Euclidean inner product of Column 2 and Column 3 is not zero, the column vectors do not form an orthonormal set and the matrix is not orthogonal.

(f) Since the norm of Column 3 is not 1, the matrix is not orthogonal.

9. The general transition matrix will be

$$\begin{bmatrix} \cos\theta & 0 & -\sin\theta \\ 0 & 1 & 0 \\ \sin\theta & 0 & \cos\theta \end{bmatrix}$$

In particular, if we rotate through $\theta = \frac{\pi}{3}$, then the transition matrix is

$$\begin{bmatrix} \frac{1}{2} & 0 & -\frac{\sqrt{3}}{2} \\ 0 & 1 & 0 \\ \frac{\sqrt{3}}{2} & 0 & \frac{1}{2} \end{bmatrix}$$

11. **(a)** See Exercise 19, above.

13. Since the row vectors (and the column vectors) of the given matrix are orthogonal, the matrix will be orthogonal provided these vectors have norm 1. A necessary and sufficient condition for this is that $a^2 + b^2 = 1/2$. Why?

15. Multiplication by the first matrix A in Exercise 24 represents a rotation and $\det(A) = 1$. The second matrix has determinant -1 and can be written as

$$\begin{bmatrix} \cos\theta & -\sin\theta \\ -\sin\theta & -\cos\theta \end{bmatrix} = \begin{bmatrix} 1 & 0 \\ 0 & -1 \end{bmatrix} \begin{bmatrix} \cos\theta & -\sin\theta \\ \sin\theta & \cos\theta \end{bmatrix}$$

Thus it represents a rotation followed by a reflection about the x-axis.

19. Note that A is orthogonal if and only if A^T is orthogonal. Since the rows of A^T are the columns of A, we need only apply the equivalence of Parts (a) and (b) to A^T to obtain the equivalence of Parts (a) and (c).

21. If A is the standard matrix associated with a rigid transformation, then Theorem 6.5.3 guarantees that A must be orthogonal. But if A is orthogonal, then Theorem 6.5.2 guarantees that $\det(A) = \pm 1$.

SUPPLEMENTARY EXERCISES 6

1. **(a)** We must find a vector $\mathbf{x} = (x_1, x_2, x_3, x_4)$ such that

$$\mathbf{x}\cdot\mathbf{u}_1 = 0, \quad \mathbf{x}\cdot\mathbf{u}_4 = 0, \quad \text{and} \quad \frac{\mathbf{x}\cdot\mathbf{u}_2}{\|\mathbf{x}\|\,\|\mathbf{u}_2\|} = \frac{\mathbf{x}\cdot\mathbf{u}_3}{\|\mathbf{x}\|\,\|\mathbf{u}_3\|}$$

The first two conditions guarantee that $x_1 = x_4 = 0$. The third condition implies that $x_2 = x_3$. Thus any vector of the form $(0, a, a, 0)$ will satisfy the given conditions provided $a \neq 0$.

(b) We must find a vector $\mathbf{x} = (x_1, x_2, x_3, x_4)$ such that $\mathbf{x} \cdot \mathbf{u}_1 = \mathbf{x} \cdot \mathbf{u}_4 = 0$. This implies that $x_1 = x_4 = 0$. Moreover, since $\|\mathbf{x}\| = \|\mathbf{u}_2\| = \|\mathbf{u}_3\| = 1$, the cosine of the angle between $\mathbf{x}$ and $\mathbf{u}_2$ is $\mathbf{x} \cdot \mathbf{u}_2$ and the cosine of the angle between $\mathbf{x}$ and $\mathbf{u}_3$ is $\mathbf{x} \cdot \mathbf{u}_3$. Thus we are looking for a vector $\mathbf{x}$ such that $\mathbf{x} \cdot \mathbf{u}_2 = 2\mathbf{x} \cdot \mathbf{u}_3$, or $x_2 = 2x_3$. Since $\|\mathbf{x}\| = 1$, we have $\mathbf{x} = (0, 2x_3, x_3, 0)$ where $4x_3^2 + x_3^2 = 1$ or $x_3 = \pm 1/\sqrt{5}$. Therefore

$$\mathbf{x} = \pm\left(0, \frac{2}{\sqrt{5}}, \frac{1}{\sqrt{5}}, 0\right)$$

7. Let

(*) $$\langle \mathbf{u}, \mathbf{v}\rangle = w_1u_1v_1 + w_2u_2v_2 + \cdots + w_nu_nv_n$$

be the weighted Euclidean inner product. Since $\langle \mathbf{v}_i, \mathbf{v}_j\rangle = 0$ whenever $i \neq j$, the vectors $\{\mathbf{v}_1, \mathbf{v}_2, \ldots, \mathbf{v}_n\}$ form an orthogonal set with respect to (*) for any choice of the constants $w_1, w_2, \ldots, w_n$. We must now choose the positive constants $w_1, w_2, \ldots, w_n$ so that $\|\mathbf{v}_k\| = 1$ for all k. But $\|\mathbf{v}_k\|^2 = kw_k$. If we let $w_k = 1/k$ for $k = 1, 2, \ldots, n$, the given vectors will then form an orthonormal set with respect to (*).

9. Let $Q = [a_{ij}]$ be orthogonal. Then $Q^{-1} = Q^T$ and $\det(Q) = \pm 1$. If C_{ij} is the cofactor of a_{ij}, then

$$Q = [(a_{ij})] = (Q^{-1})^T = \left(\frac{1}{\det(Q)}(C_{ij})^T\right)^T = \det(Q)(C_{ij})$$

so that $a_{ij} = \det(Q)C_{ij}$.

11. **(a)** The length of each "side" of this "cube" is $|k|$. The length of the "diagonal" is $\sqrt{n}|k|$. The inner product of any "side" with the "diagonal" is k^2. Therefore,

$$\cos\theta = \frac{k^2}{|k|\sqrt{n}|k|} = \frac{1}{\sqrt{n}}$$

(b) As $n \to +\infty$, $\cos\theta \to 0$, so that $\theta \to \pi/2$.

13. Recall that **u** can be expressed as the linear combination

$$\mathbf{u} = a_1\mathbf{v}_1 + \cdots + a_n\mathbf{v}_n$$

where $a_i = \langle \mathbf{u}, \mathbf{v}_i \rangle$ for $i = 1, \ldots, n$. Thus

$$\cos^2 \alpha_i = \left(\frac{\langle \mathbf{u}, \mathbf{v}_i \rangle}{\|\mathbf{u}\|\|\mathbf{v}_i\|}\right)^2$$

$$= \left(\frac{a_i}{\|\mathbf{u}\|}\right)^2 \qquad (\|\mathbf{v}_i\| = 1)$$

$$= \frac{a_i^2}{a_1^2 + a_2^2 + \cdots + a_n^2} \qquad \text{(Why?)}$$

Therefore

$$\cos^2 \alpha_1 + \cdots + \cos^2 \alpha_n = \frac{a_1^2 + a_2^2 + \cdots + a_n^2}{a_1^2 + a_2^2 + \cdots + a_n^2} = 1$$

15. Recall that A is orthogonal provided $A^{-1} = A^T$. Hence

$$\begin{aligned}\langle \mathbf{u}, \mathbf{v} \rangle &= \mathbf{v}^T A^T A \mathbf{u} \\ &= \mathbf{v}^T A^{-1} A \mathbf{u} = \mathbf{v}^T \mathbf{u}\end{aligned}$$

which is the Euclidean inner product.

EXERCISE SET 7.1

1. **(a)** Since

$$\det(\lambda I - A) = \det\begin{bmatrix} \lambda-3 & 0 \\ -8 & \lambda+1 \end{bmatrix} = (\lambda-3)(\lambda+1)$$

the characteristic equation is $\lambda^2 - 2\lambda - 3 = 0$.

(e) Since

$$\det(\lambda I - A) = \det\begin{bmatrix} \lambda & 0 \\ 0 & \lambda \end{bmatrix} = \lambda^2$$

the characteristic equation is $\lambda^2 = 0$.

3. **(a)** The equation $(\lambda I - A)\mathbf{x} = \mathbf{0}$ becomes

$$\begin{bmatrix} \lambda-3 & 0 \\ -8 & \lambda+1 \end{bmatrix}\begin{bmatrix} x_1 \\ x_2 \end{bmatrix} = \begin{bmatrix} 0 \\ 0 \end{bmatrix}$$

The eigenvalues are $\lambda = 3$ and $\lambda = -1$. Substituting $\lambda = 3$ into $(\lambda I - A)\mathbf{x} = \mathbf{0}$ yields

$$\begin{bmatrix} 0 & 0 \\ -8 & 4 \end{bmatrix}\begin{bmatrix} x_1 \\ x_2 \end{bmatrix} = \begin{bmatrix} 0 \\ 0 \end{bmatrix}$$

or

$$-8x_1 + 4x_2 = 0$$

Thus $x_1 = \frac{1}{2}s$ and $x_2 = s$ where s is arbitrary, so that a basis for the eigenspace corresponding to $\lambda = 3$ is $\begin{bmatrix} 1/2 \\ 1 \end{bmatrix}$. Of course, $\begin{bmatrix} 1 \\ 2 \end{bmatrix}$ and $\begin{bmatrix} \pi \\ 2\pi \end{bmatrix}$ are also bases.

Substituting $\lambda = -1$ into $(\lambda I - A)\mathbf{x} = \mathbf{0}$ yields

$$\begin{bmatrix} -4 & 0 \\ -8 & 0 \end{bmatrix}\begin{bmatrix} x_1 \\ x_2 \end{bmatrix} = \begin{bmatrix} 0 \\ 0 \end{bmatrix}$$

or

$$-4x_1 = 0$$
$$-8x_1 = 0$$

Hence, $x_1 = 0$ and $x_2 = s$ where s is arbitrary. In particular, if $s = 1$, then a basis for the eigenspace corresponding to $\lambda = -1$ is $\begin{bmatrix} 0 \\ 1 \end{bmatrix}$.

3. **(e)** The equation $(\lambda I - A)\mathbf{x} = \mathbf{0}$ becomes

$$\begin{bmatrix} \lambda & 0 \\ 0 & \lambda \end{bmatrix}\begin{bmatrix} x_1 \\ x_2 \end{bmatrix} = \begin{bmatrix} 0 \\ 0 \end{bmatrix}$$

Clearly, $\lambda = 0$ is the only eigenvalue. Substituting $\lambda = 0$ into the above equation yields $x_1 = s$ and $x_2 = t$ where s and t are arbitrary. In particular, if $s = t = 1$, then we find that $\begin{bmatrix} 1 \\ 0 \end{bmatrix}$ and $\begin{bmatrix} 0 \\ 1 \end{bmatrix}$ form a basis for the eigenspace associated with $\lambda = 0$.

5. **(c)** From the solution to 4(c), we have

$$\lambda^3 + 8\lambda^2 + \lambda + 8 = (\lambda + 8)(\lambda^2 + 1)$$

Since $\lambda^2 + 1 = 0$ has no real solutions, then $\lambda = -8$ is the only (real) eigenvalue.

7. **(a)** Since

$$\det(\lambda I - A) = \det\begin{bmatrix} \lambda & 0 & -2 & 0 \\ -1 & \lambda & -1 & 0 \\ 0 & -1 & \lambda+2 & 0 \\ 0 & 0 & 0 & \lambda-1 \end{bmatrix}$$

$$= \lambda^4 + \lambda^3 - 3\lambda^2 - \lambda + 2$$

$$= (\lambda - 1)^2(\lambda + 2)(\lambda + 1)$$

the characteristic equation is

$$(\lambda - 1)^2 (\lambda + 2)(\lambda + 1) = 0$$

9. **(a)** The eigenvalues are $\lambda = 1$, $\lambda = -2$, and $\lambda = -1$. If we set $\lambda = 1$, then $(\lambda I - A)\mathbf{x} = \mathbf{0}$ becomes

$$\begin{bmatrix} 1 & 0 & -2 & 0 \\ -1 & 1 & -1 & 0 \\ 0 & -1 & 3 & 0 \\ 0 & 0 & 0 & 0 \end{bmatrix}\begin{bmatrix} x_1 \\ x_2 \\ x_3 \\ x_4 \end{bmatrix} = \begin{bmatrix} 0 \\ 0 \\ 0 \\ 0 \end{bmatrix}$$

The augmented matrix can be reduced to

$$\begin{bmatrix} 1 & 0 & -2 & 0 & 0 \\ 0 & 1 & -3 & 0 & 0 \\ 0 & 0 & 0 & 0 & 0 \\ 0 & 0 & 0 & 0 & 0 \end{bmatrix}$$

Thus, $x_1 = 2s$, $x_2 = 3s$, $x_3 = s$, and $x_4 = t$ is a solution for all s and t. In particular, if we let $s = t = 1$, we see that

$$\begin{bmatrix} 2 \\ 3 \\ 1 \\ 0 \end{bmatrix} \text{ and } \begin{bmatrix} 0 \\ 0 \\ 0 \\ 1 \end{bmatrix}$$

form a basis for the eigenspace associated with $\lambda = 1$.

If we set $\lambda = -2$, then $(\lambda I - A)\mathbf{x} = \mathbf{0}$ becomes

$$\begin{bmatrix} -2 & 0 & -2 & 0 \\ -1 & -2 & -1 & 0 \\ 0 & -1 & 0 & 0 \\ 0 & 0 & 0 & -3 \end{bmatrix} \begin{bmatrix} x_1 \\ x_2 \\ x_3 \\ x_4 \end{bmatrix} = \begin{bmatrix} 0 \\ 0 \\ 0 \\ 0 \end{bmatrix}$$

The augmented matrix can be reduced to

$$\begin{bmatrix} 1 & 0 & 1 & 0 & 0 \\ 0 & 1 & 0 & 0 & 0 \\ 0 & 0 & 0 & 1 & 0 \\ 0 & 0 & 0 & 0 & 0 \end{bmatrix}$$

This implies that $x_1 = -s$, $x_2 = x_4 = 0$, and $x_3 = s$. Therefore the vector

$$\begin{bmatrix} -1 \\ 0 \\ 1 \\ 0 \end{bmatrix}$$

forms a basis for the eigenspace associated with $\lambda = -2$.

Finally, if we set $\lambda = -1$, then $(\lambda I - A)\mathbf{x} = \mathbf{0}$ becomes

$$\begin{bmatrix} -1 & 0 & -2 & 0 \\ -1 & -1 & -1 & 0 \\ 0 & -1 & 1 & 0 \\ 0 & 0 & 0 & -2 \end{bmatrix} \begin{bmatrix} x_1 \\ x_2 \\ x_3 \\ x_4 \end{bmatrix} = \begin{bmatrix} 0 \\ 0 \\ 0 \\ 0 \end{bmatrix}$$

The augmented matrix can be reduced to

$$\begin{bmatrix} 1 & 0 & 2 & 0 & 0 \\ 0 & 1 & -1 & 0 & 0 \\ 0 & 0 & 0 & 1 & 0 \\ 0 & 0 & 0 & 0 & 0 \end{bmatrix}$$

Thus, $x_1 = -2s$, $x_2 = s$, $x_3 = s$, and $x_4 = 0$ is a solution. Therefore the vector

$$\begin{bmatrix} -2 \\ 1 \\ 1 \\ 0 \end{bmatrix}$$

forms a basis for the eigenspace associated with $\lambda = -1$.

11. By Theorem 7.1.1, the eigenvalues of A are 1, 1/2, 0, and 2. Thus by Theorem 7.1.3, the eigenvalues of A^9 are $1^9 = 1$, $(1/2)^9 = 1/512$, $0^9 = 0$, and $2^9 = 512$.

13. The vectors $A\mathbf{x}$ and $\mathbf{x}$ will lie on the same line through the origin if and only if there exists a real number λ such that $A\mathbf{x} = \lambda\mathbf{x}$, that is, if and only if λ is a real eigenvalue for A and $\mathbf{x}$ is the associated eigenvector.

(a) In this case, the eigenvalues are $\lambda = 3$ and $\lambda = 2$, while associated eigenvectors are

$$\begin{bmatrix} 1 \\ 1 \end{bmatrix} \text{ and } \begin{bmatrix} 1 \\ 2 \end{bmatrix}$$

respectively. Hence the lines $y = x$ and $y = 2x$ are the only lines which are invariant under A.

(b) In this case, the characteristic equation for A is $\lambda^2 + 1 = 0$. Since A has no real eigenvalues, there are no lines which are invariant under A.

15. Let a_{ij} denote the ijth entry of A. Then the characteristic polynomial of A is $\det(\lambda I - A)$ or

$$\det \begin{bmatrix} \lambda - a_{11} & a_{12} & \cdots & -a_{1n} \\ -a_{21} & \lambda - a_{22} & \cdots & -a_{2n} \\ \vdots & \vdots & & \vdots \\ -a_{n1} & -a_{n2} & \cdots & \lambda - a_{nn} \end{bmatrix}$$

This determinant is a sum each of whose terms is the product of n entries from the given matrix. Each of these entries is either a constant or is of the form $\lambda - a_{ij}$. The only term with a λ in each factor of the product is

$$(\lambda - a_{11})(\lambda - a_{22}) \cdots (\lambda - a_{nn})$$

Therefore, this term must produce the highest power of λ in the characteristic polynomial. This power is clearly n and the coefficient of λ^n is 1.

17. The characteristic equation of A is

$$\lambda^2 - (a + d)\lambda + ad - bc = 0$$

This is a quadratic equation whose discriminant is

$$\begin{aligned}(a + d)^2 - 4ad + 4bc &= a^2 - 2ad + d^2 + 4bc \\ &= (a - d)^2 + 4bc\end{aligned}$$

The roots are

$$\lambda = \frac{1}{2}\left[(a+d) \pm \sqrt{(a-b)^2 + 4bc}\right]$$

If the discriminant is positive, then the equation has two distinct real roots; if it is zero, then the equation has one real root (repeated); if it is negative, then the equation has no real roots. Since the eigenvalues are assumed to be real numbers, the result follows.

19. As in Exercise 17, we have

$$\begin{aligned}\lambda &= \frac{a + d \pm \sqrt{(a-d)^2 + 4bc}}{2} \\ &= \frac{a + d \pm \sqrt{(c-b)^2 + 4bc}}{2} \quad \text{because } a - d = c - b \\ &= \frac{a + d \pm \sqrt{(c + b)^2}}{2} \\ &= \frac{a+d+c+d}{2} \quad \text{or} \quad \frac{a-b-c+d}{2} \\ &= a + b \text{ or } a - c\end{aligned}$$

<u>Alternate Solution:</u> Recall that if r_1 and r_2 are roots of the quadratic equation $x^2 + Bx + C = 0$, then $B = -(r_1 + r_2)$ and $C = r_1 r_2$. The converse of this result is also true. Thus the result will follow if we can show that the system of equations

$$\lambda_1 + \lambda_2 = a + d$$

$$\lambda_1 \lambda_2 = ad - bc$$

is satisfied by $\lambda_1 = a + b$ and $\lambda_2 = a - c$. This is a straightforward computation and we leave it to you.

21. Suppose that $A\mathbf{x} = \lambda\mathbf{x}$. Then

$$(A - sI)\mathbf{x} = A\mathbf{x} - sI\mathbf{x} = \lambda\mathbf{x} - s\mathbf{x} = (\lambda - s)\mathbf{x}$$

That is, $\lambda - s$ is an eigenvalue of $A - sI$ and $\mathbf{x}$ is a corresponding eigenvector.

23. **(a)** For any square matrix B, we know that $\det(B) = \det(B^T)$. Thus

$$\begin{aligned}\det(\lambda I - A) &= \det(\lambda I - A)^T \\ &= \det(\lambda I^T - A^T) \\ &= \det(\lambda I - A^T)\end{aligned}$$

from which it follows that A and A^T have the same eigenvalues because they have the same characteristic equation.

(b) Consider, for instance, the matrix $\begin{bmatrix} 2 & 1 \\ -1 & 0 \end{bmatrix}$ which has $\lambda = 1$ as a (repeated) eigenvalue. Its eigenspace is spanned by the vector $\begin{bmatrix} -1 \\ 1 \end{bmatrix}$, while the eigenspace of its transpose is spanned by the vector $\begin{bmatrix} 1 \\ 1 \end{bmatrix}$

25. **(a)** Since $p(\lambda)$ has degree 6, A is 6×6.

(b) Yes, A is invertible because $\lambda = 0$ is not an eigenvalue.

(c) A will have 3 eigenspaces corresponding to the 3 eigenvalues.

EXERCISE SET 7.2

1. The eigenspace corresponding to $\lambda = 0$ can have dimension 1 or 2. The eigenspace corresponding to $\lambda = 1$ must have dimension 1. The eigenspace corresponding to $\lambda = 2$ can have dimension 1, 2, or 3.

5. Call the matrix A. Since A is triangular, the eigenvalues are $\lambda = 3$ and $\lambda = 2$. The matrices $3I - A$ and $2I - A$ both have rank 2 and hence nullity 1. Thus A has only 2 linearly independent eigenvectors, so it is not diagonalizable.

13. The characteristic equation is $\lambda^3 - 6\lambda^2 + 11\lambda - 6 = 0$, the eigenvalues are $\lambda = 1$, $\lambda = 2$, and $\lambda = 3$, and the eigenspaces are spanned by the vectors

$$\begin{bmatrix} 1 \\ 1 \\ 1 \end{bmatrix} \quad \begin{bmatrix} 2/3 \\ 1 \\ 1 \end{bmatrix} \quad \begin{bmatrix} 1/4 \\ 3/4 \\ 1 \end{bmatrix}$$

Thus, one possibility is

$$P = \begin{bmatrix} 1 & 2 & 1 \\ 1 & 3 & 3 \\ 1 & 3 & 4 \end{bmatrix}$$

and

$$P^{-1}AP = \begin{bmatrix} 1 & 0 & 0 \\ 0 & 2 & 0 \\ 0 & 0 & 3 \end{bmatrix}$$

15. The characteristic equation is $\lambda^2(\lambda - 1) = 0$; thus $\lambda = 0$ and $\lambda = 1$ are the only eigenvalues.

The eigenspace associated with $\lambda = 0$ is spanned by the vectors $\begin{bmatrix} 1 \\ 0 \\ -3 \end{bmatrix}$ and $\begin{bmatrix} 0 \\ 1 \\ 0 \end{bmatrix}$; the eigenspace associated with $\lambda = 1$ is spanned by $\begin{bmatrix} 0 \\ 0 \\ 1 \end{bmatrix}$. Thus, one possibility is

$$P = \begin{bmatrix} 1 & 0 & 0 \\ 0 & 1 & 0 \\ -3 & 0 & 1 \end{bmatrix}$$

and hence

$$P^{-1}AP = \begin{bmatrix} 0 & 0 & 0 \\ 0 & 0 & 0 \\ 0 & 0 & 1 \end{bmatrix}$$

21. The characteristic equation of A is $(\lambda - 1)(\lambda - 3)(\lambda - 4) = 0$ so that the eigenvalues are $\lambda = 1$, 3, and 4. Corresponding eigenvectors are $[1\ 2\ 1]^T$, $[1\ 0\ -1]^T$, and $[1\ -1\ 1]^T$, respectively, so we let

$$P = \begin{bmatrix} 1 & 1 & 1 \\ 2 & 0 & -1 \\ 1 & -1 & 1 \end{bmatrix}$$

Hence

$$P^{-1} = \begin{bmatrix} 1/6 & 1/3 & 1/6 \\ 1/2 & 0 & -1/2 \\ 1/3 & -1/3 & 1/3 \end{bmatrix}$$

and therefore

$$A^n = \begin{bmatrix} 1 & 1 & 1 \\ 2 & 0 & -1 \\ 1 & -1 & 1 \end{bmatrix} \begin{bmatrix} 1^n & 0 & 0 \\ 0 & 3^n & 0 \\ 0 & 0 & 4^n \end{bmatrix} \begin{bmatrix} 1/6 & 1/3 & 1/6 \\ 1/2 & 0 & -1/2 \\ 1/3 & -1/3 & 1/3 \end{bmatrix}$$

25. **(a)** False. For instance the matrix $\begin{bmatrix} 0 & 1 \\ -1 & 2 \end{bmatrix}$, which has linearly independent column vectors, has characteristic polynomial $(\lambda - 1)^2$. Thus $\lambda = 1$ is the only eigenvalue. The corresponding eigenvectors all have the form $t\begin{bmatrix} 1 \\ 1 \end{bmatrix}$. Thus this 2×2 matrix has only 1 linearly independent eigenvector, and hence is not diagonalizable.

(b) False. Any matrix Q which is obtained from P by multiplying each entry by a nonzero number k will also work. Why?

(c) True by Theorem 7.2.2.

(d) True. Suppose that A is invertible and diagonalizable. Then there is an invertible matrix P such that $P^{-1}AP = D$ where D is diagonal. Since D is the product of invertible matrices, it is invertible, which means that each of its diagonal elements d_i is nonzero and D^{-1} is the diagonal matrix with diagonal elements $1/d_i$. Thus we have

$$(P^{-1}AP)^{-1} = D^{-1}$$

or

$$P^{-1}A^{-1}P = D^{-1}$$

That is, the same matrix P will diagonalize both A and A^{-1}.

27. **(a)** Since A is diagonalizable, there exists an invertible matrix P such that $P^{-1}AP = D$ where D is a diagonal matrix containing the eigenvalues of A along its diagonal. Moreover, it easily follows that $P^{-1}A^kP = D^k$ for k a positive integer. In addition, Theorem 7.1.3 guarantees that if λ is an eigenvalue for A, then λ^k is an eigenvalue for A^k. In other words, D^k displays the eigenvalues of A^k along its diagonal.

Therefore, the sequence

$$\begin{gathered} P^{-1}AP = D \\ P^{-1}A^2P = D^2 \\ \vdots \\ P^{-1}A^kP = D^k \\ \vdots \end{gathered}$$

will converge if and only if the sequence $A, A^2, \ldots, A^k, \ldots$ converges. Moreover, this will occur if and only if the sequences $\lambda_i, \lambda_i^2, \ldots, \lambda_i^k, \ldots$ converges for each of the n eigenvalues λ_i of A.

(b) In general, a given sequence of real numbers $a, a^2, a^3, \ldots$ will converge to 0 if and only if $-1 < a < 1$ and to 1 if $a = 1$. The sequence diverges for all other values of a.

Recall that $P^{-1} A^k P = D^k$ where D^k is a diagonal matrix containing the eigenvalues $\lambda_1^k, \lambda_2^k, \ldots, \lambda_n^k$ on its diagonal. If $|\lambda_i| < 1$ for all $i = 1, 2, \ldots, n$, then $\lim_{k\to\infty} D^k = 0$ and hence $\lim_{k\to\infty} A^k = 0$.

If $\lambda_i = 1$ is an eigenvalue of A for one or more values of i and if all of the other eigenvalues satisfy the inequality $|\lambda_j| < 1$, then $\lim_{k\to\infty} A^k$ exists and equals PD_LP^{-1} where D_L is a diagonal matrix with only 1's and 0's on the diagonal.

If A possesses one or more eigenvalues λ which do not satisfy the inequality $-1 < \lambda \leq 1$, then $\lim_{k\to\infty} A^k$ does not exist.

29. The Jordan block matrix is

$$J_n = \begin{bmatrix} 1 & 1 & 0 & \cdots & 0 & 0 \\ 0 & 1 & 1 & \cdots & 0 & 0 \\ \vdots & \vdots & \ddots & \ddots & \vdots & \vdots \\ 0 & 0 & \cdots & 1 & 1 & 0 \\ 0 & 0 & \cdots & 0 & 1 & 1 \end{bmatrix}.$$

Since this is an upper triangular matrix, we can see that the only eigenvalue is $\lambda = 1$, with algebraic multiplicity n. Solving for the eigenvectors leads to the system

$$(\lambda I - J_n)\,\mathrm{x} = \begin{bmatrix} 0 & 1 & 0 & \cdots & 0 & 0 \\ 0 & 0 & 1 & \cdots & 0 & 0 \\ \vdots & \vdots & \ddots & \ddots & \vdots & \vdots \\ 0 & 0 & \cdots & 0 & 1 & 0 \\ 0 & 0 & \cdots & 0 & 0 & 1 \end{bmatrix} \mathrm{x}.$$

EXERCISE SET 7.3

1. **(a)** The characteristic equation is $\lambda(\lambda - 5) = 0$. Thus each eigenvalue is repeated once and hence each eigenspace is 1-dimensional.

(c) The characteristic equation is $\lambda^2(\lambda - 3) = 0$. Thus the eigenspace corresponding to $\lambda = 0$ is 2-dimensional and that corresponding to $\lambda = 3$ is 1-dimensional.

(e) The characteristic equation is $\lambda^3(\lambda - 8) = 0$. Thus the eigenspace corresponding to $\lambda = 0$ is 3-dimensional and that corresponding to $\lambda = 8$ is 1-dimensional.

13. By the result of Exercise 17, Section 7.1, the eigenvalues of the symmetric 2×2 matrix $\begin{bmatrix} a & b \\ b & a \end{bmatrix}$, are $\lambda = \frac{1}{2}\left[(a+d) \pm \sqrt{(a-d)^2 + 4b^2}\right]$ Since $(a - d)^2 + 4b^2$ cannot be negative, the eigenvalues are real.

15. Yes. Notice that the given vectors are pairwise orthogonal, so we consider the equation

$$P^{-1}AP = D$$

or

$$A = PDP^{-1}$$

where the columns of P consist of the given vectors each divided by its norm and where D is the diagonal matrix with the eigenvalues of A along its diagonal. That is,

$$P = \begin{bmatrix} 0 & 1 & 0 \\ 1/\sqrt{2} & 0 & 1/\sqrt{2} \\ -1/\sqrt{2} & 0 & 1/\sqrt{2} \end{bmatrix} \text{ and } D = \begin{bmatrix} -1 & 0 & 0 \\ 0 & 3 & 0 \\ 0 & 0 & 7 \end{bmatrix}$$

From this, it follows that

$$A = PDP^{-1} = \begin{bmatrix} 3 & 0 & 0 \\ 0 & 3 & 4 \\ 0 & 4 & 3 \end{bmatrix}$$

Alternatively, we could just substitute the appropriate values for λ and $\mathbf{x}$ in the equation $A\mathbf{x} = \lambda\mathbf{x}$ and solve for the matrix A.

SUPPLEMENTARY EXERCISES 7

1. **(a)** The characteristic equation of A is $\lambda^2 - 2\cos\theta + 1 = 0$. The discriminant of this equation is $4(\cos^2\theta - 1)$, which is negative unless $\cos^2\theta = 1$. Thus A can have no real eigenvalues or eigenvectors in case $0 < \theta < \pi$.

3. **(a)** If

$$D = \begin{bmatrix} a_1 & 0 & \cdots & 0 \\ 0 & a_2 & \cdots & 0 \\ \vdots & \vdots & & \vdots \\ 0 & 0 & \cdots & a_n \end{bmatrix}$$

then $D = S^2$, where

$$S = \begin{bmatrix} \sqrt{a_1} & 0 & \cdots & 0 \\ 0 & \sqrt{a_2} & \cdots & 0 \\ \vdots & \vdots & & \vdots \\ 0 & 0 & \cdots & \sqrt{a_n} \end{bmatrix}$$

Of course, this makes sense only if $a_1 \geq 0, \ldots, a_n \geq 0$.

(b) If A is diagonalizable, then there are matrices P and D such that D is diagonal and $D = P^{-1}AP$. Moreover, if A has nonnegative eigenvalues, then the diagonal entries of D are nonnegative since they are all eigenvalues. Thus there is a matrix T, by virtue of Part (a), such that $D = T^2$. Therefore,

$$A = PDP^{-1} = PT^2P^{-1} = PTP^{-1}PTP^{-1} = (PTP^{-1})^2$$

That is, if we let $S = PTP^{-1}$, then $A = S^2$.

3. **(c)** The eigenvalues of A are $\lambda = 9$, $\lambda = 1$, and $\lambda = 4$. The eigenspaces are spanned by the vectors

$$\begin{bmatrix} 1 \\ 2 \\ 2 \end{bmatrix} \quad \begin{bmatrix} 1 \\ 0 \\ 0 \end{bmatrix} \quad \begin{bmatrix} 1 \\ 1 \\ 0 \end{bmatrix}$$

Thus, we have

$$P = \begin{bmatrix} 1 & 1 & 1 \\ 2 & 0 & 1 \\ 2 & 0 & 0 \end{bmatrix} \text{ and } P^{-1} = \begin{bmatrix} 0 & 0 & 1/2 \\ 1 & -1 & 1/2 \\ 0 & 1 & -1 \end{bmatrix}$$

while

$$D = \begin{bmatrix} 9 & 0 & 0 \\ 0 & 1 & 0 \\ 0 & 0 & 4 \end{bmatrix} \text{ and } T = \begin{bmatrix} 3 & 0 & 0 \\ 0 & 1 & 0 \\ 0 & 0 & 2 \end{bmatrix}$$

Therefore

$$S = PTP^{-1} = \begin{bmatrix} 1 & 1 & 0 \\ 0 & 2 & 1 \\ 0 & 0 & 3 \end{bmatrix}$$

5. Since $\det(\lambda I - A)$ is a sum of signed elementary products, we ask which terms involve λ^{n-1}. Obviously the signed elementary product

$$\begin{aligned} q &= (\lambda - a_{11})(\lambda - a_{22}) \cdots (\lambda - a_{nn}) \\ &= \lambda^n - (a_{11} + a_{22} + \cdots + a_{nn})\lambda^{n-1} \\ &\quad + \text{terms involving } \lambda^r \text{ where } r < n - 1 \end{aligned}$$

has a term $-$ (trace of A)λ^{n-1}. Any elementary product containing fewer than $n - 1$ factors of the form $\lambda - a_{ii}$ cannot have a term which contains λ^{n-1}. But there is no elementary product which contains exactly $n - 1$ of these factors (Why?). Thus the coefficient of λ^{n-1} is the negative of the trace of A.

7. **(b)** The characteristic equation is

$$p(\lambda) = -1 + 3\lambda - 3\lambda^2 + \lambda^3 = 0$$

Moreover,

$$A^2 = \begin{bmatrix} 0 & 0 & 1 \\ 1 & -3 & 3 \\ 3 & -8 & 6 \end{bmatrix}$$

and

$$A^3 = \begin{bmatrix} 1 & -3 & 3 \\ 3 & -8 & 6 \\ 6 & -15 & 10 \end{bmatrix}$$

It then follows that

$$p(A) = -I + 3A - 3A^2 + A^3 = 0$$

$$= P\left(\begin{bmatrix} a_0 & 0 & \cdots & 0 \\ 0 & a_0 & \cdots & 0 \\ \vdots & \vdots & \ddots & \vdots \\ 0 & 0 & \cdots & a_0 \end{bmatrix} + \begin{bmatrix} a_1\lambda_1 & 0 & \cdots & 0 \\ 0 & a_1\lambda_2 & \cdots & 0 \\ \vdots & \vdots & \ddots & \vdots \\ 0 & 0 & \cdots & a_1\lambda_n \end{bmatrix} + \begin{bmatrix} a_2\lambda_1^2 & 0 & \cdots & 0 \\ 0 & a_2\lambda_2^2 & \cdots & 0 \\ \vdots & \vdots & \ddots & \vdots \\ 0 & 0 & \cdots & a_2\lambda_n^2 \end{bmatrix}\right.$$

$$\left. + \cdots + \begin{bmatrix} a_n\lambda_1^n & 0 & \cdots & 0 \\ 0 & a_n\lambda_2^n & \cdots & 0 \\ \vdots & \vdots & \ddots & \vdots \\ 0 & 0 & \cdots & a_n\lambda_n^n \end{bmatrix}\right) P^{-1}$$

$$= P\begin{bmatrix} a_0 + a_1\lambda_1 + \cdots + a_n\lambda_1^n & 0 & \cdots & 0 \\ 0 & a_0 + a_1\lambda_2 + \cdots + a_n\lambda_2^n & \cdots & 0 \\ \vdots & \vdots & \ddots & \vdots \\ 0 & 0 & \cdots & a_0 + a_1\lambda_n + \cdots + a_n\lambda_n^n \end{bmatrix} P^{-1}$$

$$= P\begin{bmatrix} p(\lambda_1) & 0 & \cdots & 0 \\ 0 & p(\lambda_2) & \cdots & 0 \\ \vdots & \vdots & \ddots & \vdots \\ 0 & 0 & \cdots & p(\lambda_n) \end{bmatrix} P^{-1}$$

However, each λ_i is a root of the characteristic polynomial, so $p(\lambda_i) = 0$ for $i = 1, \ldots, n$. Then,

$$p(A) = P\begin{bmatrix} 0 & 0 & \cdots & 0 \\ 0 & 0 & \cdots & 0 \\ \vdots & \vdots & \ddots & \vdots \\ 0 & 0 & \cdots & 0 \end{bmatrix}P^{-1}$$

$$= \mathbf{0}.$$

Thus, a diagonalizable matrix satisfies its characteristic polynomial.

9. Since $c_0 = 0$ and $c_1 = -5$, we have $A^2 = 5A$, and, in general, $A^n = 5^{n-1}A$.

11. Call the matrix A and show that $A^2 = (c_1 + c_2 + \cdots + c_n)A = [\text{tr}(A)]A$. Thus $A^k = [\text{tr}(A)]^{k-1}A$ for $k = 2, 3, \ldots$. Now if λ is an eigenvalue of A, then λ^k is an eigenvalue of A^k, so that in case $\text{tr}(A) \neq 0$, we have that $\lambda^k[\text{tr}(A)]^{k-1} = [\lambda/\text{tr}(A)]^{k-1}\lambda$ is an eigenvalue of A for $k = 2, 3, \ldots$. Why? We know that A has at most n eigenvalues, so that this expression can take on only finitely many values. This means that either $\lambda = 0$ or $\lambda = \text{tr}(A)$. Why? In case $\text{tr}(A) = 0$, then all of the eigenvalues of A are 0. Why? Thus the only possible eigenvalues of A are zero and $\text{tr}(A)$. It is easy to check that each of these is, in fact, an eigenvalue of A.

Alternatively, we could evaluate $\det(I\lambda - A)$ by brute force. If we add Column 1 to Column 2, the new Column 2 to Column 3, the new Column 3 to Column 4, and so on, we obtain the equation

$$\det(I\lambda - A) = \det\begin{bmatrix} \lambda - c_1 & \lambda - c_1 - c_2 & \lambda - c_1 - c_2 - c_3 & \cdots & \lambda - c_1 - c_2 - \cdots - c_n \\ -c_1 & \lambda - c_1 - c_2 & \lambda - c_1 - c_2 - c_3 & \cdots & \lambda - c_1 - c_2 - \cdots - c_n \\ -c_1 & -c_1 - c_2 & \lambda - c_1 - c_2 - c_3 & \cdots & \lambda - c_1 - c_2 - \cdots - c_n \\ \vdots & \vdots & \vdots & & \vdots \\ -c_1 & -c_1 - c_2 & -c_1 - c_2 - c_3 & \cdots & \lambda - c_1 - c_2 - \cdots - c_n \end{bmatrix}$$

If we subtract Row 1 from each of the other rows and then expand by cofactors along the nth column, we have

$$\det(I-A)=\det\begin{bmatrix} \lambda-c_1 & \lambda-c_1-c_2 & \lambda-c_1-c_2-c_3 & \cdots & \lambda-\mathrm{tr}(A) \\ -\lambda & 0 & 0 & \cdots & 0 \\ -\lambda & -\lambda & 0 & \cdots & 0 \\ \vdots & \vdots & \vdots & & \vdots \\ -\lambda & -\lambda & -\lambda & \cdots & 0 \end{bmatrix}$$

$$=(-1)^{n+1}(\lambda-tr(A))\det\begin{bmatrix} -\lambda & 0 & 0 & \cdots & 0 \\ -\lambda & -\lambda & 0 & \cdots & 0 \\ -\lambda & -\lambda & -\lambda & \cdots & 0 \\ \vdots & \vdots & \vdots & & \vdots \\ -\lambda & -\lambda & -\lambda & \cdots & -\lambda \end{bmatrix}$$

$$=(-1)^{n+1}(\lambda-\mathrm{tr}(A))(-\lambda)^{n-1} \quad \text{because the above matrix is triangular}$$

$$=(-1)^{2n}(\lambda-\mathrm{tr}(A))\lambda^{n-1}$$

$$=\lambda^{n-1}(\lambda-\mathrm{tr}(A))$$

Thus $\lambda = \mathrm{tr}(A)$ and $\lambda = 0$ are the eigenvalues, with $\lambda = 0$ repeated $n - 1$ times.

17. Since every odd power of A is again A, we have that every odd power of an eigenvalue of A is again an eigenvalue of A. Thus the only possible eigenvalues of A are $\lambda = 0, \pm 1$.

EXERCISE SET 8.1

3. Since $T(-\mathbf{u}) = \|-\mathbf{u}\| = \|\mathbf{u}\| = T(\mathbf{u}) \neq -T(\mathbf{u})$ unless $\mathbf{u} = \mathbf{0}$, the function is not linear.

5. We observe that

$$T(A_1 + A_2) = (A_1 + A_2)B = A_1B + A_2B = T(A_1) + T(A_2)$$

and

$$T(cA) = (cA)B = c(AB) = cT(A)$$

Hence, T is linear.

17. **(a)** Since T_1 is defined on all of R^2, the domain of $T_2 \circ T_1$ is R^2. We have $T_2 \circ T_1(x, y) = T_2(T_1(x, y)) = T_2(2x, 3y) = (2x - 3y, 2x + 3y)$. Since the system of equations

$$2x - 3y = a$$

$$2x + 3y = b$$

can be solved for all values of a and b, the codomain is also all of R^2.

(d) Since T_1 is defined on all of R^2, the domain of $T_2 \circ T_1$ is R^2. We have

$$T_2(T_1(x, y)) = T_2(x - y, y + z, x - z) = (0, 2x)$$

Thus the codomain of $T_2 \circ T_1$ is the y-axis.

19. **(a)** We have

$$(T_1 \circ T_2)(A) = \mathrm{tr}(A^T) = \mathrm{tr}\begin{bmatrix} a & c \\ b & d \end{bmatrix} = a + d$$

(b) Since the range of T_1 is not contained in the domain of T_2, $T_2 \circ T_1$ is not well defined.

25. Since (1, 0, 0) and (0, 1, 0) form an orthonormal basis for the xy-plane, we have $T(x, y, z) = (x, 0, 0) + (0, y, 0) = (x, y, 0)$, which can also be arrived at by inspection. Then $T(T(x, y, z)) = T(x, y, 0) = (x, y, 0) = T(x, y, z)$. This says that T leaves every point in the x-y plane unchanged.

31. **(b)** We have

$$(J \circ D)(\sin x) = \int_0^x (\sin t)'dt = \sin(x) - \sin(0) = \sin(x)$$

(c) We have

$$(J \circ D)(e^x + 3) = \int_0^x (e^x + 3)'dt = e^x - 1$$

33. **(a)** True. Let $c_1 = c_2 = 1$ to establish Part (a) of the definition and let $c_2 = 0$ to establish Part (b).

(b) False. All linear transformations have this property, and, for instance there is more than one linear transformation from R^2 to R^2.

(c) True. If we let $\mathbf{u} = \mathbf{0}$, then we have $T(\mathbf{v}) = T(-\mathbf{v}) = -T(\mathbf{v})$. That is, $T(\mathbf{v}) = \mathbf{0}$ for all vectors $\mathbf{v}$ in V. But there is only one linear transformation which maps every vector to the zero vector.

(d) False. For this operator T, we have

$$T(\mathbf{v} + \mathbf{v}) = T(2\mathbf{v}) = \mathbf{v}_0 + 2\,\mathbf{v}$$

But

$$T(\mathbf{v}) + T(\mathbf{v}) = 2T(\mathbf{v}) = 2\mathbf{v}_0 + 2\mathbf{v}$$

Since $\mathbf{v}_0 \neq \mathbf{0}$, these two expressions cannot be equal.

35. Yes. Let $T : P^n \to P^m$ be the given transformation, and let $T_R : R^{n+1} \to R^{m+1}$ be the corresponding linear transformation in the sense of Section 4.4. Let $\varphi_n : P_n \to R^{n+1}$ be the function that maps a polynomial in P_n to its coordinate vector in R^{n+1}, and let $\varphi_m : P_m \to R^{m+1}$ be the function that maps a polynomial in P_m to its coordinate vector in R^{m+1}.

By Example 7, both φ_n and φ_m are linear transformations. Theorem 5.4.1 implies that a coordinate map is invertible, so φ_m^{-1} is also a linear transformation.

We have $T = \varphi_m^{-1} \circ T_R \circ \varphi_n$, so T is a composition of linear transformations. Refer to the diagram below:

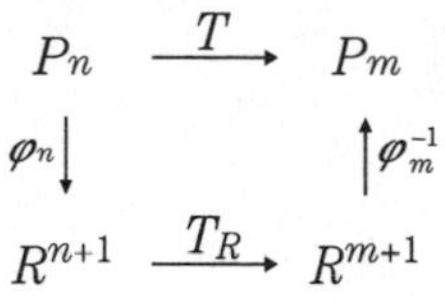

$$\begin{array}{ccc} P_n & \xrightarrow{T} & P_m \\ \varphi_n \downarrow & & \uparrow \varphi_m^{-1} \\ R^{n+1} & \xrightarrow{T_R} & R^{m+1} \end{array}$$

Thus, by Theorem 8.1.2., T is itself a linear transformation.

EXERCISE SET 8.2

1. **(a)** If $(1, -4)$ is in $R(T)$, then there must be a vector (x, y) such that $T(x, y) = (2x - y, -8x + 4y) = (1, -4)$. If we equate components, we find that $2x - y = 1$ or $y = t$ and $x = (1 + t)/2$. Thus T maps infinitely many vectors into $(1, -4)$.

(b) Proceeding as above, we obtain the system of equations

$$2x - y = 5$$

$$-8x + 4y = 0$$

Since $2x \quad y = 5$ implies that $-8x + 4y = -20$, this system has no solution. Hence $(5, 0)$ is not in $R(T)$.

3. **(b)** The vector $(1, 3, 0)$ is in $R(T)$ if and only if the following system of equations has a solution:

$$4x + y - 2z - 3w = 1$$

$$2x + y + z - 4w = 3$$

$$6x \qquad -9z + 9w = 0$$

This system has infinitely many solutions $x = (3/2)(t - 1)$, $y = 10 - 4t$, $z = t$, $w = 1$ where t is arbitrary. Thus $(1, 3, 0)$ is in $R(T)$.

5. **(a)** Since $T(x^2) = x^3 \neq 0$, the polynomial x^2 is not in $\ker(T)$.

7. **(a)** We look for conditions on x and y such that $2x - y =$ and $-8x + 4y = 0$. Since these equations are satisfied if and only if $y = 2x$, the kernel will be spanned by the vector $(1, 2)$, which is then a basis.

(c) Since the only vector which is mapped to zero is the zero vector, the kernel is $\{\mathbf{0}\}$ and has dimension zero so the basis is the empty set.

9. **(a)** Here $n = \dim(R^2) = 2$, rank$(T) = 1$ by the result of Exercise 8(a), and nullity$(T) = 1$ by Exercise 7(a). Recall that $1 + 1 = 2$.

(c) Here $n = \dim(P_2) = 3$, rank$(T) = 3$ by virtue of Exercise 8(c), and nullity$(T) = 0$ by Exercise 7(c). Thus we have $3 = 3 + 0$.

19. By Theorem 8.2.1, the kernel of T is a subspace of R^3. Since the only subspaces of R^3 are the origin, a line through the origin, a plane through the origin, or R^3 itself, the result follows. It is clear that all of these possibilities can actually occur.

21. **(a)** If

$$\begin{bmatrix} 1 & 3 & 4 \\ 3 & 4 & 7 \\ -2 & 2 & 0 \end{bmatrix} \begin{bmatrix} x \\ y \\ z \end{bmatrix} = \begin{bmatrix} 0 \\ 0 \\ 0 \end{bmatrix}$$

then $x = -t$, $y = -t$, $z = t$. These are parametric equations for a line through the origin.

(b) Using elementary column operations, we reduce the given matrix to

$$\begin{bmatrix} 1 & 0 & 0 \\ 3 & -5 & 0 \\ -2 & 8 & 0 \end{bmatrix}$$

Thus, $(1, 3, -2)^T$ and $(0, -5, 8)^T$ form a basis for the range. That range, which we can interpret as a subspace of R^3, is a plane through the origin. To find a normal to that plane, we compute

$$(1, 3, -2) \times (0, -5, 8) = (14, -8, -5)$$

Therefore, an equation for the plane is

$$14x - 8y - 5z = 0$$

Alternatively, but more painfully, we can use elementary row operations to reduce the matrix

$$\begin{bmatrix} 1 & 3 & 4 & x \\ 3 & 4 & 7 & y \\ -2 & 2 & 0 & z \end{bmatrix}$$

to the matrix

$$\begin{bmatrix} 1 & 0 & 1 & (-4x+3y)/5 \\ 0 & 1 & 1 & (3x-y)/5 \\ 0 & 0 & 0 & 14x-8y-5z \end{bmatrix}$$

Thus the vector (x, y, z) is in the range of T if and only if $14x - 8y - 5z = 0$.

23. The rank of T is at most 1, since $\dim R = 1$ and the image of T is a subspace of R. So, we know that either $\text{rank}(T) = 0$ or $\text{rank}(T) = 1$. If $\text{rank}(T) = 0$, then every matrix A is in the kernel of T, so every $n \times n$ matrix A has diagonal entries that sum to zero. This is clearly false, so we must have that $\text{rank}(T) = 1$. Thus, by the Dimension Theorem (Theorem 8.2.3), $\dim(\ker(T)) = n^2 - 1$.

27. If $f(x)$ is in the kernel of $D \circ D$, then $f''(x) = 0$ or $f(x) = ax + b$. Since these are the only eligible functions $f(x)$ for which $f''(x) = 0$ (Why?), the kernel of $D \circ D$ is the set of all functions $f(x) = ax + b$, or all straight lines in the plane. Similarly, the kernel of $D \circ D \circ D$ is the set of all functions $f(x) = ax^2 + bx + c$, or all straight lines except the y-axis and certain parabolas in the plane.

29. **(a)** Since the range of T has dimension 3 minus the nullity of T, then the range of T has dimension 2. Therefore it is a plane through the origin.

(b) As in Part (a), if the range of T has dimension 2, then the kernel must have dimension 1. Hence, it is a line through the origin.

EXERCISE SET 8.3

1. **(a)** Clearly $\ker(T) = \{(0, 0)\}$, so T is one-to-one.

(c) Since $T(x, y) = (0, 0)$ if and only if $x = y$ and $x = -y$, the kernel is $\{0, 0\}$ and T is one-to-one.

(e) Here $T(x, y) = (0, 0, 0)$ if and only if x and y satisfy the equations $x - y = 0$, $-x + y = 0$, and $2x - 2y = 0$. That is, (x, y) is in $\ker(T)$ if and only if $x = y$, so the kernel of T is this line and T is not one-to-one.

3. **(a)** Since $\det(A) = 0$, or equivalently, $\text{rank}(A) < 3$, T has no inverse.

(c) Since A is invertible, we have

$$T^{-1}\begin{bmatrix} x_1 \\ x_2 \\ x_3 \end{bmatrix} = A^{-1}\begin{bmatrix} x_1 \\ x_2 \\ x_3 \end{bmatrix} = \begin{bmatrix} \frac{1}{2} & -\frac{1}{2} & \frac{1}{2} \\ -\frac{1}{2} & \frac{1}{2} & \frac{1}{2} \\ \frac{1}{2} & \frac{1}{2} & -\frac{1}{2} \end{bmatrix}\begin{bmatrix} x_1 \\ x_2 \\ x_3 \end{bmatrix}$$

$$= \begin{bmatrix} \frac{1}{2}(\ \ x_1 - x_2 + x_3) \\ \frac{1}{2}(-x_1 + x_2 + x_3) \\ \frac{1}{2}(\ \ x_1 + x_2 - x_3) \end{bmatrix}$$

5. **(a)** The kernel of T is the line $y = -x$ since all points on this line (and only those points) map to the origin.

(b) Since the kernel is not {0, 0}, the transformation is not one-to-one.

7. **(b)** Since nullity$(T) = n - \text{rank}(T) = 1$, T is not one-to-one.

(c) Here T cannot be one-to-one since rank$(T) \leq n < m$, so nullity$(T) \geq 1$.

11. **(a)** We know that T will have an inverse if and only if its kernel is the zero vector, which means if and only if none of the numbers $a_i = 0$.

13. **(a)** By inspection, $T_1^{-1}(p(x)) = p(x)/x$, where $p(x)$ must, of course, be in the range of T_1 and hence have constant term zero. Similarly $T_2^{-1}(p(x)) = p(x - 1)$, where, again, $p(x)$ must be in the range of T_2. Therefore $(T_2 \circ T_1)^{-1}(p(x)) = p(x - 1)/x$ for appropriate polynomials $p(x)$.

17. **(a)** Since T sends the nonzero matrix $\begin{bmatrix} 0 & 1 \\ 0 & 0 \end{bmatrix}$ to the zero matrix, it is not one-to-one.

(c) Since T sends only the zero matrix to the zero matrix, it is one-to-one. By inspection, $T^{-1}(A) = T(A)$.

Alternative Solution: T can be represented by the matrix

$$T_B = \begin{bmatrix} 0 & 0 & 0 & 1 \\ 0 & -1 & 0 & 0 \\ 0 & 0 & -1 & 0 \\ 1 & 0 & 0 & 0 \end{bmatrix}$$

By direct calculation, $T_B = (T_B)^{-1}$, so $T = T^{-1}$.

19. Suppose that $\mathbf{w}_1$ and $\mathbf{w}_2$ are in $R(T)$. We must show that

$$T^{-1}(\mathbf{w}_1 + \mathbf{w}_2) = T^{-1}(\mathbf{w}_1) + T^{-1}(\mathbf{w}_2)$$

and

$$T^{-1}(k\mathbf{w}_1) = kT^{-1}(\mathbf{w}_1)$$

Because T is one-to-one, the above equalities will hold if and only if the results of applying T to both sides are indeed valid equalities. This follows immediately from the linearity of the transformation T.

21. It is easy to show that T is linear. However, T is not one-to-one, since, for instance, it sends the function $f(x) = x - 5$ to the zero vector.

25. Yes. The transformation is linear and only (0, 0, 0) maps to the zero polynomial. Clearly distinct triples in R^3 map to distinct polynomials in P_2.

27. No. T is a linear operator by Theorem 3.4.2. However, it is not one-to-one since $T(\mathbf{a}) = \mathbf{a} \times \mathbf{a} = \mathbf{0} = T(\mathbf{0})$. That is, T maps $\mathbf{a}$ to the zero vector, so if T is one-to-one, $\mathbf{a}$ must be the zero vector. But then T would be the zero transformation, which is certainly not one-to-one.

EXERCISE SET 8.4

9. **(a)** Since A is the matrix of T with respect to B, then we know that the first and second columns of A must be $[T(\mathbf{v}_1)]_B$ and $[T(\mathbf{v}_2)]_B$, respectively. That is

$$[T(\mathbf{v}_1)]_B = \begin{bmatrix} 1 \\ -2 \end{bmatrix}$$

$$[T(\mathbf{v}_2)]_B = \begin{bmatrix} 3 \\ 5 \end{bmatrix}$$

Alternatively, since $\mathbf{v}_1 = 1\mathbf{v}_1 + 0\mathbf{v}_2$ and $\mathbf{v}_2 = 0\mathbf{v}_1 + 1\mathbf{v}_2$, we have

$$[T(\mathbf{v}_1)]_B = A\begin{bmatrix} 1 \\ 0 \end{bmatrix} = \begin{bmatrix} 1 \\ -2 \end{bmatrix}$$

and

$$[T(\mathbf{v}_2)]_B = A\begin{bmatrix} 0 \\ 1 \end{bmatrix} = \begin{bmatrix} 3 \\ 5 \end{bmatrix}$$

(b) From Part (a),

$$T(\mathbf{v})_1 = \mathbf{v}_1 - 2\mathbf{v}_2 = \begin{bmatrix} 3 \\ -5 \end{bmatrix}$$

and

$$T(\mathbf{v}_2) = 3\mathbf{v}_1 + 5\mathbf{v}_2 = \begin{bmatrix} -2 \\ 29 \end{bmatrix}$$

(c) Since we already know $T(\mathbf{v}_1)$ and $T(\mathbf{v}_2)$, all we have to do is express $[x_1 \quad x_2]^T$ in terms of $\mathbf{v}_1$ and $\mathbf{v}_2$. If

$$\begin{bmatrix} x_1 \\ x_2 \end{bmatrix} = a\mathbf{v}_1 + b\mathbf{v}_2 = a\begin{bmatrix} 1 \\ 3 \end{bmatrix} + b\begin{bmatrix} -1 \\ 4 \end{bmatrix}$$

then

$$x_1 = a - b$$

$$x_2 = 3a + 4b$$

or

$$a = (4x_1 + x_2)/7$$

$$b = (-3x_1 + x_2)/7$$

Thus

$$\left(\begin{bmatrix} x_1 \\ x_2 \end{bmatrix}\right) = \frac{4x_1 + x_2}{7}\begin{bmatrix} 3 \\ -5 \end{bmatrix} + \frac{-3x_1 + x_2}{7}\begin{bmatrix} -2 \\ 29 \end{bmatrix}$$

$$= \begin{bmatrix} \dfrac{18x_1 + x_2}{7} \\ \dfrac{-107x_1 + 24x_2}{7} \end{bmatrix}$$

(d) By the above formula,

$$T\left(\begin{bmatrix} 1 \\ 1 \end{bmatrix}\right) = \begin{bmatrix} 19/7 \\ -83/7 \end{bmatrix}$$

11. **(a)** The columns of A, by definition, are $[T(\mathbf{v}_1)]_B$, $[T(\mathbf{v}_2)]_B$, and $[T(\mathbf{v}_3)]_B$, respectively.

(b) From Part (a),

$$T(\mathbf{v}_1) = \mathbf{v}_1 + 2\mathbf{v}_2 + 6\mathbf{v}_3 = 16 + 51x + 19x^2$$

$$T(\mathbf{v}_2) = 3\mathbf{v}_1 - 2\mathbf{v}_3 = -6 - 5x + 5x^2$$

$$T(\mathbf{v}_3) = -\mathbf{v}_1 + 5\mathbf{v}_2 + 4\mathbf{v}_3 = 7 + 40x + 15x^2$$

(c) Let $a_0 + a_1x + a_2x^2 = b_0\mathbf{v}_1 + b_1\mathbf{v}_2 + b_2\mathbf{v}_3$. Then

$$\begin{aligned} a_0 &= \quad\quad\;\; - b_1 + 3b_2 \\ a_1 &= 3b_0 + 3b_1 + 7b_2 \\ a_2 &= 3b_0 + 2b_1 + 2b_2 \end{aligned}$$

This system of equations has the solution

$$\begin{aligned} b_0 &= (a_0 - a_1 + 2a_2)/3 \\ b_1 &= (-5a_0 + 3a_1 - 3a_2)/8 \\ b_2 &= (a_0 + a_1 - a_2)/8 \end{aligned}$$

Thus

$$\begin{aligned} T(a_0 + a_1x + a_2x^2) &= b_0T(\mathbf{v}_1) + b_1T(\mathbf{v}_2) + b_2T(\mathbf{v}_3) \\ &= \frac{239a_0 - 161a_1 + 247a_2}{24} \\ &\quad + \frac{201a_0 - 111a_1 + 247a_2}{8}x \\ &\quad + \frac{61a_0 - 31a_1 + 107a_2}{12}x^2 \end{aligned}$$

(d) By the above formula,

$$T(1 + x^2) = 2 + 56x + 14x^2$$

13. **(a)** Since

$$T_1(1) = 2 \quad \text{and} \quad T_1(x) = -3x^2$$

$$T_2(1) = 3x \quad T_2(x) = 3x^2 \quad \text{and} \quad T_2(x_2) = 3x^3$$

$$T_2 \circ T_1(1) = 6x \quad \text{and} \quad T_2 \circ T_1(x) = -9x_3$$

we have

$$[T_1]_{B'',B} = \begin{bmatrix} 2 & 0 \\ 0 & 0 \\ 0 & -3 \end{bmatrix} \quad [T_2]_{B',B''} = \begin{bmatrix} 0 & 0 & 0 \\ 3 & 0 & 0 \\ 0 & 3 & 0 \\ 0 & 0 & 3 \end{bmatrix}$$

and

$$\left[T_2 \circ T_1\right]_{B',B} = \begin{bmatrix} 0 & 0 \\ 6 & 0 \\ 0 & 0 \\ 0 & -9 \end{bmatrix}$$

(b) We observe that here

$$[T_2 \circ T_1]_{B',B} = [T_2]_{B',B''}\,[T_1]_{B'',B}$$

15. If T is a contraction or a dilation of V, then T maps any basis $B = \{\mathbf{v}_1, \ldots, \mathbf{v}_n\}$ of V to $\{k\mathbf{v}_1, \ldots, k\mathbf{v}_n\}$ where k is a nonzero constant. Therefore the matrix of T with respect to B is

$$\begin{bmatrix} k & 0 & 0 & \cdots & 0 \\ 0 & k & 0 & \cdots & 0 \\ 0 & 0 & k & \cdots & 0 \\ \vdots & \vdots & \vdots & & \vdots \\ 0 & 0 & 0 & \cdots & k \end{bmatrix}$$

17. The standard matrix for T is just the $m \times n$ matrix whose columns are the transforms of the standard basis vectors. But since B is indeed the standard basis for R^n, the matrices are the same. Moreover, since B' is the standard basis for R^m, the resulting transformation will yield vector components relative to the standard basis, rather than to some other basis.

19. **(c)** Since $D(\mathbf{f}_1) = 2\mathbf{f}_1$, $D(\mathbf{f}_2) = \mathbf{f}_1 + 2\mathbf{f}_2$, and $D(\mathbf{f}_3) = 2\mathbf{f}_2 + 2\mathbf{f}_3$, we have the matrix

$$\begin{bmatrix} 2 & 1 & 0 \\ 0 & 2 & 2 \\ 0 & 0 & 2 \end{bmatrix}$$

EXERCISE SET 8.5

1. First, we find the matrix of T with respect to B. Since

$$T(\mathbf{u}_1) = \begin{bmatrix} 1 \\ 0 \end{bmatrix}$$

and

$$T(\mathbf{u}_2) = \begin{bmatrix} -2 \\ -1 \end{bmatrix}$$

then

$$A = [T]_B = \begin{bmatrix} 1 & -2 \\ 0 & -1 \end{bmatrix}$$

In order to find P, we note that $\mathbf{v}_1 = 2\mathbf{u}_1 + \mathbf{u}_2$ and $\mathbf{v}_2 = -3\mathbf{u}_1 + 4\mathbf{u}_2$. Hence the transition matrix from B' to B is

$$P = \begin{bmatrix} 2 & -3 \\ 1 & 4 \end{bmatrix}$$

Thus

$$P^{-1} = \begin{bmatrix} \frac{4}{11} & \frac{3}{11} \\ -\frac{1}{11} & \frac{2}{11} \end{bmatrix}$$

and therefore

$$A' = [T]_{B'} = P^{-1}[T]_B P = \frac{1}{11}\begin{bmatrix} 4 & 3 \\ -1 & 2 \end{bmatrix}\begin{bmatrix} 1 & -2 \\ 0 & -1 \end{bmatrix}\begin{bmatrix} 2 & -3 \\ 1 & 4 \end{bmatrix}$$

$$= \begin{bmatrix} -\frac{3}{11} & -\frac{56}{11} \\ -\frac{2}{11} & \frac{3}{11} \end{bmatrix}$$

3. Since $T(\mathbf{u}_1) = (1/\sqrt{2}, 1/\sqrt{2})$ and $T(\mathbf{u}_2) = (-1/\sqrt{2}, 1/\sqrt{2})$, then the matrix of T with respect to B is

$$A = [T]_B = \begin{bmatrix} 1/\sqrt{2} & -1/\sqrt{2} \\ 1/\sqrt{2} & 1/\sqrt{2} \end{bmatrix}$$

From Exercise 1, we know that

$$P = \begin{bmatrix} 2 & -3 \\ 1 & 4 \end{bmatrix} \quad \text{and} \quad P^{-1} = \frac{1}{11}\begin{bmatrix} 4 & 3 \\ -1 & 2 \end{bmatrix}$$

Thus

$$A' = [T]_{B'} = P^{-1}AP = \frac{1}{11\sqrt{2}}\begin{bmatrix} 13 & -25 \\ 5 & 9 \end{bmatrix}$$

5. Since $T(\mathbf{e}_1) = (1, 0, 0)$, $T(\mathbf{e}_2) = (0, 1, 0)$, and $T(\mathbf{e}_3) = (0, 0, 0)$, we have

$$A = [T]_B = \begin{bmatrix} 1 & 0 & 0 \\ 0 & 1 & 0 \\ 0 & 0 & 0 \end{bmatrix}$$

In order to compute P, we note that $\mathbf{v}_1 = \mathbf{e}_1$, $\mathbf{v}_2 = \mathbf{e}_1 + \mathbf{e}_2$, and $\mathbf{v}_3 = \mathbf{e}_1 + \mathbf{e}_2 + \mathbf{e}_3$. Hence,

$$P = \begin{bmatrix} 1 & 1 & 1 \\ 0 & 1 & 1 \\ 0 & 0 & 1 \end{bmatrix}$$

and

$$P^{-1} = \begin{bmatrix} 1 & -1 & 0 \\ 0 & 1 & -1 \\ 0 & 0 & 1 \end{bmatrix}$$

Thus

$$[T]_{B'} = \begin{bmatrix} 1 & -1 & 0 \\ 0 & 1 & -1 \\ 0 & 0 & 1 \end{bmatrix} \begin{bmatrix} 1 & 0 & 0 \\ 0 & 1 & 0 \\ 0 & 0 & 0 \end{bmatrix} \begin{bmatrix} 1 & 1 & 1 \\ 0 & 1 & 1 \\ 0 & 0 & 1 \end{bmatrix} = \begin{bmatrix} 1 & 0 & 0 \\ 0 & 1 & 1 \\ 0 & 0 & 0 \end{bmatrix}$$

7. Since

$$T(\mathbf{p}_1) = 9 + 3x = \frac{2}{3}\mathbf{p}_1 + \frac{1}{2}\mathbf{p}_2$$

and

$$T(\mathbf{p}_2) = 12 + 2x = -\frac{2}{9}\mathbf{p}_1 + \frac{4}{3}\mathbf{p}_2$$

we have

$$[T]_B = \begin{bmatrix} \frac{2}{3} & -\frac{2}{9} \\ \frac{1}{2} & \frac{4}{3} \end{bmatrix}$$

We note that $\mathbf{q}_1 = -\frac{2}{9}\mathbf{p}_1 + \frac{1}{3}\mathbf{p}_2$ and $\mathbf{q}_2 = \frac{7}{9}\mathbf{p}_1 - \frac{1}{6}\mathbf{p}_2$. Hence

$$P = \begin{bmatrix} -\frac{2}{9} & \frac{7}{9} \\ \frac{1}{3} & -\frac{1}{6} \end{bmatrix}$$

and

$$P^{-1} = \begin{bmatrix} \frac{3}{4} & \frac{7}{2} \\ \frac{3}{2} & 1 \end{bmatrix}$$

Therefore

$$[T]_{B'} = \begin{bmatrix} \frac{3}{4} & \frac{7}{2} \\ \frac{3}{2} & 1 \end{bmatrix} \begin{bmatrix} \frac{2}{3} & -\frac{2}{9} \\ \frac{1}{2} & \frac{4}{3} \end{bmatrix} \begin{bmatrix} -\frac{2}{9} & \frac{7}{9} \\ \frac{1}{3} & -\frac{1}{6} \end{bmatrix} = \begin{bmatrix} 1 & 1 \\ 0 & 1 \end{bmatrix}$$

9. (a) If A and C are similar $n \times n$ matrices, then there exists an invertible $n \times n$ matrix P such that $A = P^{-1}CP$. We can interpret P as being the transition matrix from a basis B' for R^n to a basis B. Moreover, C induces a linear transformation $T : R^n \to R^n$ where $C = [T]_B$. Hence $A = [T]_{B'}$. Thus A and C are matrices for the same transformation with respect to different bases. But from Theorem 8.2.2, we know that the rank of T is equal to the rank of C and hence to the rank of A.

Alternate Solution: We observe that if P is an invertible $n \times n$ matrix, then P represents a linear transformation of R^n *onto* R^n. Thus the rank of the transformation represented by the matrix CP is the same as that of C. Since P^{-1} is also invertible, its null space contains only the zero vector, and hence the rank of the transformation represented by the matrix $P^{-1}\,CP$ is also the same as that of C. Thus the ranks of A and C are equal. Again we use the result of Theorem 8.2.2 to equate the rank of a linear transformation with the rank of a matrix which represents it.

Second Alternative: Since the assertion that similar matrices have the same rank deals only with matrices and not with transformations, we outline a proof which involves only matrices. If $A = P^{-1}\,CP$, then P^{-1} and P can be expressed as products of elementary matrices. But multiplication of the matrix C by an elementary matrix is equivalent to performing an elementary row or column operation on C. From Section 5.5, we know that such operations do not change the rank of C. Thus A and C must have the same rank.

11. **(a)** The matrix for T relative to the standard basis B is

$$[T]_B = \begin{bmatrix} 1 & -1 \\ 2 & 4 \end{bmatrix}$$

The eigenvalues of $[T]_B$ are $\lambda = 2$ and $\lambda = 3$, while corresponding eigenvectors are $(1, -1)$ and $(1, -2)$, respectively. If we let

$$P = \begin{bmatrix} 1 & 1 \\ -1 & -2 \end{bmatrix} \quad \text{then} \quad P^{-1} = \begin{bmatrix} 2 & 1 \\ -1 & -1 \end{bmatrix}$$

and

$$P^{-1}[T]_B P = \begin{bmatrix} 2 & 0 \\ 0 & 3 \end{bmatrix}$$

is diagonal. Since P represents the transition matrix from the basis B' to the standard basis B, we have

$$B' = \left\{ \begin{bmatrix} 1 \\ -1 \end{bmatrix}, \begin{bmatrix} 1 \\ -2 \end{bmatrix} \right\}$$

as a basis which produces a diagonal matrix for $[T]_{B'}$.

13. **(a)** The matrix of T with respect to the standard basis for P_2 is

$$A = \begin{bmatrix} 5 & 6 & 2 \\ 0 & -1 & -8 \\ 1 & 0 & -2 \end{bmatrix}$$

The characteristic equation of A is

$$\lambda^3 - 2\lambda^2 - 15\lambda + 36 = (\lambda - 3)^2(\lambda + 4) = 0$$

and the eigenvalues are therefore $\lambda = -4$ and $\lambda = 3$.

(b) If we set $\lambda = -4$, then $(\lambda I - A)\mathbf{x} = \mathbf{0}$ becomes

$$\begin{bmatrix} -9 & -6 & -2 \\ 0 & -3 & 8 \\ -1 & 0 & -2 \end{bmatrix} \begin{bmatrix} x_1 \\ x_2 \\ x_3 \end{bmatrix} = \begin{bmatrix} 0 \\ 0 \\ 0 \end{bmatrix}$$

The augmented matrix reduces to

$$\begin{bmatrix} 1 & 0 & 2 & 0 \\ 0 & 1 & -8/3 & 0 \\ 0 & 0 & 0 & 0 \end{bmatrix}$$

and hence $x_1 = -2s$, $x_2 = \frac{8}{3}s$, and $x_3 = s$. Therefore the vector

$$\begin{bmatrix} -2 \\ 8/3 \\ 1 \end{bmatrix}$$

is a basis for the eigenspace associated with $\lambda = -4$. In P^2, this vector represents the polynomial $-2 + \frac{8}{3}x + x^2$.

If we set $\lambda = 3$ and carry out the above procedure, we find that $x_1 = 5\,s$, $x_2 = -2s$, and $x_3 = s$. Thus the polynomial $5 - 2x + x^2$ is a basis for the eigenspace associated with $\lambda = 3$.

15. If $\mathbf{v}$ is an eigenvector of T corresponding to λ, then $\mathbf{v}$ is a nonzero vector which satisfies the equation $T(\mathbf{v}) = \lambda\mathbf{v}$ or $(\lambda I - T)\mathbf{v} = \mathbf{0}$. Thus $\lambda I - T$ maps $\mathbf{v}$ to $\mathbf{0}$, or $\mathbf{v}$ is in the kernel of $\lambda I - T$.

17. Since $C[\mathbf{x}]_B = D[\mathbf{x}]_B$ for all $\mathbf{x}$ in V, we can, in particular, let $\mathbf{x} = \mathbf{v}_i$ for each of the basis vectors $\mathbf{v}_1, \ldots, \mathbf{v}_n$ of V. Since $[\mathbf{v}_i]_B = \mathbf{e}_i$ for each i where $\{\mathbf{e}_1, \ldots, \mathbf{e}_n\}$ is the standard basis for R^n, this yields $C\mathbf{e}_i = D\mathbf{e}_i$ for $i = 1, \ldots, n$. But $C\mathbf{e}_i$ and $D\mathbf{e}_i$ are just the ith columns of C and D, respectively. Since corresponding columns of C and D are all equal, we have $C = D$.

19. **(a)** False. Every matrix is similar to itself, since $A = I^{-1}AI$.

(b) True. Suppose that $A = P^{-1}BP$ and $B = Q^{-1}CQ$. Then

$$A = P^{-1}(Q^{-1}CQ)P = (P^{-1}Q^{-1})C(QP) = (QP)^{-1}C(QP)$$

Therefore A and C are similar.

(c) True. By Table 1, A is invertible if and only if B is invertible, which guarantees that A is singular if and only if B is singular.

Alternatively, if $A = P^{-1} BP$, then $B = PAP^{-1}$. Thus, if B is singular, then so is A. Otherwise, B would be the product of 3 invertible matrices.

(d) True. If $A = P^{-1} BP$, then $A^{-1}=(P^{-1} BP)^{-1} = P^{-1} B^{-1}(P^{-1})^{-1} = P^{-1}B^{-1}P$, so A^{-1} and B^{-1} are similar.

25. First, we need to prove that for any square matrices A and B, the trace satisfies $\text{tr}(A) = \text{tr}(B)$. Let

$$A = \begin{bmatrix} a_{11} & a_{12} & \cdots & a_{1n} \\ a_{21} & a_{22} & \cdots & a_{2n} \\ \vdots & \vdots & \ddots & \vdots \\ a_{n1} & a_{n2} & \cdots & a_{nn} \end{bmatrix} \quad \text{and} \quad B = \begin{bmatrix} b_{11} & b_{12} & \cdots & b_{1n} \\ b_{21} & b_{22} & \cdots & b_{2n} \\ \vdots & \vdots & \ddots & \vdots \\ b_{n1} & b_{n2} & \cdots & b_{nn} \end{bmatrix}$$

Then,

$$[AB]_{11} = a_{11}b_{11} + a_{12}b_{21} + a_{13}b_{31} + \cdots + a_{1n}b_{n1} = \sum_{j=1}^{n} a_{1j}b_{j1}$$

$$[AB]_{22} = a_{21}b_{12} + a_{22}b_{22} + a_{23}b_{32} + \cdots + a_{2n}b_{n2} = \sum_{j=1}^{n} a_{2j}b_{j2}$$

$$\vdots$$

$$[AB]_{nn} = a_{n1}b_{1n} + a_{n2}b_{2n} + a_{n3}b_{3n} + \cdots + a_{nn}b_{nn} = \sum_{j=1}^{n} a_{nj}b_{jn}.$$

Thus,

$$\begin{aligned} \text{tr}(AB) &= [AB]_{11} + [AB]_{22} + \cdots + [AB]_{nn} \\ &= \sum_{j=1}^{n} a_{1j}b_{j1} + \sum_{j=1}^{n} a_{2j}b_{j2} + \cdots + \sum_{j=1}^{n} a_{nj}b_{jn} \\ &= \sum_{k=1}^{n}\sum_{j=1}^{n} a_{kj}b_{kj}. \end{aligned}$$

Reversing the order of summation and the order of multiplication, we have

$$\begin{aligned}
\operatorname{tr}(AB) &= \sum_{j=1}^{n}\sum_{k=1}^{n} b_{jk}a_{jk} \\
&= \sum_{k=1}^{n} b_{1k}a_{k1} + \sum_{k=1}^{n} b_{2k}a_{k2} + \cdots + \sum_{k=1}^{n} b_{nk}a_{kn} \\
&= [BA]_{11} + [BA]_{22} + \cdots + [BA]_{nn} \\
&= \operatorname{tr}(BA).
\end{aligned}$$

Now, we show that the trace is a similarity invariant. Let $B = P^{-1}AP$. Then

$$\begin{aligned}
\operatorname{tr}(B) &= \operatorname{tr}(P^{-1}AP) \\
&= \operatorname{tr}((P^{-1}A)P) \\
&= \operatorname{tr}(P(P^{-1}A)) \\
&= \operatorname{tr}(PP^{-1})A) \\
&= \operatorname{tr}(IA) \\
&= \operatorname{tr}(A).
\end{aligned}$$

EXERCISE SET 8.6

1. **(a)** This transformation is onto because for any ordered pair (a, b) in R^2, $T(b, a) = (a, b)$.

(b) We use a counterexample to show that this transformation is not onto. Since there is no pair (x, y) that satisfies $T(x, y) = (1, 0)$, T is not onto.

(c) This transformation is onto. For any ordered pair (a, b) in R^2, $T\left(\frac{a+b}{2}, \frac{a-b}{2}\right) = (a, b)$.

(d) This is not an onto transformation. For example, there is no pair (x, y) that satisfies $T(x, y) = (1, 1, 0)$.

(e) The image of this transformation is all vectors in R^3 of the form $(a, -a, 2a)$. Thus, the image of T is a one-dimensional subspace of R^3 and cannot be all of R^3. In particular, there is no vector (x, y) that satisfies $T(x, y) = (1, 1, 0)$, and this transformation is not onto.

(f) This is an onto transformation. For any point (a, b) in R^2, there are an infinite number of points that map to it. One such example is $T\left(\frac{a+b}{2}, \frac{a-b}{2}, 0\right) = (a, b)$.

3. **(a)** We find that $\text{rank}(A) = 2$, so the image of T is a two-dimensional subspace of R^3. Thus, T is not onto.

(b) We find that $\text{rank}(A) = 3$, so the image of T is all of R^3. Thus, T is not onto.

(c) We find that $\text{rank}(A) = 3$, so the image of T is all of R^3. Thus, T is onto.

(d) We find that $\text{rank}(A) = 3$, so the image of T is all of R^3. Thus, T is onto.

5. **(a)** The transformation T is not a bijection because it is not onto. There is no $p(x)$ in $P_2(x)$ so that $xp(x) = 1$.

(b) The transformation $T(A) = A^T$ is one-to-one, onto, and linear, so it is a bijection.

(c) By Theorem 8.6.1, there is no bijection between R^4 and R^3, so T cannot be a bijection. In particular, it fails being one-to-one. As an example, $T(1, 1, 2, 2) = T(1, 1, 0, 0) = (1, 1, 0)$.

(d) Because dim P_3 = 4 and dim R^3 = 3, Theorem 8.6.1 states that there is no bijection between P_3 and R^3, so T cannot be a bijection. In particular, it fails being one-to-one. As an example, $T(x + x^2 + x^3) = T(1 + x + x^2 + x^3) = (1, 1, 1)$.

7. Assume there exists a surjective (onto) linear transformation $T : V \rightarrow W$, where dim $W >$ dim V. Let m = dim V and n = dim W, with $m < n$. Then, the matrix A_T of the transformation is an $n \times m$ matrix, with $m < n$. The maximal rank of A_T is m, so the dimension of the image of T is at most m. Since the dimension of the image of T is smaller than the dimension of the codomain R^n, T is not onto. Thus, there cannot be a surjective transformation from V onto W if dim $V <$ dim W.

If n = dim $W \leq$ dim $V = m$, then the matrix A_T of the transformation is an $n \times m$ matrix with maximim possible rank n. If rank$(A_T) = n$, then T is a surjective transformation.

Thus, it is only possible for $T : V \rightarrow W$ to be a surjective linear transformation if dim $W \leq$ dim V.

9. Let $T : V \rightarrow R^n$ be defined by $T(\mathbf{v}) = (\mathbf{v})_S$, where $S = \{\mathbf{u}_1, \mathbf{u}_2, \ldots, \mathbf{u}_n\}$ is a basis of V. We know from Example 7 in Section 8.1 that the coordinate map is a linear transformation. Let $(a_1, a_2, \ldots, a_n)$ be any point in R^n. Then, for the vector $\mathbf{v} = a_1\mathbf{u}_1 + a_2\mathbf{u}_2 + \cdots + a_n\mathbf{u}_n$, we have

$$T(\mathbf{v}) = T(a_1\mathbf{u}_1 + a_2\mathbf{u}_2 + \cdots + a_n\mathbf{u}_n) = (a_1, a_2, \ldots, a_n)$$

so T is onto.

Also, let $\mathbf{v}_1 = a_1\mathbf{u}_1 + a_2\mathbf{u}_2 + \cdots + a_n\mathbf{u}_n$ and $\mathbf{v}_2 = b_1\mathbf{u}_1 + b_2\mathbf{u}_2 + \cdots + b_n\mathbf{u}_n$. If $T(\mathbf{v}_1) = T(\mathbf{v}_2)$, then $(\mathbf{v}_1)_S = (\mathbf{v}_2)_S$, and thus $(a_1, a_2, \ldots, a_n) = (b_1, b_2, \ldots, b_n)$. It follows that $a_1 = b_1$, $a_2 = b_2, \ldots, a_n = b_n$, and thus

$$\mathbf{v}_1 = a_1\mathbf{u}_1 + a_2\mathbf{u}_2 + \cdots + a_n\mathbf{u}_n = \mathbf{v}_2.$$

So, T is one-to-one and is thus an isomorphism.

11. Let V = Span{1, sin x, cos x, sin $2x$, cos $2x$}. Differentiation is a linear transformation (see Example 11, Section 8.1). In this case, D maps functions in V into other functions in V. To construct the matrix of the linear transformation with respect to the basis B = {1, sin x, cos x, sin $2x$, cos $2x$}, we look at coordinate vectors of the derivatives of the basis vectors:

$$D(1) = 0 \quad D(\sin x) = \cos x \quad D(\cos x) = -\sin x \quad D(\sin 2x) = 2\cos 2x$$
$$D(\cos 2x) = 2\sin 2x$$

The coordinate matrices are:

$$[D(1)]_B = \begin{bmatrix} 0 \\ 0 \\ 0 \\ 0 \\ 0 \end{bmatrix} \quad [D(\sin x)]_B = \begin{bmatrix} 0 \\ 0 \\ 1 \\ 0 \\ 0 \end{bmatrix} \quad [D(\cos x)]_B = \begin{bmatrix} 0 \\ -1 \\ 0 \\ 0 \\ 0 \end{bmatrix} \quad [D(\sin 2x)]_B = \begin{bmatrix} 0 \\ 0 \\ 0 \\ 0 \\ 2 \end{bmatrix} \quad [D(\cos 2x)]_B = \begin{bmatrix} 0 \\ 0 \\ 0 \\ -2 \\ 0 \end{bmatrix}.$$

Thus, the matrix of the transformation is

$$A_D = \begin{bmatrix} 0 & 0 & 0 & 0 & 0 \\ 0 & 0 & -1 & 0 & 0 \\ 0 & 1 & 0 & 0 & 0 \\ 0 & 0 & 0 & 0 & -2 \\ 0 & 0 & 0 & 2 & 0 \end{bmatrix}$$

Then, differentiation of a function in V can be accomplished by matrix multiplication by the formula

$$[D(\mathbf{f})]_B = A_D[\mathbf{f}]_B.$$

The final vector, once transformed back to V from coordinates in R^5, will be the desired derivative.

For example,

$$[D(3 - 4\sin x + \sin 2x + 5\cos 2x)]_B = A_D\begin{bmatrix} 3 \\ -4 \\ 0 \\ 1 \\ 5 \end{bmatrix} = \begin{bmatrix} 0 & 0 & 0 & 0 & 0 \\ 0 & 0 & -1 & 0 & 0 \\ 0 & 1 & 0 & 0 & 0 \\ 0 & 0 & 0 & 0 & -2 \\ 0 & 0 & 0 & 2 & 0 \end{bmatrix}\begin{bmatrix} 3 \\ -4 \\ 0 \\ 1 \\ 5 \end{bmatrix} = \begin{bmatrix} 0 \\ 0 \\ -4 \\ -10 \\ 2 \end{bmatrix}.$$

Thus, $D(3 - 4\sin x + \sin 2x + 5\cos 2x) = -4\cos x - 10\sin 2x + 2\cos 2x$.

SUPPLEMENTARY EXERCISES 8

3. By the properties of an inner product, we have

$$\begin{aligned} T(\mathbf{v} + \mathbf{w}) &= \langle \mathbf{v} + \mathbf{w}, \mathbf{v}_0 \rangle \mathbf{v}_0 \\ &= (\langle \mathbf{v}, \mathbf{v}_0 \rangle + \langle \mathbf{w}, \mathbf{v}_0 \rangle)\mathbf{v}_0 \\ &= \langle \mathbf{v}, \mathbf{v}_0 \rangle \mathbf{v}_0 + \langle \mathbf{w}, \mathbf{v}_0 \rangle \mathbf{v}_0 \\ &= T(\mathbf{v}) + T(\mathbf{w}) \end{aligned}$$

and

$$T(k\mathbf{v}) = \langle k\mathbf{v}, \mathbf{v}_0 \rangle \mathbf{v}_0 = k\langle \mathbf{v}, \mathbf{v}_0 \rangle \mathbf{v}_0 = kT(\mathbf{v})$$

Thus T is a linear operator on V.

5. **(a)** The matrix for T with respect to the standard basis is

$$A = \begin{bmatrix} 1 & 0 & 1 & 1 \\ 2 & 1 & 3 & 1 \\ 1 & 0 & 0 & 1 \end{bmatrix}$$

We first look for a basis for the range of T; that is, for the space of vectors B such that $A\mathbf{x} = \mathbf{b}$. If we solve the system of equations

$$\begin{aligned} x \qquad + z + w &= b_1 \\ 2x + y + 3x + w &= b_2 \\ x \qquad\qquad + w &= b_3 \end{aligned}$$

we find that $= z = b_1 - b_3$ and that any one of x, y, or w will determine the other two. Thus, $T(\mathbf{e}_3)$ and any two of the remaining three columns of A is a basis for $\mathrm{R}(T)$.

Alternate Solution :We can use the method of Section 5.5 to find a basis for the column space of A by reducing A^T to row-echelon form. This yields

$$\begin{bmatrix} 1 & 2 & 1 \\ 0 & 1 & 0 \\ 0 & 0 & 1 \\ 0 & 0 & 0 \end{bmatrix}$$

so that the three vectors

$$\begin{bmatrix} 1 \\ 2 \\ 1 \end{bmatrix} \quad \begin{bmatrix} 0 \\ 1 \\ 0 \end{bmatrix} \quad \begin{bmatrix} 0 \\ 0 \\ 1 \end{bmatrix}$$

form a basis for the column space of T and hence for its range.

Second Alternative: Note that since $\text{rank}(A) = 3$, then $R(T)$ is a 3-dimensional subspace of R^3 and hence is all of R^3. Thus the standard basis for R^3 is also a basis for $R(T)$.

To find a basis for the kernel of T, we consider the solution space of $A\mathbf{x} = \mathbf{0}$. If we set $b_1 = b_2 = b_3 = 0$ in the above system of equations, we find that $z = 0$, $x = -w$, and $y = w$. Thus the vector $(-1, 1, 0, 1)$ forms a basis for the kernel.

7. (a) We know that T can be thought of as multiplication by the matrix

$$[T]_B = \begin{bmatrix} 1 & 1 & 2 & -2 \\ 1 & -1 & -4 & 6 \\ 1 & 2 & 5 & -6 \\ 3 & 2 & 3 & -2 \end{bmatrix}$$

where reduction to row-echelon form easily shows that $\text{rank}([T]_B) = 2$. Therefore the rank of T is 2 and the nullity of T is $4 - 2 = 2$.

(b) Since $[T]_B$ is not invertible, T is not one-to-one.

9. **(a)** If $A = P^{-1} BP$, then

$$\begin{aligned} A^T &= (P^{-1} BP)^T \\ &= P^T B^T (P^{-1})^T \\ &= ((P^T)^{-1})^{-1} B^T (P^{-1})^T \\ &= ((P^{-1})^T)^T B (P^{-1})^T \end{aligned}$$

Therefore A^T and B^T are similar. You should verify that if P is invertible, then so is P^T and that $(P^T)^{-1} = (P^{-1})^T$.

11. If we let $X = \begin{bmatrix} a & b \\ c & d \end{bmatrix}$, then we have

$$\begin{aligned} T\left(\begin{bmatrix} a & b \\ c & d \end{bmatrix}\right) &= \begin{bmatrix} a+c & b+d \\ 0 & 0 \end{bmatrix} + \begin{bmatrix} b & b \\ d & d \end{bmatrix} \\ &= \begin{bmatrix} a+b+c & 2b+d \\ d & d \end{bmatrix} \end{aligned}$$

The matrix X is in the kernel of T if and only if $T(X) = \mathbf{0}$, i.e., if and only if

$$\begin{aligned} a + b + c \quad &= 0 \\ 2b \quad + d &= 0 \\ d &= 0 \end{aligned}$$

Hence

$$X = \begin{bmatrix} a & 0 \\ -a & 0 \end{bmatrix}$$

The space of all such matrices X is spanned by the matrix $\begin{bmatrix} 1 & 0 \\ -1 & 0 \end{bmatrix}$, and therefore has dimension 1. Thus the nullity is 1. Since the dimension of M_{22} is 4, the rank of T must therefore be 3.

Alternate Solution. Using the computations done above, we have that the matrix for this transformation with respect to the standard basis in M_{22} is

$$\begin{bmatrix} 1 & 1 & 1 & 0 \\ 0 & 2 & 0 & 1 \\ 0 & 0 & 0 & 1 \\ 0 & 0 & 0 & 1 \end{bmatrix}$$

Since this matrix has rank 3, the rank of T is 3, and therefore the nullity must be 1.

13. The standard basis for M_{22} is the set of matrices

$$\begin{bmatrix} 1 & 0 \\ 0 & 0 \end{bmatrix}, \begin{bmatrix} 0 & 1 \\ 0 & 0 \end{bmatrix}, \begin{bmatrix} 0 & 0 \\ 1 & 0 \end{bmatrix}, \begin{bmatrix} 0 & 0 \\ 0 & 1 \end{bmatrix}$$

If we think of the above matrices as the vectors

$$[1 \quad 0 \quad 0 \quad 0]^T, \quad [0 \quad 1 \quad 0 \quad 0]^T, \quad [0 \quad 0 \quad 1 \quad 0]^T, \quad [0 \quad 0 \quad 0 \quad 1]^T$$

then L takes these vectors to

$$[1 \quad 0 \quad 0 \quad 0]^T, \quad [0 \quad 0 \quad 1 \quad 0]^T, \quad [0 \quad 1 \quad 0 \quad 0]^T, \quad [0 \quad 0 \quad 0 \quad 1]^T$$

Therefore the desired matrix for L is

$$\begin{bmatrix} 1 & 0 & 0 & 0 \\ 0 & 0 & 1 & 0 \\ 0 & 1 & 0 & 0 \\ 0 & 0 & 0 & 1 \end{bmatrix}$$

15. The transition matrix P from B' to B is

$$P = \begin{bmatrix} 1 & 1 & 1 \\ 0 & 1 & 1 \\ 0 & 0 & 1 \end{bmatrix}$$

Therefore, by Theorem 8.5.2, we have

$$[T]_{B'} = P^{-1}[T]_B P = \begin{bmatrix} -4 & 0 & 9 \\ 1 & 0 & -2 \\ 0 & 1 & 1 \end{bmatrix}$$

Alternate Solution: We compute the above result more directly. It is easy to show that $\mathbf{u}_1 = \mathbf{v}_1$, $\mathbf{u}_2 = -\mathbf{v}_1 + \mathbf{v}_2$, and $\mathbf{u}_3 = -\mathbf{v}_2 + \mathbf{v}_3$. So

$$T(\mathbf{v}_1) = T(\mathbf{u}_1) = -3\mathbf{u}_1 + \mathbf{u}_2 = -4\mathbf{v}_1 + \mathbf{v}_2$$

$$T(\mathbf{v}_2) = T(\mathbf{u}_1 + \mathbf{u}_2) = T(\mathbf{u}_1) + T(\mathbf{u}_2)$$

$$= \mathbf{u}_1 + \mathbf{u}_2 + \mathbf{u}_3 = \mathbf{v}_3$$

$$T(\mathbf{v}_3) = T(\mathbf{u}_1 + \mathbf{u}_2 + \mathbf{u}_3) = T(\mathbf{u}_1) + T(\mathbf{u}_2) + T(\mathbf{u}_3)$$

$$= 8\mathbf{u}_1 - \mathbf{u}_2 + \mathbf{u}_3$$

$$= 9\mathbf{v}_1 - 2\mathbf{v}_2 + \mathbf{v}_3$$

17. Since

$$T\left(\begin{bmatrix}1\\0\\0\end{bmatrix}\right) = \begin{bmatrix}1\\0\\1\end{bmatrix},\ T\left(\begin{bmatrix}0\\1\\0\end{bmatrix}\right) = \left(\begin{bmatrix}-1\\1\\0\end{bmatrix}\right),\ \text{and } T\left(\begin{bmatrix}0\\0\\1\end{bmatrix}\right) = \begin{bmatrix}1\\0\\-1\end{bmatrix}$$

we have

$$[T]_B = \begin{bmatrix}1 & -1 & 1\\0 & 1 & 0\\1 & 0 & -1\end{bmatrix}$$

In fact, this result can be read directly from $[T(X)]_B$.

19. **(a)** Recall that $D(\mathbf{f} + \mathbf{g}) = (f(x) + g(x))'' = f''(x) + g''(x)$ and $D(c\mathbf{f}) = (cf(x))'' = cf''(x)$.

(b) Recall that $D(\mathbf{f}) = \mathbf{0}$ if and only if $f'(x) = a$ for some constant a if and only if $f(x) = ax + b$ for constants a and b. Since the functions $f(x) = x$ and $g(x) = 1$ are linearly independent, they form a basis for the kernel of D.

(c) Since $D(\mathbf{f}) = f(x)$ if and only if $f''(x) = f(x)$ if and only if $f(x) = ae^x + be^{-x}$ for a and b arbitrary constants, the functions $f(x) = e^x$ and $g(x) = e^{-x}$ span the set of all such functions. This is clearly a subspace of $C^2(-\infty, \infty)$ (Why?), and to show that it has dimension 2, we need only check that e^x and e^{-x} are linearly independent functions. To this end, suppose that there exist constants c_1 and c_2 such that $c_1e^x + c_2e^{-x} = 0$. If we let $x = 0$ and $x = 1$, we obtain the equations $c_1 + c_2 = 0$ and $c_1e + c_2e^{-1} = 0$. These imply that $c_1 = c_2 = 0$, so e^x and e^{-x} are linearly independent.

21. **(a)** We have

$$T(p(x)+q(x)) = \begin{bmatrix} p(x_1)+q(x_1) \\ p(x_2)+q(x_2) \\ p(x_3)+q(x_3) \end{bmatrix} = \begin{bmatrix} p(x_1) \\ p(x_2) \\ p(x_3) \end{bmatrix} + \begin{bmatrix} q(x_1) \\ q(x_2) \\ q(x_3) \end{bmatrix} = T(p(x)) + T(q(x))$$

and

$$T(kp(x)) = \begin{bmatrix} kp(x_1) \\ kp(x_2) \\ kp(x_3) \end{bmatrix} = k\begin{bmatrix} p(x_1) \\ p(x_2) \\ p(x_3) \end{bmatrix} = kT(p(x))$$

(b) Since T is defined for quadratic polynomials only, and the numbers x_1, x_2, and x_3 are distinct, we can have $p(x_1) = p(x_2) = p(x_3) = 0$ if and only if p is the zero polynomial. (Why?) Thus $\ker(T) = \{\mathbf{0}\}$, so T is one-to-one.

(c) We have

$$T(a_1P_1(x) + a_2P_2(x) + a_3P_3(x)) = a_1T(P_1(x)) + a_2T(P_2(x)) + a_3T(P_3(x))$$

$$= a_1\begin{bmatrix} 1 \\ 0 \\ 0 \end{bmatrix} + a_2\begin{bmatrix} 0 \\ 1 \\ 0 \end{bmatrix} + a_3\begin{bmatrix} 0 \\ 0 \\ 1 \end{bmatrix}$$

$$= \begin{bmatrix} a_1 \\ a_2 \\ a_3 \end{bmatrix}$$

(d) From the above calculations, we see that the points must lie on the curve.

23. Since

$$D(x^k) = \begin{cases} 0 & \text{if } k = 0 \\ kx^{k-1} & \text{if } k = 1, 2, \ldots, n \end{cases}$$

then

$$[D(x^k)]_B = \begin{cases} (0, \ldots, 0) & \text{if } k = 0 \\ (0, \ldots, k, \ldots, 0) & \text{if } k = 1, 2, \ldots, n \end{cases}$$

$\uparrow$

k^{th} component

where the above vectors all have $n + 1$ components. Thus the matrix of D with respect to B is

$$\begin{bmatrix} 0 & 1 & 0 & 0 & \cdots & 0 \\ 0 & 0 & 2 & 0 & \cdots & 0 \\ 0 & 0 & 0 & 3 & \cdots & 0 \\ \vdots & \vdots & \vdots & \vdots & & \vdots \\ 0 & 0 & 0 & 0 & \cdots & n \\ 0 & 0 & 0 & 0 & \cdots & 0 \end{bmatrix}$$

25. Let B_n and B_{n+1} denote the bases for P_n and P_{n+1}, respectively. Since

$$J(x^k) = \frac{x^{k+1}}{k+1} \quad \text{for } k = 0, \ldots, n$$

we have

$$[J(x^k)]B_{n+1} = \left(0, \ldots, \frac{1}{k+1}, \ldots, 0\right) \quad (n + 2 \text{ components})$$

$\uparrow$

$(k + 2)^{nd}$ component

where $[x^k]_{B_n} = [0, \ldots, 1, \ldots, 0]^T$ with the entry 1 as the $(k + 1)^{st}$ component out of a total of $n + 1$ components. Thus the matrix of J with respect to B_{n+1} is

$$\begin{bmatrix} 0 & 0 & 0 & \cdots & 0 \\ 1 & 0 & 0 & \cdots & 0 \\ 0 & 1/2 & 0 & \cdots & 0 \\ 0 & 0 & 1/3 & \cdots & 0 \\ \vdots & \vdots & \vdots & & \vdots \\ 0 & 0 & 0 & \cdots & 1/(n+1) \end{bmatrix}$$

with $n + 2$ rows and $n + 1$ columns.

EXERCISE SET 9.1

1. **(a)** The system is of the form $\mathbf{y}' = A\mathbf{y}$ where

$$A = \begin{bmatrix} 1 & 4 \\ 2 & 3 \end{bmatrix}$$

The eigenvalues of A are $\lambda = 5$ and $\lambda = -1$ and the corresponding eigenspaces are spanned by the vectors

$$\begin{bmatrix} 1 \\ 1 \end{bmatrix} \text{ and } \begin{bmatrix} -2 \\ 1 \end{bmatrix}$$

respectively. Thus if we let

$$P = \begin{bmatrix} 1 & -2 \\ 1 & 1 \end{bmatrix}$$

we have

$$D = P^{-1}\, AP = \begin{bmatrix} 5 & 0 \\ 0 & -1 \end{bmatrix}$$

Let $\mathbf{y} = P\mathbf{u}$ and hence $\mathbf{y}' = P\mathbf{u}'$. Then

$$\mathbf{u}' = \begin{bmatrix} 5 & 0 \\ 0 & -1 \end{bmatrix} \mathbf{u}$$

or

$$u'_1 = 5u_1$$

$$u'_2 = -u_2$$

Therefore

$$u_1 = c_1e^{5x}$$

$$u_2 = c_2e^{-x}$$

Thus the equation $\mathbf{y} = P\mathbf{u}$ is

$$\begin{bmatrix} y_1 \\ y_1 \end{bmatrix} = \begin{bmatrix} 1 & -2 \\ 1 & 1 \end{bmatrix} \begin{bmatrix} c_1e^{5x} \\ c_2e^{-x} \end{bmatrix} = \begin{bmatrix} c_1e^{5x} - 2c_2e^{-x} \\ c_1e^{5x} + c_2e^{-x} \end{bmatrix}$$

or

$$y_1 = c_1e^5x - 2c_2e^{-x}$$

$$y_2 = c_1e^{5x} + c_2e^{-x}$$

1. (b) If $y_1(0) = y_2(0) = 0$, then

$$c_1 - 2c_2 = 0$$

$$c_1 + c_2 = 0$$

so that $c_1 = c_2 = 0$. Thus $y_1 = 0$ and $y_2 = 0$.

3. (a) The system is of the form $\mathbf{y}' = A\mathbf{y}$ where

$$A = \begin{bmatrix} 4 & 0 & 1 \\ -2 & 1 & 0 \\ -2 & 0 & 1 \end{bmatrix}$$

The eigenvalues of A are $\lambda = 1$, $\lambda = 2$, and $\lambda = 3$ and the corresponding eigenspaces are spanned by the vectors

$$\begin{bmatrix} 0 \\ 1 \\ 0 \end{bmatrix} \begin{bmatrix} -1/2 \\ 1 \\ 1 \end{bmatrix} \begin{bmatrix} -1 \\ 1 \\ 1 \end{bmatrix}$$

respectively. Thus, if we let

$$P = \begin{bmatrix} 0 & -1/2 & -1 \\ 1 & 1 & 1 \\ 0 & 1 & 1 \end{bmatrix}$$

then

$$D = P^{-1}AP \begin{bmatrix} 1 & 0 & 0 \\ 0 & 2 & 0 \\ 0 & 0 & 3 \end{bmatrix}$$

Let $\mathbf{y} = P\mathbf{u}$ and hence $\mathbf{y}' = P\mathbf{u}'$. Then

$$\mathbf{u}' = \begin{bmatrix} 1 & 0 & 0 \\ 0 & 2 & 0 \\ 0 & 0 & 3 \end{bmatrix} \mathbf{u}$$

so that

$$u'_1 = u_1$$

$$u'_2 = 2u_2$$

$$u'_3 = 3u_3$$

Therefore

$$u_1 = c_1e^x$$

$$u_2 = c_2e^{2x}$$

$$u_3 = c_3e^{3x}$$

Thus the equation $\mathbf{y} = P\mathbf{u}$ is

$$\begin{bmatrix} y_1 \\ y_2 \\ y_3 \end{bmatrix} = \begin{bmatrix} 0 & -1/2 & -1 \\ 1 & 1 & 1 \\ 0 & 1 & 1 \end{bmatrix} \begin{bmatrix} c_1e^x \\ c_2e^{2x} \\ c_3e^{3x} \end{bmatrix}$$

or

$$y_1 = -\frac{1}{2}c_2e^{2x} - c_3e^{3x}$$

$$y_2 = c_1e^x + c_2e^{2x} + c_3e^{3x}$$

$$y_3 = c_2e^{2x} + c_3e^{3x}$$

Note: If we use

$$\begin{bmatrix} 0 \\ 1 \\ 0 \end{bmatrix} \begin{bmatrix} -1 \\ 2 \\ 2 \end{bmatrix} \begin{bmatrix} 1 \\ -1 \\ -1 \end{bmatrix}$$

as basis vectors for the eigenspaces, then

$$P = \begin{bmatrix} 0 & -1 & 1 \\ 1 & 2 & -1 \\ 0 & 2 & -1 \end{bmatrix}$$

and

$$y_1 = -c_2e^{2x} + c_3e^{3x}$$

$$y_2 = c_1e^{x} + 2c_2e^{2x} - c_3e^{3x}$$

$$y_3 = 2c_2e^{2x} - c_3e^{3x}$$

There are, of course, infinitely many other ways of writing the answer, depending upon what bases you choose for the eigenspaces. Since the numbers c_1, c_2, and c_3 are arbitrary, the "different" answers do, in fact, represent the same functions.

3. (b) If we set $x = 0$, then the initial conditions imply that

$$-\frac{1}{2}c_2 - c_3 = -1$$

$$c_1 + c_2 + c_3 = 1$$

$$c_2 + c_3 = 0$$

or, equivalently, that $c_1 = 1$, $c_2 = -2$, and $c_3 = 2$. If we had used the "different" solution we found in Part (a), then we would have found that $c_1 = 1$, $c_2 = -1$, and $c_3 = -2$. In either case, when we substitute these values into the appropriate equations, we find that

$$y_1 = e^{2x} - 2e^{3x}$$

$$y_2 = e^{x} - 2e^{2x} + 2e^{3x}$$

$$y_3 = -2e^{2x} + 2e^{3x}$$

5. Following the hint, let $y = f(x)$ be a solution to $y' = ay$, so that $f'(x) = af(x)$. Now consider the function $g(x) = f(x)e^{-ax}$. Observe that

$$\begin{aligned} g'(x) &= f'(x)e^{-ax} - af(x)e^{-ax} \\ &= af(x)e^{-ax} - af(x)e^{-ax} \\ &= 0 \end{aligned}$$

Thus $g(x)$ must be a constant; say $g(x) = c$. Therefore,

$$f(x)e^{-ax} = c$$

or

$$f(x) = ce^{ax}$$

That is, every solution of $y' = ay$ has the form $y = ce^{ax}$.

. . . .

7. If $y_1 = y$ and $y_2 = y'$, then $y_1' = y_2$ and $y_2' = y'' = y' + 6y = y_2 + 6y_1$. That is,

$$\begin{aligned} y_1' &= \quad\quad y_2 \\ y_2' &= 6y_1 + y_2 \end{aligned}$$

or $\mathbf{y}' = A\mathbf{y}$ where

$$A = \begin{bmatrix} 0 & 1 \\ 6 & 1 \end{bmatrix}$$

The eigenvalues of A are $\lambda = -2$ and $\lambda = 3$ and the corresponding eigenspaces are spanned by the vectors

$$\begin{bmatrix} -1 \\ 2 \end{bmatrix} \text{ and } \begin{bmatrix} 1 \\ 3 \end{bmatrix}$$

respectively. Thus, if we let

$$P = \begin{bmatrix} -1 & 1 \\ 2 & 3 \end{bmatrix}$$

then

$$P^{-1}AP = \begin{bmatrix} -2 & 0 \\ 0 & 3 \end{bmatrix}$$

Let $\mathbf{y} = P\mathbf{u}$ and hence $\mathbf{y}' = P\mathbf{u}'$. Then

$$\mathbf{u}' = \begin{bmatrix} -2 & 0 \\ 0 & 3 \end{bmatrix} \mathbf{u}$$

or

$$y_1 = -c_1e^{-2x} + c_2e^{3x}$$
$$y_2 = 2c_1e^{-2x} + 3c_2e^{3x}$$

Therefore

$$u_1 = c_1e^{-2x}$$
$$u_2 = c_2e^{3x}$$

Thus the equation $\mathbf{y} = P\mathbf{u}$ is

$$\begin{bmatrix} y_1 \\ y_2 \end{bmatrix} = \begin{bmatrix} -1 & 1 \\ 2 & 3 \end{bmatrix} \begin{bmatrix} c_1e^{-2x} \\ c_2e^{3x} \end{bmatrix}$$

or

$$y_1 = -c_1e^{-2x} + c_2e^{3x}$$
$$y_2 = 2c_1e^{-2x} + 3c_2e^{3x}$$

Note that $y_1' = y_2$, as required, and, since $y_1 = y$, then

$$y = -c_1e^{-2x} + c_2e^{3x}$$

Since c_1 and c_2 are arbitrary, any answer of the form $y = ae^{-2x} + be^{3x}$ is correct.

9. If we let $y_1 = y$, $y_2 = y'$, and $y_3 = y''$, then we obtain the system

$$y_1' = y_2$$
$$y_2' = y_3$$
$$y_3' = 6y_1 - 11y_2 + 6y_3$$

The associated matrix is therefore

$$A = \begin{bmatrix} 0 & 1 & 0 \\ 0 & 0 & 1 \\ 6 & -11 & 6 \end{bmatrix}$$

The eigenvalues of A are $\lambda = 1$, $\lambda = 2$, and $\lambda = 3$ and the corresponding eigenvectors are

$$\begin{bmatrix} 1 \\ 1 \\ 1 \end{bmatrix}, \begin{bmatrix} 1 \\ 2 \\ 4 \end{bmatrix} \text{ and } \begin{bmatrix} 1 \\ 3 \\ 9 \end{bmatrix}$$

The solution is, after some computation,

$$y = c_1e^x + c_2e^{2x} + c_3e^{3x}$$

11. Consider $y' = Ay$, where $A = \begin{bmatrix} a_{11} & a_{12} \\ a_{21} & a_{22} \end{bmatrix}$, with a_{ij} real. Solving the system

$$\det(\lambda I - A) = \begin{vmatrix} \lambda - a_{11} & -a_{12} \\ -a_{21} & \lambda - a_{21} \end{vmatrix} = 0$$

yields the quadratic equation

$$\lambda^2 - (a_{11} + a_{22})\lambda + a_{11}a_{22} - a_{21}a_{12} = 0, \text{ or}$$
$$\lambda^2 - (\text{Tr}A)\lambda + \det A = 0.$$

Let λ_1, λ_2 be the solutions of the characteristic equation. Using the quadratic formula yields

$$\lambda_1, \lambda_2 = \frac{Tr\,A \pm \sqrt{Tr^2 A - 4\det A}}{2}$$

Now the solutions $y_1(t)$ and $y_2(t)$ to the system $y' = Ay$ will approach zero as $t \to \infty$ if and only if $\text{Re}(\lambda_1, \lambda_2) < 0$. (Both are < 0)

Case I: $Tr^2A - 4 \det A < 0$.

In this case $R_e(\lambda_1) = R_e(\lambda_2) = \frac{TrA}{2}$. Thus $y_1(t), y_2(t) \to 0$ if and only if $TrA < 0$.

Case II: $Tr^2A - 4 \det A = 0$. Then $\lambda_1 = \lambda_2$, and $Re(\lambda_1, \lambda_2) = \frac{TrA}{2}$, so $y_1, y_2 \to 0$ if and only if $TrA < 0$.

Case III: $Tr^2A - 4 \det A > 0$. Then λ_1, λ_2 are both real.

Subcase 1: $\det A > 0$.

Then $|Tr\ A| > \sqrt{Tr^2A - 4\det A} > 0$

If $TrA > 0$, then both $(\lambda_1, \lambda_2) > 0$, so $y_1, y_2 \not\to 0$.

If $TrA < 0$, then both $(\lambda_1, \lambda_2) < 0$, so $y_1, y_2 \to 0$. $TrA = 0$ is not possible in this case.

Subcase 2: $\det A < 0$

Then $\sqrt{Tr^2A - 4\det A} > |Tr\ A| \geq 0$

If $TrA > 0$, then one root (say λ_1) is positive, the other is negative, so $y_1, y_2 \not\to 0$.

If $TrA = 0$, then again $\lambda_1 > 0, \lambda_2 < 0$, so $y_1, y_2 \not\to 0$.

Subcase 3: $\det A = 0$. Then $\lambda_1 = 0$ or $\lambda_2 = 0$, so $y_1, y_2 \not\to 0$.

13. The system $\begin{cases} y_1' = 2y_1 + y_2 + t \\ y_2' = y_1 + 2y_2 + 2t \end{cases}$.

can be put into the form

$$\begin{bmatrix} y_1' \\ y_2' \end{bmatrix} = \begin{bmatrix} 2 & 1 \\ 1 & 2 \end{bmatrix} \begin{bmatrix} y_1 \\ y_2 \end{bmatrix} + t \begin{bmatrix} 1 \\ 2 \end{bmatrix} = \mathrm{A}\bar{\mathrm{y}} + \bar{\mathrm{f}}$$

The eigenvalues and eigenvectors of A are:

$$\lambda_1 = 1,\ \mathrm{x}_1 = \begin{pmatrix} 1 \\ -1 \end{pmatrix}\quad \lambda_2 = 3,\ \mathrm{x}_2 = \begin{pmatrix} 1 \\ 1 \end{pmatrix}$$

Solving:

$$\begin{bmatrix} y_1 \\ y_2 \end{bmatrix} = c_1 e^{t} \begin{bmatrix} 1 \\ -1 \end{bmatrix} + c_2 e^{3t} \begin{bmatrix} 1 \\ 1 \end{bmatrix} + t \begin{bmatrix} 0 \\ -1 \end{bmatrix} + \begin{bmatrix} 1/3 \\ -2/3 \end{bmatrix}$$

EXERCISE SET 9.2

1. **(a)** Since $T(x, y) = (-y, -x)$, the standard matrix is

$$A = \begin{bmatrix} -1 & 0 \\ 0 & -1 \end{bmatrix}$$

(b)

$$A = \begin{bmatrix} -1 & 0 \\ 0 & -1 \end{bmatrix}$$

(c) Since $T(x, y) = (x, 0)$, the standard matrix is

$$\begin{bmatrix} 1 & 0 \\ 0 & 0 \end{bmatrix}$$

(d)

$$A = \begin{bmatrix} 0 & 0 \\ 0 & 1 \end{bmatrix}$$

3. **(b)** Since $T(x, y, z) = (x, -y, z)$, the standard matrix is

$$\begin{bmatrix} 0 & -1 & 0 \\ 1 & 0 & 0 \\ 0 & 0 & 1 \end{bmatrix}$$

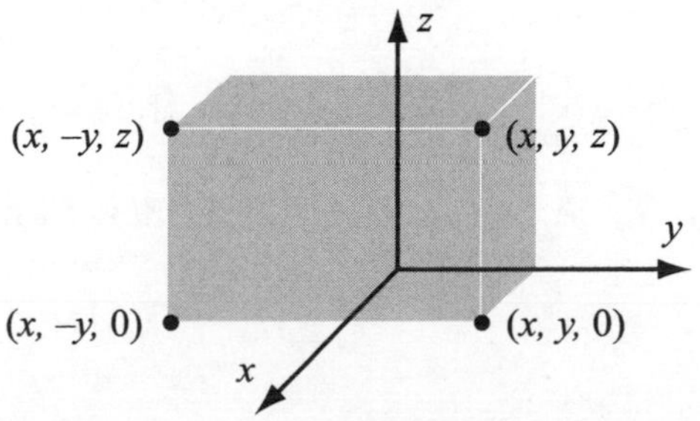

(x, −y, z)
(x, y, z)
(x, −y, 0)
(x, y, 0)
x
y
z

5. **(a)** This transformation leaves the z-coordinate of every point fixed. However it sends (1, 0, 0) to (0, 1, 0) and (0, 1, 0) to (–1, 0, 0). The standard matrix is therefore

$$\begin{bmatrix} 0 & -1 & 0 \\ 1 & 0 & 0 \\ 0 & 0 & 1 \end{bmatrix}$$

(c) This transformation leaves the y-coordinate of every point fixed. However it sends (1, 0, 0) to (0, 0, –1) and (0, 0, 1) to (1, 0, 0). The standard matrix is therefore

$$\begin{bmatrix} 0 & 0 & 1 \\ 0 & 1 & 0 \\ -1 & 0 & 0 \end{bmatrix}$$

13. **(a)** $\begin{bmatrix} 1 & 0 \\ 0 & 5 \end{bmatrix}\begin{bmatrix} 1/2 & 0 \\ 0 & 1 \end{bmatrix} = \begin{bmatrix} 1/2 & 0 \\ 0 & 5 \end{bmatrix}$

(c) $\begin{bmatrix} -1 & 0 \\ 0 & -1 \end{bmatrix}\begin{bmatrix} 0 & 1 \\ 1 & 0 \end{bmatrix} = \begin{bmatrix} 0 & -1 \\ -1 & 0 \end{bmatrix}$

15. **(b)** The matrices which represent compressions along the x- and y- axes are $\begin{bmatrix} k & 0 \\ 0 & 1 \end{bmatrix}$ and $\begin{bmatrix} 1 & 0 \\ 0 & k \end{bmatrix}$, respectively, where $0 < k < 1$. But

$$\begin{bmatrix} k & 0 \\ 0 & 1 \end{bmatrix}^{-1} = \begin{bmatrix} 1/k & 0 \\ 0 & 1 \end{bmatrix}$$

and

$$\begin{bmatrix} 1 & 0 \\ 0 & k \end{bmatrix}^{-1} = \begin{bmatrix} 1 & 0 \\ 0 & 1/k \end{bmatrix}$$

Since $0 < k < 1$ implies that $1/k > 1$, the result follows.

(c) The matrices which represent reflections about the x- and y- axes are $\begin{bmatrix} 1 & 0 \\ 0 & -1 \end{bmatrix}$ and $\begin{bmatrix} -1 & 0 \\ 0 & 1 \end{bmatrix}$, respectively. Since these matrices are their own inverses, the result follows.

17. **(a)** The matrix which represents this shear is $\begin{bmatrix} 1 & 3 \\ 0 & 1 \end{bmatrix}$; its inverse is $\begin{bmatrix} 1 & -3 \\ 0 & 1 \end{bmatrix}$. Thus, points$(x', y')$ on the image line must satisfy the

equations

$$\begin{aligned} x &= x' - 3y' \\ y &= \quad\;\; y' \end{aligned}$$

where $y = 2x$. Hence $y' = 2x' - 6y'$, or $2x' - 7y' = 0$. That is, the equation of the image line is $2x - 7y = 0$.

Alternatively, we could note that the transformation leaves (0, 0) fixed and sends (1, 2) to (7, 2). Thus (0, 0) and (7, 2) determine the image line which has the equation $2x - 7y = 0$.

(c) The reflection and its inverse are both represented by the matrix $\begin{bmatrix} 0 & 1 \\ 1 & 0 \end{bmatrix}$. Thus the point (x', y') on the image line must satisfy the equations

$$\begin{aligned} x &= y' \\ y &= x' \end{aligned}$$

where $y = 2x$. Hence $x' = 2y'$, so the image line has the equation $x - 2y = 0$.

(e) The rotation can be represented by the matrix $\begin{bmatrix} 1/2 & -\sqrt{3}/2 \\ \sqrt{3}/2 & 1/2 \end{bmatrix}$. This sends the origin to itself and the point (1, 2) to the point $((1 - 2\sqrt{3})/2, (2 + \sqrt{3})/2)$. Since both (0, 0) and (1, 2) lie on the line $y = 2x$, their images determine the image of the line under the required rotation. Thus, the image line is represented by the equation $(2 + \sqrt{3})x + (2\sqrt{3} - 1)y = 0$.

Alternatively, we could find the inverse of the matrix, $\begin{bmatrix} 1/2 & \sqrt{3}/2 \\ -\sqrt{3}/2 & 1/2 \end{bmatrix}$, and proceed as we did in Parts (a) and (c).

21. We use the notation and the calculations of Exercise 20. If the line $Ax + By + C = 0$ passes through the origin, then $C = 0$, and the equation of the image line reduces to $(dA - cB)x + (-bA + aB)y = 0$. Thus it also must pass through the origin.

The two lines $A_1x + B_1y + C_1 = 0$ and $A_2x + B_2y + C_2 = 0$ are parallel if and only if $A_1B_2 = A_2B_1$. Their image lines are parallel if and only if

$$(dA_1 - cB_1)(-bA_2 + aB_2) = (dA_2 - cB_2)(-bA_1 + aB_1)$$

or

$$bcA_2B_1 + adA_1B_2 = bcA_1B_2 + adA_2B_1$$

or

$$(ad - bc)(A_1B_2 - A_2B_1) = 0$$

or

$$A_1B_2 - A_2B_1 = 0$$

Thus the image lines are parallel if and only if the given lines are parallel.

23. **(a)** The matrix which transforms (x, y, z) to $(x + kz, y + kz, z)$ is

$$\begin{bmatrix} 1 & 0 & k \\ 0 & 1 & k \\ 0 & 0 & 1 \end{bmatrix}$$

EXERCISE SET 9.3

1. We have

$$\begin{bmatrix} a \\ b \end{bmatrix} = \left(\begin{bmatrix} 1 & 0 \\ 1 & 1 \\ 1 & 2 \end{bmatrix}^T \begin{bmatrix} 1 & 0 \\ 1 & 1 \\ 1 & 2 \end{bmatrix} \right)^{-1} \begin{bmatrix} 1 & 0 \\ 1 & 1 \\ 1 & 2 \end{bmatrix}^T \begin{bmatrix} 0 \\ 2 \\ 7 \end{bmatrix}$$

$$= \begin{bmatrix} 3 & 3 \\ 3 & 5 \end{bmatrix}^{-1} \begin{bmatrix} 9 \\ 16 \end{bmatrix} = \begin{bmatrix} \frac{5}{6} & -\frac{1}{2} \\ -\frac{1}{2} & \frac{1}{2} \end{bmatrix} \begin{bmatrix} 9 \\ 16 \end{bmatrix}$$

$$= \begin{bmatrix} -1/2 \\ 7/2 \end{bmatrix}$$

Thus the desired line is $y = -1/2 + (7/2)x$.

3. Here

$$M = \begin{bmatrix} 1 & 2 & 4 \\ 1 & 3 & 9 \\ 1 & 5 & 25 \\ 1 & 6 & 36 \end{bmatrix}$$

and

$$\begin{bmatrix} a \\ b \\ c \end{bmatrix} = (M^T M)^{-1} MT \begin{bmatrix} 0 \\ -10 \\ -48 \\ -76 \end{bmatrix}$$

$$= \begin{bmatrix} 4 & 16 & 74 \\ 16 & 74 & 376 \\ 74 & 376 & 2018 \end{bmatrix}^{-1} \begin{bmatrix} -134 \\ -726 \\ -4026 \end{bmatrix}$$

$$= \begin{bmatrix} \frac{221}{10} & -\frac{62}{5} & \frac{3}{2} \\ -\frac{62}{5} & \frac{649}{90} & -\frac{8}{9} \\ \frac{3}{2} & -\frac{8}{9} & \frac{1}{9} \end{bmatrix} \begin{bmatrix} -134 \\ -726 \\ -4026 \end{bmatrix}$$

$$= \begin{bmatrix} 2 \\ 5 \\ -3 \end{bmatrix}$$

Thus the desired quadratic is $y = 2 + 5x - 3x^2$.

5. The two column vectors of M are linearly independent if and only if neither is a nonzero multiple of the other. Since all of the entries in the first column are equal, the columns are linearly independent if and only if the second column has at least two different entries, or if and only if at least two of the numbers x_i are distinct.

EXERCISE SET 9.4

1. **(a)** Since $f(x) = 1 + x$, we have

$$a_0 = \frac{1}{\pi}\int_0^{2\pi}(1+x)dx = 2+2\pi$$

Using Example 1 and some simple integration, we obtain

$$a_k = \frac{1}{\pi}\int_0^{2\pi}(1+x)\cos(kx)dx = 0$$

$$k = 1, 2, \ldots$$

$$b_k = \frac{1}{\pi}\int_0^{2\pi}(1+x)\sin(kx)dx = -\frac{2}{k}$$

Thus, the least squares approximation to $1 + x$ on $[0, 2\pi]$ by a trigonometric polynomial of order ≤ 2 is

$$1 + x \simeq (1 + \pi) - 2\sin x - \sin 2x$$

3. **(a)** The space W of continuous functions of the form $a + be^x$ over $[0, 1]$ is spanned by the functions $\mathbf{u}_1 = 1$ and $\mathbf{u}_2 = e^x$. First we use the Gram-Schmidt process to find an orthonormal basis $\{\mathbf{v}_1, \mathbf{v}_2\}$ for W.

Since $\langle \mathbf{f}, \mathbf{g}\rangle = \int_0^1 f(x)g(x)dx$, then $\|\mathbf{u}_1\| = 1$ and hence

$$\mathbf{v}_1 = 1$$

Thus

$$\mathbf{v}_2 = \frac{e^x - (e^x, 1)1}{\left\|e^x - (e^x, 1)1\right\|} = \frac{e^x - e + 1}{\alpha}$$

where α is the constant

$$\alpha = \|e^x - e + 1\| = \left[\int_0^1 (e^x - e + 1)^2 dx\right]^{1/2}$$
$$= \left[\frac{(3-e)(e-1)}{2}\right]^{1/2}$$

Therefore the orthogonal projection of x on W is

$$\text{proj}_W x = \langle x, 1\rangle 1 + \left\langle x, \frac{e^x - e + 1}{\alpha}\right\rangle \frac{e^x - e + 1}{\alpha}$$
$$= \int_0^1 x\,dx + \frac{e^x - e + 1}{\alpha}\int_0^1 \frac{x(e^x - e + 1)}{\alpha}\,dx$$
$$= \frac{1}{2} + \frac{e^x - e + 1}{\alpha}\left[\frac{3-e}{2\alpha}\right]$$
$$= \frac{1}{2} + \left[\frac{1}{e-1}\right](e^x - e + 1)$$
$$= -\frac{1}{2} + \left[\frac{1}{e-1}\right]e^x$$

(b) The mean square error is

$$\int_0^1 \left[x - (-\frac{1}{2} + \left[\frac{1}{e-1}\right]e^x\right]^2 dx = \frac{13}{12} + \frac{1+e}{2(1-e)} = \frac{1}{2} + \frac{3-e}{2(1-e)}$$

The answer above is deceptively short since a great many calculations are involved.

To shortcut some of the work, we derive a different expression for the mean square error (m.s.e.). By definition,

$$\text{m.s.e.} = \int_a^b [f(x) - g(x)]^2\,dx$$
$$= \|\mathbf{f} - \mathbf{g}\|^2$$
$$= \langle \mathbf{f} - \mathbf{g}, \mathbf{f} - \mathbf{g}\rangle$$
$$= \langle \mathbf{f}, \mathbf{f} - \mathbf{g}\rangle - \langle \mathbf{g}, \mathbf{f} - \mathbf{g}\rangle$$

Recall that $\mathbf{g} = \text{proj}_W\mathbf{f}$, so that $\mathbf{g}$ and $\mathbf{f} - \mathbf{g}$ are orthogonal. Therefore,

$$\begin{aligned} \text{m.s.e.} &= \langle \mathbf{f}, \mathbf{f} - \mathbf{g} \rangle \\ &= \langle \mathbf{f}, \mathbf{f} \rangle - \langle \mathbf{f} \mathbf{g} \rangle \end{aligned}$$

But $\mathbf{g} = \langle \mathbf{f}, \mathbf{v}_1 \rangle \mathbf{v}_1 + \langle \mathbf{f}, \mathbf{v}_2 \rangle \mathbf{v}_2$, so that

$$(*) \qquad \text{m.s.e.} = \langle \mathbf{f}, \mathbf{f} \rangle - \langle \mathbf{f}, \mathbf{v}_1 \rangle^2 - \langle \mathbf{f}, \mathbf{v}_2 \rangle^2$$

Now back to the problem at hand. We know $\langle \mathbf{f}, \mathbf{v}_1 \rangle$ and $\langle \mathbf{f}, \mathbf{v}_2 \rangle$ from Part (a). Thus, in this case,

$$\begin{aligned} \text{m.s.e.} &= \int_0^1 x^2 dx - \left(\frac{1}{2}\right)^2 - \left(\frac{3-e}{2\alpha}\right)^2 \\ &= \frac{1}{12} - \frac{3-e}{2(e-1)} \approx .0014 \end{aligned}$$

Clearly the formula (*) above can be generalized. If W is an n-dimensional space with orthonormal basis $\{\mathbf{v}_1, \mathbf{v}_2, \ldots, \mathbf{v}_n\}$, then

$$(**) \qquad \text{m.s.e.} = \|\mathbf{f}\|^2 - \langle \mathbf{f}, \mathbf{v}_1 \rangle^2 - \cdots - \langle \mathbf{f}, \mathbf{v}_n \rangle^2$$

5. **(a)** The space W of polynomials of the form $a_0 + a_1 x + a_2 x^2$ over $[-1, 1]$ has the basis $\{1, x, x_2\}$. Using the inner product $\langle \mathbf{u}, \mathbf{v} \rangle = \int_{-1}^1 u(x)v(x)dx$ and the Gram-Schmidt process, we obtain the orthonormal basis

$$\left(\frac{1}{\sqrt{2}}, \sqrt{\frac{3}{2}}x, \frac{1}{2}\sqrt{\frac{5}{2}}(3x^2 - 1)\right)$$

(See Exercise 29, Section 6.3.) Thus

$$\begin{aligned} \langle \sin \pi x, \mathbf{v}_1 \rangle &= \frac{1}{\sqrt{2}} \int_{-1}^1 \sin(\pi x) dx = 0 \\ \langle \sin \pi x, \mathbf{v}_2 \rangle &= \sqrt{\frac{3}{2}} \int_{-1}^1 x \sin(\pi x) dx = \frac{2}{\pi}\sqrt{\frac{3}{2}} \\ \langle \sin \pi x, \mathbf{v}_3 \rangle &= \frac{1}{2}\sqrt{\frac{5}{2}} \int_{-1}^1 (3x^2 - 1)\sin(\pi x) dx = 0 \end{aligned}$$

Therefore,

$$\sin \pi x \approx \frac{3}{\pi} x$$

(b) From Part (a) and (**) in the solution to 3(b), we have

$$\text{m.s.e.} = \int_{-1}^{1} \sin^2(\pi x) dx - \frac{6}{\pi^2} = 1 - \frac{6}{\pi^2} \approx .39$$

9. Let $f(x) = \begin{cases} 1, & 0 \le x < \pi \\ 0, & \pi \le x \le 2\pi \end{cases}$

(note a slight correction to definition of $f(x)$ as stated on pg. 479.)

Then

$$a_0 = \frac{1}{\pi}\int_0^{2\pi} f(x)\, dx = \frac{1}{\pi}\int_0^{\pi} dx = 1$$

$$a_k = \frac{1}{\pi}\int_0^{2\pi} f(x)\cos kx\, dx = \frac{1}{\pi}\int_0^{\pi} \cos kx\, dx = 0$$

$$k = 1, 2, \cdots$$

$$b_k = \frac{1}{\pi}\int_0^{2\pi} f(x)\sin kx\, dx = \frac{1}{\pi}\int_0^{\pi} \sin kx\, dx$$

$$= \frac{1}{k\pi}\left[1-(-1)^k\right].$$

So the Fourier Series is $\frac{1}{2} + \sum_{k=1}^{\infty} \frac{1}{k\pi}\left[1-(-1)^k\right]\sin kx$

EXERCISE SET 9.5

1. **(d)** Since one of the terms in this expression is the product of 3 rather than 2 variables, the expression is not a quadratic form.

5. **(b)** The quadratic form can be written as

$$[x_1 \quad x_2]\begin{bmatrix} 7 & 1/2 \\ 1/2 & 4 \end{bmatrix}\begin{bmatrix} x_1 \\ x_2 \end{bmatrix} = x^T Ax$$

The characteristic equation of A is $(\lambda - 7)(\lambda - 4) - 1/4 = 0$ or

$$4\lambda^2 - 44\lambda + 111 = 0$$

which gives us

$$\lambda = \frac{11 \pm \sqrt{10}}{2}$$

If we solve for the eigenvectors, we find that for $\lambda = \dfrac{11+\sqrt{10}}{2}$

$$x_1 = (3 + \sqrt{10})x_2$$

and $\lambda = \dfrac{11-\sqrt{10}}{2}$

$$x_1 = (3 - \sqrt{10})x_2$$

Therefore the normalized eigenvectors are

$$\begin{bmatrix} \dfrac{3+\sqrt{10}}{\sqrt{20+6\sqrt{10}}} \\ \dfrac{1}{\sqrt{20+6\sqrt{10}}} \end{bmatrix} \text{ and } \begin{bmatrix} \dfrac{3-\sqrt{10}}{\sqrt{20-6\sqrt{10}}} \\ \dfrac{1}{\sqrt{20-6\sqrt{10}}} \end{bmatrix}$$

or, if we simplify,

$$\begin{bmatrix} \dfrac{1}{\sqrt{20-6\sqrt{10}}} \\ \dfrac{1}{\sqrt{20+6\sqrt{10}}} \end{bmatrix} \quad \text{and} \quad \begin{bmatrix} \dfrac{-1}{\sqrt{20+6\sqrt{10}}} \\ \dfrac{1}{\sqrt{20-6\sqrt{10}}} \end{bmatrix}$$

Thus the maximum value of the given form with its constraint is

$$\frac{11+\sqrt{10}}{2} \quad \text{at} \quad x_1 = \frac{1}{\sqrt{20-6\sqrt{10}}} \quad \text{and} \quad x_2 = \frac{1}{\sqrt{20+6\sqrt{10}}}$$

The minimum value is

$$\frac{11-\sqrt{10}}{2} \quad \text{at} \quad x_1 = \frac{-1}{\sqrt{20+6\sqrt{10}}} \quad \text{and} \quad x_2 = \frac{1}{\sqrt{20-6\sqrt{10}}}$$

7. **(b)** The eigenvalues of this matrix are the roots of the equation $\lambda^2 - 10\lambda + 24 = 0$. They are $\lambda = 6$ and $\lambda = 4$ which are both positive. Therefore the matrix is positive definite.

9. **(b)** The characteristic equation of this matrix is $\lambda^3 - 3\lambda + 2 = (\lambda + 1)^2(\lambda - 2)$. Since two of the eigenvalues are negative, the matrix is not positive definite.

11. **(a)** Since $x_1^2 + x_2^2 > 0$ unless $x_1 = x_2 = 0$, the form is positive definite.

(c) Since $(x_1 - x_2)^2 \geq 0$, the form is positive semidefinite. It is not positive definite because it can equal zero whenever $x_1 = x_2$ even when $x_1 = x_2 \neq 0$.

(e) If $|x_1| > |x_2|$, then the form has a positive value, but if $|x_1| < |x_2|$, then the form has a negative value. Thus it is indefinite.

13. **(a)** By definition,

$$
\begin{aligned}
T(\mathbf{x}+\mathbf{y}) &= (\mathbf{x}+\mathbf{y})^T A(\mathbf{x}+\mathbf{y}) \\
&= (\mathbf{x}^T+\mathbf{y}^T)A(\mathbf{x}+\mathbf{y}) \\
&= \mathbf{x}^T A\mathbf{x} + \mathbf{x}^T A\mathbf{y} + \mathbf{y}^T A\mathbf{x} + \mathbf{y}^T A\,\mathbf{y} \\
&= T(\mathbf{x}) + \mathbf{x}^T A\mathbf{y} + (\mathbf{x}^T A^T T\mathbf{y})^T + T(\mathbf{y}) \\
&= T(\mathbf{x}) + x^T A\mathbf{y} + \mathbf{x}^T A^T\mathbf{y} + T(\mathbf{y})
\end{aligned}
$$

(The transpose of a 1×1 matrix is itself.)

$$= T(\mathbf{x}) + 2\mathbf{x}^T A\mathbf{y} + T(\mathbf{y})$$

(Assuming that A is symmetric, $A^T = A$.)

(b) We have

$$
\begin{aligned}
T(k\mathbf{x}) &= (k\mathbf{x})^T A(k\mathbf{x}) \\
&= k^2\mathbf{x}^T A\mathbf{x} \qquad \text{(Every term has a factor of } k^2.) \\
&= k^2 T(\mathbf{x})
\end{aligned}
$$

(c) The transformation is not linear because $T(k\mathbf{x}) \neq kT(\mathbf{x})$ unless $k = 0$ or 1 by Part (b).

15. If we expand the quadratic form, it becomes

$$
\begin{aligned}
&c_1^2x_1^2 + c_2^2x_2^2 + \cdots + c_n^2x_n^2 + 2c_1c_2x_1x_2 + 2c_1c_3x_1x_3 + \\
&\cdots + 2c_1c_nx_1x_n + 2c_2c_3x_2x_3 + \cdots + 2c_{n-1}c_nx_{n-1}x_n
\end{aligned}
$$

Thus we have

$$
A = \begin{bmatrix}
c_1^2 & c_1c_2 & c_1c_3 & \cdots & c_1c_n \\
c_1c_2 & c_2^2 & c_1c_3 & \cdots & c_2c_n \\
c_1c_3 & c_2c_3 & c_3^2 & \cdots & c_3c_n \\
\cdots & \cdots & \cdots & \cdots & \cdots \\
c_1c_n & c_2c_n & c_3c_n & \cdots & c_n^2
\end{bmatrix}
$$

and the quadratic form is given by $\mathrm{x}^T A\mathbf{x}$ where $\mathbf{x} = [x_1\ x_2\ \cdots\ x_n]^T$.

17. To show that $\lambda_n \le \mathbf{x}^T A\mathbf{x}$ if $\|\mathbf{x}\| = 1$, we use the equation from the proof dealing with λ_1 and the fact that λ_n is the smallest eigenvalue. This gives

$$\begin{aligned}\mathbf{x}^T A\mathbf{x} = \langle \mathbf{x}, A\mathbf{x}\rangle &= \lambda_1 \langle \mathbf{x}, \mathbf{v}_1\rangle^2 + \lambda_2 \langle \mathbf{x}, \mathbf{v}_2\rangle^2 + \cdots + \lambda_n \langle \mathbf{x}, \mathbf{v}_n\rangle^2 \\ &\ge \lambda_n \langle \mathbf{x}, \mathbf{v}_1\rangle^2 + \lambda_n \langle \mathbf{x}, \mathbf{v}_2\rangle^2 + \cdots + \lambda_n \langle \mathbf{x}, \mathbf{v}_n\rangle^2 \\ &= \lambda_n (\langle \mathbf{x}, \mathbf{v}_1\rangle^2 + \cdots + \langle \mathbf{x}, \mathbf{v}_n\rangle^2) \\ &= \lambda_n\end{aligned}$$

Now suppose that $\mathbf{x}$ is an eigenvector of A corresponding to λ_n. As in the proof dealing with λ_1, we have

$$\mathbf{x}^T A\mathbf{x} = \langle \mathbf{x}, A\mathbf{x}\rangle = \langle \mathbf{x}, \lambda_n\mathbf{x}\rangle = \lambda_n \langle \mathbf{x}, \mathbf{x}\rangle = \lambda_n\|\mathbf{x}\|^2 = \lambda_n$$

EXERCISE SET 9.6

1. **(a)** The quadratic form $\mathbf{x}^T A\mathbf{x}$ can be written as

$$\begin{bmatrix} x_1 & x_2 \end{bmatrix}\begin{bmatrix} 2 & -1 \\ -1 & 2 \end{bmatrix}\begin{bmatrix} x_1 \\ x_2 \end{bmatrix}$$

The characteristic equation of A is $\lambda^2 - 4\lambda + 3 = 0$. The eigenvalues are $\lambda = 3$ and $\lambda = 1$. The corresponding eigenspaces are spanned by the vectors

$$\begin{bmatrix} 1 \\ -1 \end{bmatrix} \text{ and } \begin{bmatrix} 1 \\ 1 \end{bmatrix}$$

respectively. These vectors are orthogonal. If we normalize them, we can use the result to obtain a matrix P such that the substitution $\mathbf{x} = P\mathbf{y}$ or

$$\begin{bmatrix} x_1 \\ x_2 \end{bmatrix} = \begin{bmatrix} 1/\sqrt{2} & 1/\sqrt{2} \\ -1/\sqrt{2} & 1/\sqrt{2} \end{bmatrix}\begin{bmatrix} y_1 \\ y_2 \end{bmatrix}$$

will eliminate the cross-product term. This yields the new quadratic form

$$\begin{bmatrix} y_1 & y_2 \end{bmatrix}\begin{bmatrix} 3 & 0 \\ 0 & 1 \end{bmatrix}\begin{bmatrix} y_1 \\ y_2 \end{bmatrix}$$

or $3y_1^2 + y_2^2$.

7. **(a)** If we complete the squares, then the equation $9x^2 + 4y^2 - 36x - 24y + 36 = 0$ becomes

$$9(x^2 - 4x + 4) + 4(y^2 - 6y + 9) = -36 + 9(4) + 4(9)$$

or

$$9(x-2)^2 + 4(y-3)^2 = 36$$

or

$$\frac{(x')^2}{4} + \frac{(y')^2}{9} = 1$$

This is an ellipse.

(c) If we complete the square, then $y^2 - 8x - 14y + 49 = 0$ becomes

$$y^2 - 14y + 49 = 8x$$

or

$$(y-7)^2 = 8x$$

This is the parabola $(y')^2 = 8x'$.

(e) If we complete the squares, then $2x^2 - 3y^2 + 6x + 2y = -41$ becomes

$$2\left(x^2 + 3x + \frac{9}{4}\right) - 3\left(y^2 - \frac{20}{3}y + \frac{100}{9}\right) = -41 + \frac{9}{2} - \frac{100}{3}$$

or

$$2\left(x + \frac{3}{2}\right)^2 - 3\left(y - \frac{10}{3}\right)^2 = -\frac{419}{6}$$

or

$$12(x')^2 - 18(y')^2 = -419$$

This is a hyperbola.

9. The matrix form for the conic $9x^2 - 4xy + 6y^2 - 10x - 20y = 5$ is

$$\mathbf{x}^T A\mathbf{x} + K\mathbf{x} = 5$$

where

$$A = \begin{bmatrix} 9 & -2 \\ -2 & 6 \end{bmatrix} \text{ and } \mathbf{K} = [-10 \quad -20]$$

The eigenvalues of A are $\lambda_1 = 5$ and $\lambda_2 = 1$ and the eigenspaces are spanned by the vectors

$$\begin{bmatrix} 1 \\ 2 \end{bmatrix} \quad \text{and} \quad \begin{bmatrix} -2 \\ 1 \end{bmatrix}$$

Thus we can let

$$P = \begin{bmatrix} \frac{1}{\sqrt{5}} & \frac{-2}{\sqrt{5}} \\ \frac{2}{\sqrt{5}} & \frac{1}{\sqrt{5}} \end{bmatrix}$$

Note that $\det(P) = 1$. If we let $\mathbf{x} = P\mathbf{x}'$, then

$$(\mathbf{x}')^T (P^T AP)\mathbf{x}' + KP\mathbf{x}' = 5$$

where

$$P^T AP = \begin{bmatrix} 5 & 0 \\ 0 & 10 \end{bmatrix} \quad \text{and} \quad KP = \begin{bmatrix} -10\sqrt{5} & 0 \end{bmatrix}$$

Thus we have the equation

$$5(x')^2 + 10(y')^2 - 10\sqrt{5}x' = 5$$

If we complete the square, we obtain the equation

$$5\left((x')^2 - 2\sqrt{5}x' + 5\right) + 10\,(y')^2 = 5 + 25$$

or

$$(x'')^2 + 2(y'')^2 = 6$$

where $x'' = x' - \sqrt{5}$ and $y'' = y'$. This is the ellipse

$$\frac{(x'')^2}{6}+\frac{(y'')^2}{3}=1$$

Of course we could also rotate to obtain the same ellipse in the form $2(x'')^2 + (y'')^2 = 6$, which is just the other standard position.

11. The matrix form for the conic $2x^2 - 4xy - y^2 - 4x - 8y = -14$ is

$$\mathbf{x}^T A\mathbf{x} + K\mathbf{x} = -14$$

where

$$A=\begin{bmatrix} 2 & -2 \\ -2 & -1 \end{bmatrix} \quad \text{and} \quad K=\begin{bmatrix} -4 & -8 \end{bmatrix}$$

The eigenvalues of A are $\lambda_1 = 3$, $\lambda_2 = -2$ and the eigenspaces are spanned by the vectors

$$\begin{bmatrix} -2 \\ 1 \end{bmatrix} \quad \text{and} \quad \begin{bmatrix} 1 \\ 2 \end{bmatrix}$$

Thus we can let

$$P=\begin{bmatrix} \frac{2}{\sqrt{5}} & \frac{1}{\sqrt{5}} \\ \frac{-1}{\sqrt{5}} & \frac{2}{\sqrt{5}} \end{bmatrix}$$

Note that $\det(P) = 1$. If we let $\mathbf{x} = P\mathbf{x}'$, then

$$(\mathbf{x}')^T (P^T AP)\mathbf{x}' + KP\mathbf{x}' = -14$$

where

$$P^T AP=\begin{bmatrix} 3 & 0 \\ 0 & -2 \end{bmatrix} \quad \text{and} \quad KP=\begin{bmatrix} 0 & -4\sqrt{5} \end{bmatrix}$$

Thus we have the equation

$$3(x')^2 - 2(y')^2 - 4\sqrt{5}y' = -14$$

If we complete the square, then we obtain

$$3(x')^2 - 2\left((y')^2 + 2\sqrt{5}y' + 5\right) = -14 - 10$$

or

$$3(x'')^2 - 2(y'')^2 = -24$$

where $x'' = x'$ and $y'' = y' + \sqrt{5}$. This is the hyperbola

$$\frac{(y'')^2}{12} - \frac{(x'')^2}{8} = 1$$

We could also rotate to obtain the same hyperbola in the form $2(x'')^2 - 3(y'')^2 = 24$.

15. **(a)** The equation $x^2 - y^2 = 0$ can be written as $(x - y)(x + y) = 0$. Thus it represents the two intersecting lines $x \pm y = 0$.

(b) The equation $x^2 + 3y^2 + 7 = 0$ can be written as $x^2 + 3y^2 = -7$. Since the left side of this equation cannot be negative, then there are no points (x, y) which satisfy the equation.

(c) If $8x^2 + 7y^2 = 0$, then $x = y = 0$. Thus the graph consists ofthe single point $(0, 0)$.

(d) This equation can be rewritten as $(x - y)^2 = 0$. Thus it represents the single line $y = x$.

(e) The equation $9x^2 + 12xy + 4y^2 - 52 = 0$ can be written as $(3x + 2y)^2 = 52$ or $3x + 2y = \pm\sqrt{52}$. Thus its graph is the two parallel lines $3x + 2y \pm 2\sqrt{13} = 0$.

(f) The equation $x^2 + y^2 - 2x - 4y = -5$ can be written as $x^2 - 2x + 1 + y^2 - 4y + 4 = 0$ or $(x - 1)^2 + (y - 2)^2 = 0$. Thus it represents the point $(1, 2)$.

EXERCISE SET 9.7

5. **(a)** ellipse $36x^2 + 9y^2 = 32$

(b) ellipse $2x^2 + 6y^2 = 21$

(c) hyperbola $6x^2 - 3y^2 = 8$

(d) ellipse $9x^2 + 4y^2 = 1$

(e) ellipse $16x^2 + y^2 = 16$

(f) hyperbola $3y^2 - 7x^2 = 1$

(g) circle $x^2 + y^2 = 24$

7. **(a)** If we complete the squares, the quadratic becomes

$$9(x^2 - 2x + 1) + 36(y^2 - 4y + 4) + 4(z^2 - 6z + 9)$$

$$= -153 + 9 + 144 + 36$$

or

$$9(x - 1)^2 + 36(y - 2)^2 + 4(z - 3)^2 = 36$$

or

$$\frac{(x')^2}{4} + \frac{(y')^2}{1} - \frac{(z')}{9} = 1$$

This is an ellipsoid.

(c) If we complete the square, the quadratic becomes

$$3(x^2 + 14z + 49) - 3y^2 - z^2 = 144 + 147$$

or

$$3(x + 7)^2 - 3y^2 - z^2 = 3$$

or

$$\frac{(x')^2}{1} + \frac{(y')^2}{1} - \frac{(z')^2}{3} = 1$$

This is a hyperboloid of two sheets.

7. **(e)** If we complete the squares, the quadric becomes

$$(x^2 + 2x + 1) + 16(y^2 - 2y + 1) - 16z = 15 + 1 + 16$$

or

$$(x + 1)^2 + 16(y - 1)^2 - 16(z + 2) = 0$$

or

$$\frac{(x')^2}{4} + \frac{(y')^2}{1} - \frac{(z')}{9} = 1$$

This is an elliptic paraboloid.

(g) If we complete the squares, the quadric becomes

$$(x^2 - 2x + 1) + (y^2 + 4y + 4) + (z^2 - 6z + 9) = 11 + 1 + 4 + 9$$

or

$$(x - 1)^2 + (y + 2)^2 + (z - 3)^2 = 25$$

or

$$\frac{(x')^2}{1} + \frac{(y')^2}{1} - \frac{(z')^2}{3} = 1$$

This is a sphere.

9. The matrix form for the quadric is $\mathbf{x}^T A\mathbf{x} + K\mathbf{x} = -9$ where

$$A = \begin{bmatrix} 0 & 1 & 1 \\ 1 & 0 & 1 \\ 1 & 1 & 0 \end{bmatrix} \text{ and } K = \begin{bmatrix} -6 & -6 & -4 \end{bmatrix}$$

The eigenvalues of A are $\lambda_1 = \lambda_2 = -1$ and $\lambda_3 = 2$, and the vectors

$$\mathbf{u}_1 = \begin{bmatrix} -1 \\ 0 \\ 1 \end{bmatrix} \quad \mathbf{u}_2 = \begin{bmatrix} -1 \\ 1 \\ 0 \end{bmatrix} \quad \text{and} \quad \mathbf{u}_3 = \begin{bmatrix} 1 \\ 1 \\ 1 \end{bmatrix}$$

span the corresponding eigenspaces. Note that $\mathbf{u}_1 \cdot \mathbf{u}_3 = \mathbf{u}_2 \cdot \mathbf{u}_3 = 0$ but that $\mathbf{u}_1 \cdot \mathbf{u}_2 \neq 0$. Hence, we must apply the Gram-Schmidt process to $\{\mathbf{u}_1, \mathbf{u}_2\}$. We must also normalize $\mathbf{u}_3$. This gives the orthonormal set

$$\begin{bmatrix} \frac{-1}{\sqrt{2}} \\ 0 \\ \frac{1}{\sqrt{2}} \end{bmatrix} \quad \begin{bmatrix} \frac{-1}{\sqrt{6}} \\ \frac{5}{\sqrt{6}} \\ \frac{-1}{\sqrt{6}} \end{bmatrix} \quad \begin{bmatrix} \frac{1}{\sqrt{3}} \\ \frac{1}{\sqrt{3}} \\ \frac{-1}{\sqrt{3}} \end{bmatrix}$$

Thus we can let

$$P = \begin{bmatrix} \frac{-1}{\sqrt{2}} & \frac{-1}{\sqrt{6}} & \frac{1}{\sqrt{3}} \\ 0 & \frac{2}{\sqrt{6}} & \frac{1}{\sqrt{3}} \\ \frac{1}{\sqrt{2}} & \frac{-1}{\sqrt{6}} & \frac{-1}{\sqrt{3}} \end{bmatrix}$$

Note that $\det(P) = 1$,

$$P^T AP = \begin{bmatrix} -1 & 0 & 0 \\ 0 & -1 & 0 \\ 0 & 0 & 2 \end{bmatrix} \text{ and } KP = \begin{bmatrix} -\sqrt{2} & \frac{-2}{\sqrt{6}} & \frac{-16}{\sqrt{3}} \end{bmatrix}$$

Therefore the transformation $\mathbf{x} = P\mathbf{x}'$ reduces the quadric to

$$-(x')^2 - (y')^2 + 2(z')^2 - \sqrt{2}x' - \frac{2}{\sqrt{6}}y' - \frac{16}{\sqrt{3}}z' = -9$$

If we complete the squares, this becomes

$$\left[(x')^2 + \sqrt{2}x' + \frac{1}{2}\right] + \left[(y')^2 + \frac{2}{\sqrt{6}}y' + \frac{1}{6}\right]$$
$$-2\left[(z')^2 - \frac{8}{\sqrt{3}}z' + \frac{16}{3}\right] = 9 + \frac{1}{2} + \frac{1}{6} - \frac{32}{3}$$

Letting $x'' = x' + \frac{1}{\sqrt{2}}$, $y'' = y' + \frac{1}{16}$, and $z'' = z' - \frac{4}{\sqrt{3}}$ yields

$$(x'')^2 + (y'')^2 - 2(z'')^2 = -1$$

This is the hyperboloid of two sheets

$$\frac{(z'')^2}{1/2} - \frac{(x'')^2}{1} - \frac{(y'')^2}{1} = 1$$

11. The matrix form for the quadric is $\mathbf{x}^T A\mathbf{x} + K\mathbf{x} - 31 =$ where

$$A = \begin{bmatrix} 0 & 1 & 0 \\ 1 & 0 & 0 \\ 0 & 0 & 0 \end{bmatrix} \quad \text{and} \quad K = \begin{bmatrix} -6 & 10 & 1 \end{bmatrix}$$

The eigenvalues of A are $\lambda_1 = 1$, $\lambda_2 = -1$, and $\lambda_3 = 0$, and the corresponding eigenspaces are spanned by the orthogonal vectors

$$\begin{bmatrix} 1 \\ 1 \\ 0 \end{bmatrix} \quad \begin{bmatrix} -1 \\ 1 \\ 0 \end{bmatrix} \quad \begin{bmatrix} 0 \\ 0 \\ 1 \end{bmatrix}$$

Thus, we let $\mathbf{x} = P\mathbf{x}'$ where

$$P = \begin{bmatrix} \frac{1}{\sqrt{2}} & \frac{-1}{\sqrt{2}} & 0 \\ \frac{1}{\sqrt{2}} & \frac{1}{\sqrt{2}} & 0 \\ 0 & 0 & 1 \end{bmatrix}$$

Note that $\det(P) = 1$,

$$P^T AP = \begin{bmatrix} 1 & 0 & 0 \\ 0 & -1 & 0 \\ 0 & 0 & 0 \end{bmatrix} \quad \text{and} \quad KP = \begin{bmatrix} 2\sqrt{2} & 8\sqrt{2} & 1 \end{bmatrix}$$

Therefore, the equation of the quadric is reduced to

$$(x')^2 - (y')^2 + 2\sqrt{2}x' + 8\sqrt{2}y' + z' - 31 = 0$$

If we complete the squares, this becomes

$$\left[(x')^2 + 2\sqrt{2}x' + 2\right] - \left[(y')^2 - 8\sqrt{2}y' + 32\right] + z' = 31 + 2 - 32$$

Letting $x'' = x' + \sqrt{2}$, $y'' = y' - 4\sqrt{2}$, and $z'' = z' - 1$ yields

$$(x'')^2 - (y'')^2 + z'' = 0$$

This is a hyperbolic paraboloid.

13. We know that the equation of a general quadric Q can be put into the standard matrix form $\mathbf{x}^T A\mathbf{x} + K\mathbf{x} + j = 0$ where

$$A = \begin{bmatrix} a & d & e \\ d & b & f \\ e & f & c \end{bmatrix} \quad \text{and} \quad K = \begin{bmatrix} g & h & i \end{bmatrix}$$

Since A is a symmetric matrix, then A is orthogonally diagonalizable by Theorem 7.3.1. Thus, by Theorem 7.2.1, A has 3 linearly independent eigenvectors. Now let T be the matrix whose column vectors are the 3 linearly independent eigenvectors of A. It follows from the proof of Theorem 7.2.1 and the discussion immediately following that theorem, that $T^{-1}AT$ will be a diagonal matrix whose diagonal entries are the eigenvalues λ_1, λ_2, and λ_3 of A. Theorem 7.3.2 guarantees that these eigenvalues are real.

As noted immediately after Theorem 7.3.2, we can, if necessary, transform the matrix T to a matrix S whose column vectors form an orthonormal set. To do this, orthonormalize the basis of each eigenspace before using its elements as column vectors of S.

Furthermore, by Theorem 6.5.1, we know that S is orthogonal. It follows from Theorem 6.5.2 that $\det(S) = \pm 1$.

In case $\det(S) = -1$, we interchange two columns in S to obtain a matrix P such that $\det(P) = 1$. If $\det(S) = 1$, we let $P = S$. Thus, P represents a rotation. Note that P is orthogonal, so that $P^{-1} = P^T$, and also, that P orthogonally diagonalizes A. In fact,

$$P^T AP = \begin{bmatrix} \lambda_1 & 0 & 0 \\ 0 & \lambda_2 & 0 \\ 0 & 0 & \lambda_3 \end{bmatrix}$$

Hence, if we let $\mathbf{x} = P\mathbf{x}'$, then the equation of the quadric Q becomes

$$(\mathbf{x}')^T (P^T AP)\mathbf{x}' + KP\mathbf{x}' + j = 0$$

or

$$\lambda_1(x')^2 + \lambda_2(y')^2 + \lambda_3(z')^2 + g'x' + h'y' + i'z' + j = 0$$

where

$$[g' \; h' \; i'] = KP$$

Thus we have proved Theorem 9.7.1.

EXERCISE SET 9.8

1. If $AB = C$ where A is $m \times n$, B is $n \times p$, and C is $m \times p$, then C has mp entries, each of the form

$$c_{ij} = a_{i1}b_{1j} + a_{i2}b_{2j} + \cdots + a_{in}b_{nj}$$

Thus we need n multiplications and $n - 1$ additions to compute each of the numbers c_{ij}. Therefore we need mnp multiplications and $m(n - 1)p$ additions to compute C.

5. Following the hint, we have

$$S_n = 1 + \quad 2 \quad + \quad 3 \quad + \cdots + n$$

$$S_n = n + (n - 1) + (n - 2) + \cdots + 1$$

or

$$2S_n = (n + 1) + (n + 1) + (n + 1) + \cdots + (n + 1)$$

Thus

$$S_n = \frac{n(n+1)}{2}$$

7. **(a)** By direct computation,

$$(k + 1)^3 - k^3 = k^3 + 3k^2 + 3k + 1 - k^3 = 3k^2 + 3k + 1$$

(b) The sum "telescopes". That is,

$$\begin{aligned}
&[2^3 - 1^3] + [3^3 - 2^3] + [4^3 - 3^3] + \cdots + [(n + 1)^3 - n^3] \\
&\quad = 2^3 - 1^3 + 3^3 - 2^3 + 4^3 - 3^3 + \cdots + (n + 1)^3 - n^3 \\
&\quad = (n + 1)^3 - 1
\end{aligned}$$

(c) By Parts (a) and (b), we have

$$3(1)^2 + 3(1) + 1 + 3(2)^2 + 3(2) + 1 + 3(3)^2 + 3(3) + 1 + \cdots + 3n^2 + 3n + 1$$

$$= 3(1^2 + 2^2 + 3^2 + \cdots + n^2) + 3(1 + 2 + 3 + \cdots + n) + n$$

$$= (n + 1)^3 - 1$$

(d) Thus, by Part (c) and exercise 6, we have

$$1^2+2^2+3^2+\cdots+n^2=\frac{1}{3}\left[\,(n+1)^3-1-3\frac{n(n+1)}{2}-n\,\right]$$

$$=\frac{(n+1)^3}{3}-\frac{1}{3}-\frac{n(n+1)}{2}-\frac{n}{3}$$

$$=\frac{2(n+1)^3-2-3n(n+1)-2n}{6}$$

$$=\frac{(n+1)[2(n+1)^2-3n-2]}{6}$$

$$=\frac{(n+1)(2n^2+4n+2-3n-2)}{6}$$

$$=\frac{(n+1)(2n^2+n)}{6}$$

$$=\frac{n(n+1)(2n+1)}{6}$$

9. Since R is a row-echelon form of an invertible $n \times n$ matrix, it has ones down the main diagonal and nothing but zeros below. If, as usual, we let $\mathbf{x} = [x_1\ x_2\ \cdots\ x_n]^T$ and $\mathbf{b} = [b_1\ b_2\ \cdots\ b_n]^T$, then we have $x_n = b_n$ with no computations. However, since $x_{n-1} = b_{n-1} - cx_n$ for some number c, it will require one multiplication and one addition to find x_{n-1}. In general,

$$x_i = b_i - \text{some linear combination of } x_{i+1}, x_{i+2}, \cdots, x_n$$

Therefore it will require two multiplications and two additions to find x_{n-2}, three of each to find x_{n-3}, and finally, n_{-1} of each to find x_1. That is, it will require

$$1+2+3+\cdots+(n-1)=\frac{(n-1)n}{1}$$

multiplications and the same number of additions to solve the system by back substitution.

11. To solve a linear system whose coefficient matrix is an invertible $n \times n$ matrix, A, we form the $n \times (n + 1)$ matrix $[A|b]$ and reduce A to I_n. Thus we first divide Row 1 by a_{11}, using n multiplications (ignoring the multiplication whose result must be one and assuming that $a_{11} \neq 0$ since no row interchanges are required). We then subtract a_{i1} times Row 1 from Row i for $i = 2, \ldots, n$ to reduce the first column to that of I_n. This requires $n(n - 1)$ multiplications and the same number of additions (again ignoring the operations whose results we already know). The total number of multiplications so far is n^2 and the total number of additions is $n(n - 1)$.

To reduce the second column to that of I_n, we repeat the procedure, starting with Row 2 and ignoring Column 1. Thus $n - 1$ multiplications assure us that there is a one on the main diagonal, and $(n - 1)^2$ multiplications and additions will make all $n - 1$ of the remaining column entries zero. This requires $n(n - 1)$ new multiplications and $(n - 1)^2$ new additions.

In general, to reduce Column i to the ith column of I_n, we require $n + 1 - i$ multiplications followed by $(n + 1 - i)(n - 1)$ multiplications and additions, for a total of $n(n + 1 - i)$ multiplications and $(n + 1 - i)(n - 1)$ additions.

If we add up all these numbers, we find that we need

$$\begin{aligned} n^2 + n(n-1) + n(n-2) + \cdots + n(2) + n(1) &= n(n + (n-1) + \cdots + 2 + 1) \\ &= \frac{n^2(n+1)}{2} \\ &= \frac{n^3}{2} + \frac{n^2}{2} \end{aligned}$$

multiplications and

$$\begin{aligned} &n(n-1) + (n-1)^2 + (n-2)(n-1) + \cdots + 2(n-1) + (n-1) \\ &\quad = (n-1)(n + (n-1) + \cdots + 2 + 1) \\ &\quad = \frac{(n-1)(n)(n+1)}{2} \\ &\quad = \frac{n^3}{2} - \frac{n}{2} \end{aligned}$$

additions to compute the reduction.

EXERCISE SET 9.9

1. The system in matrix form is

$$\begin{bmatrix} 3 & -6 \\ -2 & -5 \end{bmatrix}\begin{bmatrix} x_1 \\ x_2 \end{bmatrix}=\begin{bmatrix} 3 & 0 \\ -2 & 1 \end{bmatrix}\begin{bmatrix} 1 & -2 \\ 0 & 1 \end{bmatrix}\begin{bmatrix} x_1 \\ x_2 \end{bmatrix}=\begin{bmatrix} 0 \\ 1 \end{bmatrix}$$

This reduces to two matrix equations

$$\begin{bmatrix} 1 & -2 \\ 0 & 1 \end{bmatrix}\begin{bmatrix} x_1 \\ x_2 \end{bmatrix}=\begin{bmatrix} y_1 \\ y_2 \end{bmatrix}$$

and

$$\begin{bmatrix} 3 & 0 \\ -2 & 1 \end{bmatrix}\begin{bmatrix} y_1 \\ y_2 \end{bmatrix}=\begin{bmatrix} 0 \\ 1 \end{bmatrix}$$

The second matrix equation yields the system

$$\begin{aligned} 3y_1 \qquad &= 0 \\ -2y_1 + y_2 &= 1 \end{aligned}$$

which has $y_1 = 0$, $y_2 = 1$ as its solution. If we substitute these values into the first matrix equation, we obtain the system

$$\begin{aligned} x_1 - 2x_2 &= 0 \\ x_2 &= 1 \end{aligned}$$

This yields the final solution $x_1 = 2$, $x_2 = 1$.

3. To reduce the matrix of coefficients to a suitable upper triangular matrix, we carry out the following operations:

$$\begin{bmatrix} 2 & 8 \\ -1 & -1 \end{bmatrix} \to \begin{bmatrix} 1 & 4 \\ -1 & -1 \end{bmatrix} \to \begin{bmatrix} 1 & 4 \\ 0 & 3 \end{bmatrix} \to \begin{bmatrix} 1 & 4 \\ 0 & 1 \end{bmatrix} = U$$

These operations involve multipliers 1/2, 1, and 1/3. Thus the corresponding lower triangular matrix is

$$L = \begin{bmatrix} 2 & 0 \\ -1 & 3 \end{bmatrix}$$

We therefore have the two matrix equations

$$\begin{bmatrix} 1 & 4 \\ 0 & 1 \end{bmatrix} \begin{bmatrix} x_1 \\ x_2 \end{bmatrix} = \begin{bmatrix} y_1 \\ y_2 \end{bmatrix}$$

and

$$\begin{bmatrix} 2 & 0 \\ -1 & 3 \end{bmatrix} \begin{bmatrix} y_1 \\ y_2 \end{bmatrix} = \begin{bmatrix} -2 \\ -2 \end{bmatrix}$$

The second matrix equation yields the system

$$\begin{aligned} 2y_1 \quad\quad &= -2 \\ -y_1 + 3y_2 &= -2 \end{aligned}$$

which has $y_1 = -1$, $y_2 = -1$ as its solution. If we substitute these values into the first matrix equation, we obtain the system

$$\begin{aligned} x_1 + 4x_2 &= -1 \\ x_2 &= -1 \end{aligned}$$

This yields the final solution $x_1 = 3$, $x_2 = -1$.

5. To reduce the matrix of coefficients to a suitable upper triangular matrix, we carry out the following operations:

$$\begin{bmatrix} 2 & -2 & -2 \\ 0 & -2 & 2 \\ -1 & 5 & 2 \end{bmatrix} \to \begin{bmatrix} 1 & -1 & -1 \\ 0 & -2 & 2 \\ -1 & 5 & 2 \end{bmatrix} \to \begin{bmatrix} 1 & -1 & -1 \\ 0 & -2 & 2 \\ 0 & 4 & 1 \end{bmatrix} \to$$

$$\begin{bmatrix} 1 & -1 & -1 \\ 0 & 1 & -1 \\ 0 & 4 & 1 \end{bmatrix} \to \begin{bmatrix} 1 & -1 & -1 \\ 0 & 1 & -1 \\ 0 & 0 & 5 \end{bmatrix} \to \begin{bmatrix} 1 & -1 & -1 \\ 0 & 1 & -1 \\ 0 & 0 & 1 \end{bmatrix} = U$$

These operations involve multipliers of 1/2, 0 (for the 2 row), 1, –1/2, –4, and 1/5. Thus the corresponding lower triangular matrix is

$$L = \begin{bmatrix} 2 & 0 & 0 \\ 0 & -2 & 0 \\ -1 & 4 & 5 \end{bmatrix}$$

We therefore have the matrix equations

$$\begin{bmatrix} 1 & -1 & -1 \\ 0 & 1 & -1 \\ 0 & 0 & 1 \end{bmatrix} \begin{bmatrix} x_1 \\ x_2 \\ x_3 \end{bmatrix} = \begin{bmatrix} y_1 \\ y_2 \\ y_3 \end{bmatrix}$$

and

$$\begin{bmatrix} 2 & 0 & 0 \\ 0 & -2 & 0 \\ -1 & 4 & 5 \end{bmatrix} \begin{bmatrix} y_1 \\ y_2 \\ y_3 \end{bmatrix} = \begin{bmatrix} -4 \\ -2 \\ 6 \end{bmatrix}$$

The second matrix equation yields the system

$$\begin{aligned} 2y_1 \quad\quad\quad\quad &= -4 \\ -2y_2 \quad\quad &= -2 \\ -y_1 + 4y_2 + 5y_3 &= 6 \end{aligned}$$

which has $y_1 = -2$, $y_2 = 1$ and $y_3 = 0$ as its solution. If we substitute these values into the first matrix equation, we obtain the system

$$\begin{aligned} x_1 - x_2 - x_3 &= -2 \\ x_2 - x_3 &= 1 \\ x_3 &= 0 \end{aligned}$$

This yields the final solution $x_1 = -1$, $x_2 = 1$, $x_3 = 0$.

11. **(a)** To reduce A to row-echelon form, we carry out the following operations:

$$\begin{bmatrix} 2 & 1 & -1 \\ -2 & -1 & 2 \\ 2 & 1 & 0 \end{bmatrix} \to \begin{bmatrix} 1 & 1/2 & -1/2 \\ -2 & -1 & 2 \\ 2 & 1 & 0 \end{bmatrix} \to \begin{bmatrix} 1 & 1/2 & -1/2 \\ 0 & 0 & 1 \\ 2 & 1 & 0 \end{bmatrix}$$

$$\to \begin{bmatrix} 1 & 1/2 & -1/2 \\ 0 & 0 & 1 \\ 0 & 0 & 1 \end{bmatrix} \to \begin{bmatrix} 1 & 1/2 & -1/2 \\ 0 & 0 & 1 \\ 0 & 0 & 0 \end{bmatrix} = U$$

This involves multipliers 1/2, 2, –2, 1 (for the 2 diagonal entry), and –1. Where no multiplier is needed in the second entry of the last row, we use the multiplier 1, thus obtaining the lower triangular matrix

$$L = \begin{bmatrix} 2 & 0 & 0 \\ -2 & 1 & 0 \\ 2 & 1 & 1 \end{bmatrix}$$

In fact, if we compute LU, we see that it will equal A no matter what entry we choose for the lower right-hand corner of L.

If we stop just before we reach row-echelon form, we obtain the matrices

$$U = \begin{bmatrix} 1 & 1/2 & -1/2 \\ 0 & 0 & 1 \\ 0 & 0 & 1 \end{bmatrix} \qquad L = \begin{bmatrix} 2 & 0 & 0 \\ -2 & 1 & 0 \\ 2 & 0 & 1 \end{bmatrix}$$

which will also serve.

(b) We have that $A = LU$ where

$$L = \begin{bmatrix} 2 & 0 & 0 \\ -2 & 1 & 0 \\ 2 & 1 & 1 \end{bmatrix} \qquad U = \begin{bmatrix} 1 & 1/2 & -1/2 \\ 0 & 0 & 1 \\ 0 & 0 & 1 \end{bmatrix}$$

If we let

$$L_1 = \begin{bmatrix} 1 & 0 & 0 \\ -1 & 1 & 0 \\ 1 & 1 & 1 \end{bmatrix} \quad \text{and} \quad D = \begin{bmatrix} 2 & 0 & 0 \\ 0 & 1 & 0 \\ 0 & 1 & 1 \end{bmatrix}$$

then $A = L_1DU$ as desired. (See the matrices at the very end of Section 9.9.)

(c) Let $U_2 = DU$ and note that this matrix is upper triangular. Then $A = L_1U_2$ is of the required form.

13. **(a)** If A has such an LU-decomposition, we can write it as

$$\begin{bmatrix} a & b \\ c & d \end{bmatrix} = \begin{bmatrix} 1 & 0 \\ w & 1 \end{bmatrix} \begin{bmatrix} x & y \\ 0 & z \end{bmatrix} = \begin{bmatrix} x & y \\ wx & yw + z \end{bmatrix}$$

This yields the system of equations

$$x = a \qquad y = b \qquad wx = c \qquad yw + z = d$$

Since $a \neq 0$, this has the unique solution

$$x = a \qquad y = b \qquad w = c/a \qquad z = (ad - bc)/a$$

The uniqueness of the solution guarantees the uniqueness of the LU-decomposition.

(b) By the above,

$$\begin{bmatrix} a & b \\ c & d \end{bmatrix} = \begin{bmatrix} 1 & 0 \\ c/a & 1 \end{bmatrix} \begin{bmatrix} a & b \\ 0 & (ad - bc)/a \end{bmatrix}$$

15. We have that $L = E_1^{-1} E_2^{-1} \cdots E_k^{-1}$ where each of the matrices E_i is an elementary matrix which does not involve interchanging rows. By Exercise 27 of Section 2.1, we know that if E is an invertible lower triangular matrix, then E^{-1} is also lower triangular. Now the matrices E_i are all lower triangular and invertible by their construction. Therefore for $i = 1, \ldots, k$ we have that E_i^{-1} is lower triangular. Hence L, as the product of lower triangular matrices, must also be lower triangular.

17. Let A be any $n \times n$ matrix. We know that A can be reduced to row-echelon form and that this may require row interchanges. If we perform these interchanges (if any) first, we reduce A to the matrix

$$E_k \cdots E_1 A = B$$

where E_i is the elementary matrix corresponding to the ith such interchange. Now we know that B has an LU-decomposition, call it LU where U is a row-echelon form of A. That is,

$$E_k \cdots E_1 A = LU$$

where each of the matrices E_i is elementary and hence invertible. (In fact, $E_i^{-1} = E_i$ for all E_i. Why?) If we let

$$P = (E_k \cdots E_1)^{-1} = E_1^{-1} \cdots E_k^{-1} \qquad \text{if } k > 0$$

and $P = I$ if no row interchanges are required, then we have $A = PLU$ as desired.

19. Assume $A = PLU$, where P is a permutation matrix. Then note $P^{-1} = P$. To solve $AX = B$, where $A = PLU$, set $C = P^{-1}B = PB$ and $Y = UX$.

First solve $LY = C$ for Y.

Then solve $UX = Y$ for X.

If $A = \begin{bmatrix} 3 & -1 & 0 \\ 3 & -1 & 1 \\ 0 & 2 & 1 \end{bmatrix}$, then $A = PLU$,

with $P = \begin{bmatrix} 1 & 0 & 0 \\ 0 & 0 & 1 \\ 0 & 1 & 0 \end{bmatrix}, L = \begin{bmatrix} 3 & 0 & 0 \\ 0 & 2 & 0 \\ 3 & 0 & 1 \end{bmatrix}, U = \begin{bmatrix} 1 & -1/3 & 0 \\ 0 & 1 & 1/2 \\ 0 & 0 & 1 \end{bmatrix}$

To solve

$$\begin{pmatrix} 3 & -1 & 0 \\ 3 & -1 & 1 \\ 0 & 2 & 1 \end{pmatrix} \begin{pmatrix} x_1 \\ x_2 \\ x_3 \end{pmatrix} = \begin{pmatrix} 0 \\ 1 \\ 0 \end{pmatrix}, \text{ or } \overrightarrow{AX} = \vec{e}_2,$$

$$\text{set } C = P\vec{e}_2 \quad \begin{pmatrix} 1 & 0 & 0 \\ 0 & 0 & 1 \\ 0 & 1 & 0 \end{pmatrix} \begin{pmatrix} 0 \\ 1 \\ 0 \end{pmatrix} = \begin{pmatrix} 0 \\ 0 \\ 1 \end{pmatrix}$$

$$\text{Solve } LY = C, \text{ or } \begin{bmatrix} 3 & 0 & 0 \\ 0 & 2 & 0 \\ 3 & 0 & 1 \end{bmatrix} \begin{bmatrix} y_1 \\ y_2 \\ y_3 \end{bmatrix} = \begin{bmatrix} 0 \\ 0 \\ 1 \end{bmatrix}, \text{ to get} \begin{bmatrix} y_1 \\ y_2 \\ y_3 \end{bmatrix} = \begin{bmatrix} 0 \\ 0 \\ 1 \end{bmatrix}$$

$$\text{Solve } UX = Y, \text{ or } \begin{bmatrix} 1 & -1/3 & 0 \\ 0 & 1 & 1/2 \\ 0 & 0 & 1 \end{bmatrix} \begin{bmatrix} x_1 \\ x_2 \\ x_3 \end{bmatrix} = \begin{bmatrix} 0 \\ 0 \\ 1 \end{bmatrix}$$

$$\text{so } \begin{bmatrix} x_1 \\ x_2 \\ x_3 \end{bmatrix} = \begin{bmatrix} -1/6 \\ -1/2 \\ 1 \end{bmatrix}$$

EXERCISE SET 10.1

3. **(b)** Since two complex numbers are equal if and only if both their real and imaginary parts are equal, we have

$$x + y = 3$$

and

$$x - y = 1$$

Thus $x = 2$ and $y = 1$.

5. **(a)** Since complex numbers obey all the usual rules of algebra, we have

$$z = 3 + 2i - (1 - i) = 2 + 3i$$

(c) Since $(i - z) + (2z - 3i) = -2 + 7i$, we have

$$i + (-z + 2z) - 3i = -2 + 7i$$

or

$$z = -2 + 7i - i + 3i = -2 + 9i$$

7. **(b)** $-2z = 6 + 8i$

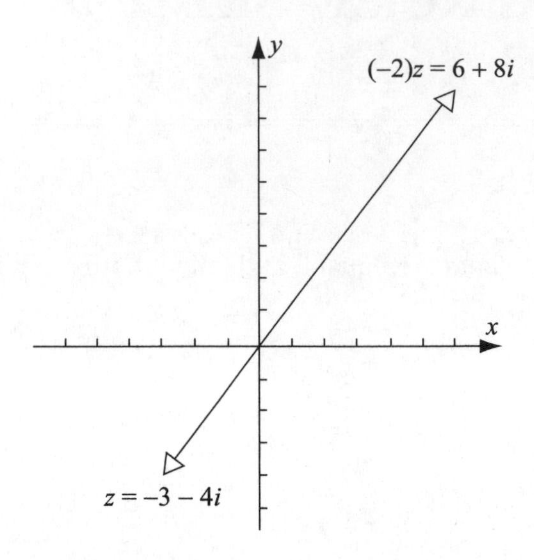

9. **(c)**

$$z_1z_2 = \frac{1}{6}(2+4i)(1-5i) = \frac{2}{6}(1+2i)(1-5i) = \frac{1}{3}(1-3i+10) = \frac{11}{3} - i$$

$$z_1^2 = \frac{4}{9}(1+2i)^2 = \frac{4}{9}(-3+4i)$$

$$z_2^2 = \frac{1}{4}(1-5i)^2 = \frac{1}{4}(-24-10i) = -6 - \frac{5}{2}i$$

11. Since $(4 - 6i)^2 = 2^2(2 - 3i)^2 = 4(-5 - 12i) = -4(5 + 12i)$, then

$$(1 + 2i)(4 - 6i)^2 = -4(1 + 2i)(5 + 12i) = -4(-19 + 22i) = 76 - 88i$$

13. Since $(1 - 3i)^2 = -8 - 6i = -2(4 + 3i)$, then

$$(1 - 3i)^3 = -2(1 - 3i)(4 + 3i) = -2(13 - 9i)$$

15.

Since $(2+i)\left(\frac{1}{2}+\frac{3}{4}i\right) = \frac{1}{4}+2i$, then

$$\left[(2+i)\left(\frac{1}{2}+\frac{3}{4}i\right)\right]^2 = \left(\frac{1}{4}+2i\right)^2 = -\frac{63}{16}+i$$

17. Since $i^2 = -1$ and $i^3 = -i$, then $1 + i + i^2 + i^3 = 0$. Thus $(1 + i + i^2 + i^3)^{100} = 0$.

19. **(a)**

$$A+3iB = \begin{bmatrix} 1 & i \\ -i & 3 \end{bmatrix} + \begin{bmatrix} 6i & -3+6i \\ 3+9i & 12i \end{bmatrix}$$

$$= \begin{bmatrix} 1+6i & -3+7i \\ 3+8i & 3+12i \end{bmatrix}$$

(d)

$$A^2 = \begin{bmatrix} 2 & 4i \\ -4i & 10 \end{bmatrix} \text{ and } B^2 = \begin{bmatrix} 11+i & 12+6i \\ 18-6i & 23+i \end{bmatrix}$$

Hence

$$B^2 - A^2 = \begin{bmatrix} 9+i & 12+2i \\ 18-2i & 13+i \end{bmatrix}$$

21. **(a)** Let $z = x + iy$. Then

$$\operatorname{Im}(iz) = \operatorname{Im}[i(x + iy)] = \operatorname{Im}(-y + ix) = x$$

$$= \operatorname{Re}(x + iy) = \operatorname{Re}(z)$$

23. **(a)** We know that $i^1 = i$, $i^2 = -1$, $i^3 = -i$, and $i^4 = 1$. We also know that $i^{m+n} = i^m i^n$ and $i^{mn} = (i^m)^n$ where m and n are positive integers. The proof can be broken into four cases:

1. $n = 1, 5, 9, \ldots$ or $n = 4k + 1$

2. $n = 2, 6, 10, \ldots$ or $n = 4k + 2$

3. $n = 3, 7, 11, \ldots$ or $n = 4k + 3$

4. $n = 4, 8, 12, \ldots$ or $n = 4k + 4$

where $k = 0, 1, 2, \ldots$. In each case, $i^n = i^{4k+\ell}$ for some integer ℓ between 1 and 4. Thus

$$i^n = i^{4k} i^{\ell} = (i^4)^k \, i^{\ell} = 1^{\,k} i^{\ell} = i^{\ell}$$

This completes the proof.

(b) Since $2509 = 4 \cdot 627 + 1$, Case 1 of Part (a) applies, and hence $i^{2509} = i$.

25. Observe that $zz_1 = zz_2 \Leftrightarrow zz_1 - zz_2 = 0 \Leftrightarrow z(z_1 - z_2) = 0$. Since $z \neq 0$ by hypothesis, it follows from Exercise 24 that $z_1 - z_2 = 0$, *i.e.*, that $z_1 = z_2$.

27. **(a)** Let $z_1 = x_1 + iy_1$ and $z_2 = x_2 + iy_2$. Then

$$\begin{aligned} z_1 z_2 &= (x_1 + iy_1)(x_2 + iy_2) \\ &= (x_1x_2 - y_1y_2) + i(x_1y_2 + x_2y_1) \\ &= (x_2x_1 - y_2y_1) + i(y_2x_1 + y_1x_2) \\ &= (x_2 + iy_2)(x_1 + iy_1) \\ &= z_2 z_1 \end{aligned}$$

EXERCISE SET 10.2

3. (a) We have

$$z\bar{z} = (2 - 4i)(2 + 4i) = 20$$

On the other hand,

$$|z|^2 = 2^2 + (-4)^2 = 20$$

(b) We have

$$z\bar{z} = (-3 + 5i)(-3 - 5i) = 34$$

On the other hand,

$$|z|^2 = (-3)^2 + 5^2 = 34$$

5. (a) Equation (5) with $z_1 = 1$ and $z_2 = i$ yields

$$\frac{1}{i} = \frac{1(-i)}{1} = -i$$

(c)

$$\frac{1}{z} = \frac{7}{-i} = \frac{7(i)}{1} = 7i$$

7. Equation (5) with $z_1 = i$ and $z_2 = 1 + i$ gives

$$\frac{i}{1+i} = \frac{i(1-i)}{2} = \frac{1}{2}+\frac{1}{2}i$$

9. Since $(3 + 4i)^2 = -7 + 24i$, we have

$$\frac{1}{(3+4i)^2} = \frac{-7-24i}{(-7)^2+(-24)^2} = \frac{-7-24i}{625}$$

11. Since

$$\frac{\sqrt{3}+i}{\sqrt{3}-i} = \frac{(\sqrt{3}+i)^2}{4} = \frac{2+2\sqrt{3}i}{4} = \frac{1}{2}+\frac{\sqrt{3}}{2}i$$

then

$$\frac{\sqrt{3}+i}{(1-i)(\sqrt{3}-i)} = \frac{\frac{1}{2}+\frac{\sqrt{3}}{2}i}{1-i} = \frac{\left(\frac{1}{2}+\frac{\sqrt{3}}{2}i\right)(1+i)}{2} = \frac{1-\sqrt{3}}{4}+\left(\frac{1+\sqrt{3}}{4}\right)i$$

13. We have

$$\frac{i}{1-i} = \frac{i(1+i)}{2} = -\frac{1}{2}+\frac{1}{2}i$$

and

$$(1-2i)(1+2i) = 5$$

Thus

$$\frac{i}{(1-i)(1-2i)(1+2i)} = \frac{-\frac{1}{2}+\frac{1}{2}i}{5} = -\frac{1}{10}+\frac{1}{10}i$$

15. **(a)** If $iz = 2 - i$, then

$$z = \frac{2-i}{i} = \frac{(2-i)(-i)}{1} = -1-2i$$

17. **(a)** The set of points satisfying the equation $|z| = 2$ is the set of all points representing vectors of length 2. Thus, it is a circle of radius 2 and center at the origin.

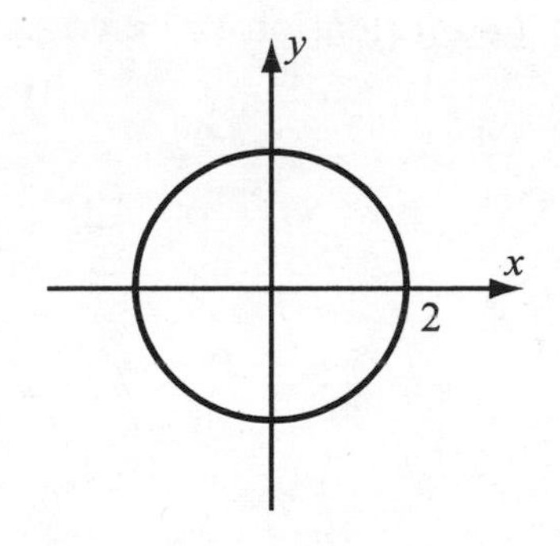

Analytically, if $z = x + iy$, then

$$|z| = 2 \Leftrightarrow \sqrt{x^2 + y^2} = 2$$
$$\Leftrightarrow x^2 + y^2 = 4$$

which is the equation of the above circle.

(c) The values of z which satisfy the equation $|z - i| = |z + i|$ are just those z whose distance from the point i is equal to their distance from the point $-i$. Geometrically, then, z can be any point on the real axis.

We now show this result analytically. Let $z = x + iy$. Then

$$|z - i| = |z + i| \Leftrightarrow |z - i|^2 = |z + i|^2$$
$$\Leftrightarrow |x + i(y - 1)|^2 = |x + i(y + 1)|^2$$
$$\Leftrightarrow x^2 + (y - 1)^2 = x^2 + (y + 1)^2$$
$$\Leftrightarrow -2y = 2y$$
$$\Leftrightarrow y = 0$$

19. **(a)** $\text{Re}(\overline{iz}) = \text{Re}(\bar{i}\,\bar{z}) = \text{Re}[(-i)(x - iy)] = \text{Re}(-y - ix) = -y$

(c) $\text{Re}(i\bar{z}) = \text{Re}[i(x - iy)] = \text{Re}(y + ix) = y$

21. **(a)** Let $z = x + iy$. Then

$$\frac{1}{2}(z + \bar{z}) = \frac{1}{2}[(x + iy) + (x - iy)] = \frac{1}{2}(2x) = x = \text{Re}(z)$$

23. **(a)** Equation (5) gives

$$\begin{aligned}\frac{z_1}{z_2} &= \frac{1}{|z_2|^2} z_1 \bar{z}_2 \\ &= \frac{1}{x_2^2 + y_2^2}(x_1 + iy_1)\,(x_2 - iy_2) \\ &= \frac{1}{x_2^2 + y_2^2}\left[(x_1x_2 + y_1y_2) + i(x_2y_1 - x_1y_2)\right]\end{aligned}$$

Thus

$$\operatorname{Re}\left(\frac{z_1}{z_2}\right) = \frac{x_1x_2 + y_1y_2}{x_2^2 + y_2^2}$$

25. $|z| = \sqrt{x^2 + y^2} = \sqrt{x^2 + (-y)^2} = |\bar{z}|$

27. **(a)** $\overline{z^2} = \overline{zz} = \bar{z}\,\bar{z} = (\bar{z})^2$

(b) We use mathematical induction. In Part (a), we verified that the result holds when $n = 2$. Now, assume that $(\bar{z})^n = \overline{z^n}$. Then

$$\begin{aligned}(\bar{z})^{n+1} &= (\bar{z})^n \bar{z} \\ &= \overline{z^n}\,\bar{z} \\ &= \overline{z^{n+1}}\end{aligned}$$

and the result is proved.

35. **(a)**

$$A^{-1} = \frac{1}{i^2 + 2}\begin{bmatrix} i & 2 \\ -1 & i \end{bmatrix} = \begin{bmatrix} i & 2 \\ -1 & i \end{bmatrix}$$

It is easy to verify that $AA^{-1} = A^{-1}A = I$.

39. **(a)**

$$\left[\begin{array}{ccc|ccc} 1 & 1+i & 0 & 1 & 0 & 0 \\ 0 & 1 & i & 0 & 1 & 0 \\ -i & 1-2i & 2 & 0 & 0 & 1 \end{array}\right]$$

$$\left[\begin{array}{ccc|ccc} 1 & 1+i & 0 & 1 & 0 & 0 \\ 0 & 1 & i & 0 & 1 & 0 \\ 0 & -i & 2 & i & 0 & 1 \end{array}\right] \quad \boxed{R_3 \to R_3 + iR_1}$$

$$\left[\begin{array}{ccc|ccc} 1 & 1+i & 0 & 1 & 0 & 0 \\ 0 & 1 & i & 0 & 1 & 0 \\ 0 & 0 & 1 & i & i & 1 \end{array}\right] \quad \boxed{R_3 \to R_3 + iR_2}$$

$$\left[\begin{array}{ccc|ccc} 1 & 1+i & 0 & 1 & 0 & 0 \\ 0 & 1 & 0 & 1 & 2 & -i \\ 0 & 0 & 1 & i & i & 1 \end{array}\right] \quad \boxed{R_2 \to R_2 - iR_3}$$

$$\left[\begin{array}{ccc|ccc} 1 & 0 & 0 & -i & -2-2i & -1+i \\ 0 & 1 & 0 & 1 & 2 & -i \\ 0 & 0 & 1 & i & i & 1 \end{array}\right] \quad \boxed{R_2 \to R_1 - (1+i)R_2}$$

Thus

$$A^{-1} = \begin{bmatrix} -i & -2-2i & -1+i \\ 1 & 2 & -i \\ i & i & 1 \end{bmatrix}$$

41. **(a)** We have $|z_1 - z_2| = \sqrt{(a_1 - a_2)^2 + (b_1 - b_2)^2}$, which is just the distance between the two numbers z_1 and z_2 when they are considered as points in the complex plane.

(b) Let $z_1 = 12$, $z_2 = 6 + 2i$, and $z_3 = 8 + 8i$. Then

$$|z_1 - z_2|^2 = 6^2 + (-2)^2 = 40$$

$$|z_1 - z_3|^2 = 4^2 + (-8)^2 = 80$$

$$|z_2 - z_3|^2 = (-2)^2 + (-6)^2 = 40$$

Since the sum of the squares of the lengths of two sides is equal to the square of the third side, the three points determine a right triangle.

EXERCISE SET 10.3

1. **(a)** If $z = 1$, then $\arg z = 2k\pi$ where $k = 0, \pm 1, \pm 2, \cdots$. Thus, $\operatorname{Arg}(1) = 0$.

(c) If $z = -i$, then $\arg z = \dfrac{3\pi}{2} + 2k\pi$ where $k = 0, \pm 1, \pm 2, \cdots$. Thus, $\operatorname{Arg}(-i) = -\pi/2$.

(e) If $z = -1 + \sqrt{3}i$, then $\arg z = \dfrac{2\pi}{3} + 2k\pi$ where $k = 0, \pm 1, \pm 2, \cdots$. Thus, Arg $(-1 + \sqrt{3}i) = \dfrac{2\pi}{3}$.

3. **(a)** Since $|2i| = 2$ and $\operatorname{Arg}(2i) = \pi/2$, we have

$$2i = 2\left[\cos\left(\frac{\pi}{2}\right) + i\sin\left(\frac{\pi}{2}\right)\right]$$

(c) Since $|5 + 5i| = \sqrt{50} = 5\sqrt{2}$ and $\operatorname{Arg}(5 + 5i) = \pi/4$, we have

$$5 + 5i = 5\sqrt{2}\left[\cos\left(\frac{\pi}{4}\right) + i\sin\left(\frac{\pi}{4}\right)\right]$$

(e) Since $|-3-3i| = \sqrt{18} = 3\sqrt{2}$ and $\operatorname{Arg}(-3-3i) = -\dfrac{3\pi}{4}$, we have

$$-3-3i = 3\sqrt{2}\left[\cos\left(-\frac{3\pi}{4}\right) + i\sin\left(-\frac{3\pi}{4}\right)\right]$$

5. We have $|z_1| = 1$, $\text{Arg}(z_1) = \dfrac{\pi}{2}$, $|z_2| = 2$, $\text{Arg}(z_2) = -\dfrac{\pi}{3}$, $|z_3| = 2$, and $\text{Arg}(z_3) = \dfrac{\pi}{6}$. So

$$\left|\frac{z_1 z_2}{z_3}\right| = \frac{|z_1||z_2|}{|z_3|} = 1$$

and

$$\text{Arg}\left(\frac{z_1 z_2}{z_3}\right) = \text{Arg}(z_1) + \text{Arg}(z_2) - \text{Arg}(z_3) = 0$$

Therefore

$$\frac{z_1 z_2}{z_3} = \cos(0) + i\sin(0) = 1$$

7. We use Formula (10).

(a) We have $r = 1$, $\theta = -\dfrac{\pi}{2}$, and $n = 2$. Thus

$$(-i)^{1/2} = \cos\left(-\frac{\pi}{4} + k\pi\right) + i\sin\left(-\frac{\pi}{4} + k\pi\right) \qquad k = 0, 1$$

Thus, the two square roots of $-i$ are:

$$\cos\left(-\frac{\pi}{4}\right) + i\sin\left(-\frac{\pi}{4}\right) = \frac{1}{\sqrt{2}} - \frac{1}{\sqrt{2}}i$$

$$\cos\left(\frac{3\pi}{4}\right) + i\sin\left(\frac{3\pi}{4}\right) = -\frac{1}{\sqrt{2}} + \frac{1}{\sqrt{2}}i$$

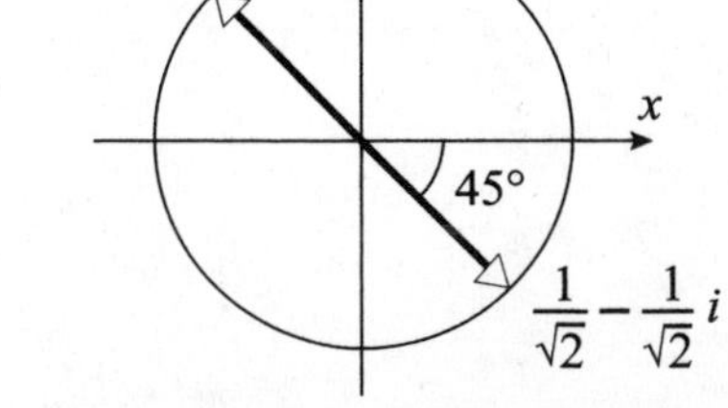

(c) We have $r = 27$, $\theta = \pi$, and $n = 3$. Thus

$$(-27)^{1/3} = 3\left[\cos\left(\frac{\pi}{3}+\frac{2k\pi}{3}\right)+i\sin\left(\frac{\pi}{3}+\frac{2k\pi}{3}\right)\right] \qquad k=0,1,2$$

Therefore, the three cube roots of –27 are:

$$3\left[\cos\left(\frac{\pi}{3}\right) + i\sin\left(\frac{\pi}{3}\right)\right] = \frac{3}{2} + \frac{3\sqrt{3}}{2}i$$

$$3\left[\cos(\pi)+i\sin(\pi)\right] = -3$$

$$3\left[\cos\left(\frac{5\pi}{3}\right) + i\sin\left(\frac{5\pi}{3}\right)\right] = \frac{3}{2} - \frac{3\sqrt{3}}{2}i$$

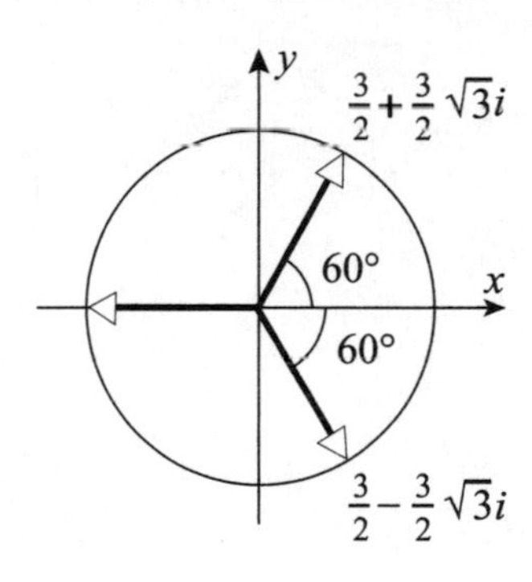

7. **(e)** Here $r = 1$, $\theta = \pi$, and $n = 4$. Thus

$$(-1)^{1/4} = \cos\left(\frac{\pi}{4} + \frac{k\pi}{2}\right) + i\sin\left(\frac{\pi}{4} + \frac{k\pi}{2}\right) \qquad k=0,\ 1,\ 2,\ 3$$

Therefore the four fourth roots of –1 are:

$$\cos\frac{\pi}{4} + i\sin\frac{\pi}{4} = \frac{1}{\sqrt{2}} + \frac{1}{\sqrt{2}}i$$
$$\cos\frac{3\pi}{4} + i\sin\frac{3\pi}{4} = -\frac{1}{\sqrt{2}} + \frac{1}{\sqrt{2}}i$$
$$\cos\frac{5\pi}{4} + i\sin\frac{5\pi}{4} = -\frac{1}{\sqrt{2}} - \frac{1}{\sqrt{2}}i$$
$$\cos\frac{7\pi}{4} + i\sin\frac{7\pi}{4} = \frac{1}{\sqrt{2}} - \frac{1}{\sqrt{2}}i$$

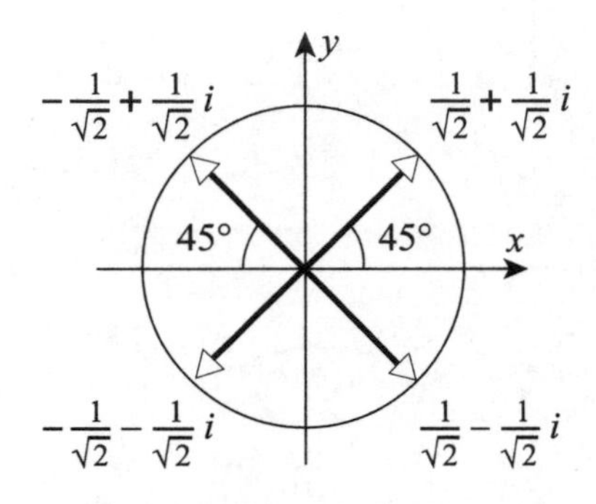

9. We observe that $w = 1$ is one sixth root of 1. Since the remaining 5 must be equally spaced around the unit circle, any two roots must be separated from one another by an angle of $\frac{2\pi}{6} = \frac{\pi}{3} = 60°$. We show all six sixth roots in the diagram.

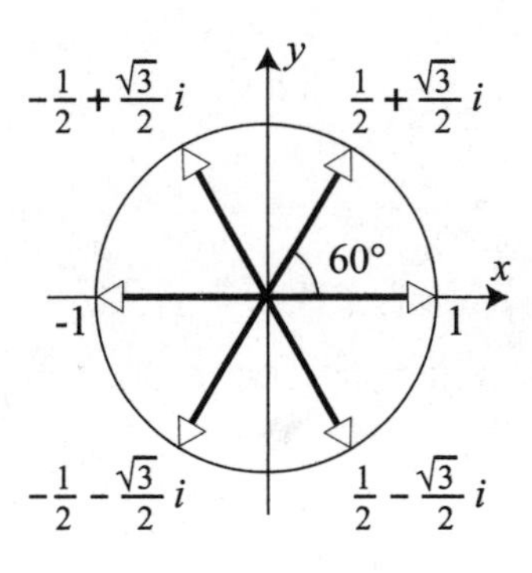

11. We have $z^4 = 16 \Leftrightarrow z = 16^{1/4}$. The fourth roots of 16 are 2, $2i$, –2, and $-2i$.

15. **(a)** Since

$$z - 3e^{i\pi} - 3[\cos(\pi) + i\sin(\pi)] = -3$$

then $\mathrm{Re}(z) = -3$ and $\mathrm{Im}(z) = 0$.

(c) Since

$$\bar{z} = \sqrt{2}e^{i\pi/2} = \sqrt{2}\left[\cos\left(\frac{\pi}{2}\right) + i\sin\left(\frac{\pi}{2}\right)\right] = \sqrt{2}i$$

then $z = -\sqrt{2}i$ and hence $\mathrm{Re}(z) = 0$ and $\mathrm{Im}(z) = -\sqrt{2}$.

17. Case 1. Suppose that $n = 0$. Then

$$(\cos\theta + i\sin\theta)^n = 1 = \cos(0) + i\sin(0)$$

So Formula (7) is valid if $n = 0$.

Case 2. In order to verify that Formula (7) holds if n is a negative integer, we first let $n = -1$. Then

$$\begin{aligned}(\cos\theta + i\sin\theta)^{-1} &= \frac{1}{\cos\theta + i\sin\theta}\\ &= \cos\theta - i\sin\theta\\ &= \cos(-\theta) + i\sin(-\theta)\end{aligned}$$

Thus, Formula (7) is valid if $n = -1$.

Now suppose that n is a positive integer (and hence that $-n$ is a negative integer). Then

$$\begin{aligned}(\cos\theta + i\sin\theta)^{-n} &= \left[\left(\cos\theta + i\sin\theta\right)^{-1}\right]^n\\ &= [\cos(-\theta) + i\sin(-\theta)]^n\\ &= \cos(-n\theta) + i\sin(-n\theta) \qquad \text{[By Formula (7)]}\end{aligned}$$

This completes the proof.

19. We have $z_1 = r_1 e^{i\theta_1}$ and $z_2 = r_2 e^{i\theta_2}$. But (see Exercise 17)

$$\frac{1}{z_2} = \frac{1}{r_2 e^{i\theta_2}} = \frac{1}{r_2} e^{-i\theta_2}$$

If we replace z_2 by $1/z_2$ in Formula (3), we obtain

$$\begin{aligned}\frac{z_1}{z_2} &= z_1\left(\frac{1}{z_2}\right)\\ &= \frac{r_1}{r_2}\left[\cos\left(\theta_1 + (-\theta_2)\right) + i\sin\left(\theta_1 + (-\theta_2)\right)\right]\\ &= \frac{r_1}{r_2}\left[\cos\left(\theta_1 - \theta_2\right) + i\sin\left(\theta_1 - \theta_2\right)\right]\end{aligned}$$

which is Formula (5).

21. If

$$e^{i\theta} = \cos\theta + i\sin\theta$$

then replacing θ with $-\theta$ yields

$$e^{-i\theta} = \cos(-\theta) + i\sin(-\theta) = \cos\theta - i\sin\theta$$

If we then compute $e^{i\theta} + e^{-i\theta}$ and $e^{i\theta} - e^{-i\theta}$, the results will follow.

23. Let $z = r(\cos\theta + i\sin\theta)$. Formula (5) guarantees that $1/z = r^{-1}(\cos(-\theta) + i\sin(-\theta))$ since $z \neq 0$. Applying Formula (6) for n a positive integer to the above equation yields

$$z^{-n} = \left(\frac{1}{z}\right)^n = r^{-n}\left(\cos(-n\theta) + i\sin(-n\theta)\right)$$

which is just Formula (6) for $-n$ a negative integer.

EXERCISE SET 10.4

1. **(a)** $\mathbf{u} - \mathbf{v} = (2i - (-i), 0 - i, -1 - (1 + i), 3 - (-1))$

$$= (3i, -i, -2 - i, 4)$$

(c) $-\mathbf{w} + \mathbf{v} = (-(1 + i) - i, i + i, -(-1 + 2i) + (1 + i), 0 + (-1))$

$$= (-1 - 2i, 2i, 2 - i, -1)$$

(e) $-i\mathbf{v} = (-1, 1, 1 - i, i)$ and $2i\mathbf{w} = (-2 + 2i, 2, -4 - 2i, 0)$. Thus

$$-i\mathbf{v} + 2i\mathbf{w} = (-3 + 2i, 3, -3 - 3i, i)$$

3. Consider the equation $c_1\mathbf{u}_1 + c_2\mathbf{u}_2 + c_3\mathbf{u}_3 = (-3 + i, 3 + 2i, 3 - 4i)$. The augmented matrix for this system of equations is

$$\begin{bmatrix} 1-i & 2i & 0 & -3+i \\ i & 1+i & 2i & 3+2i \\ 0 & 1 & 2-i & 3-4i \end{bmatrix}$$

The row-echelon form for the above matrix is

$$\begin{bmatrix} 1 & -1+i & 0 & -2-i \\ 0 & 1 & \frac{1}{2}+\frac{1}{2}i & \frac{3}{2}+\frac{1}{2}i \\ 0 & 0 & 1 & 1-i \end{bmatrix}$$

Hence, $c_3 = 2 - i$, $c_2 = \frac{3}{2} + \frac{1}{2}i - \left(\frac{1}{2} + \frac{1}{2}i\right)c_3 = 0$, and $c_1 = -2 - i$.

5. **(a)** $\|\mathbf{v}\| = \sqrt{|1|^2 + |i|^2} = \sqrt{2}$

(c) $\|\mathbf{v}\| = \sqrt{|2i|^2 + |0|^2 + |2i+1|^2 + |(-1)|^2} = \sqrt{4+0+5+1} = \sqrt{10}$

9. **(a)** $\mathbf{u} \cdot \mathbf{v} = (-i)(-3i) + (3i)(-2i) = -3 + 6 = 3.$

(c) $\mathbf{u} \cdot \mathbf{v} = (1 - i)(4 - 6i) + (1 + i)(5i) + (2i)(-1 - i) + (3)(-i)$

$= (-2 - 10i) + (-5 + 5i) + (2 - 2i) + (-3i)$

$= -5 - 10i$

11. Let V denote the set and let

$$\mathbf{u} = \begin{bmatrix} u & 0 \\ 0 & \bar{u} \end{bmatrix} \quad \text{and} \quad \mathbf{v} = \begin{bmatrix} v & 0 \\ 0 & \bar{v} \end{bmatrix}$$

We check the axioms listed in the definition of a vector space (see Section 5.1).

(1) $\mathbf{u}+\mathbf{v} = \begin{bmatrix} u+v & 0 \\ 0 & \bar{u}+\bar{v} \end{bmatrix} = \begin{bmatrix} u+v & 0 \\ 0 & \overline{u+v} \end{bmatrix}$

So $\mathbf{u} + \mathbf{v}$ belongs to V.

(2) Since $u + v = v + u$ and $\bar{u}+\bar{v} = \bar{v}+\bar{u}$, it follows that $\mathbf{u} + \mathbf{v} = \mathbf{v} + \mathbf{u}$.

(3) Axiom (3) follows by a routine, if tedious, check.

(4) The matrix $\begin{bmatrix} 0 & 0 \\ 0 & 0 \end{bmatrix}$ serves as the zero vector.

(5) Let $-\mathbf{u} = \begin{bmatrix} -u & 0 \\ 0 & -\bar{u} \end{bmatrix} = \begin{bmatrix} -u & 0 \\ 0 & -\bar{u} \end{bmatrix}$. Then $\mathbf{u} + (-\mathbf{u}) = \mathbf{0}$.

(6) Since $k\mathbf{u} = \begin{bmatrix} ku & 0 \\ 0 & k\bar{u} \end{bmatrix}$, $k\mathbf{u}$ will be in V if and only if $k\bar{u} = \overline{ku}$, which is true if and only if k is real or $u = 0$. Thus Axiom (6) fails.

(7)–(9) These axioms all hold by virtue of the properties of matrix addition and scalar multiplication. However, as seen above, the closure property of scalar multiplication may fail, so the vectors need not be in V.

(10) Clearly $1\mathbf{u} = \mathbf{u}$.

Thus, this set is not a vector space because Axiom (6) fails.

13. Suppose that $T(\mathbf{x}) = A\mathbf{x} = \mathbf{0}$. It is easy to show that the reduced row echelon form for A is

$$\begin{bmatrix} 1 & 0 & (1+3i)/2 \\ 0 & 1 & (1+i)/2 \\ 0 & 0 & 0 \end{bmatrix}$$

Hence, $x_1 = (-(1 + 3i)/2)x_3$ and $x_2 = (-(1 + i)/2)x_3$ where x_3 is an arbitrary complex number. That is,

$$\mathbf{x} = \begin{bmatrix} -(1+3i)/2 \\ -(1+i)/2 \\ 1 \end{bmatrix}$$

spans the kernel of T and hence T has nullity one.

Alternatively, the equation $A\mathbf{x} = 0$ yields the system

$$\begin{aligned} ix_1 - ix_2 - x_3 &= 0 \\ x_1 - ix_2 + (1 + i)x_3 &= 0 \\ 0 + (1 - i)x_2 + x_3 &= 0 \end{aligned}$$

The third equation implies that $x_3 = -(1 - i)x_2$. If we substitute this expression for x_3 into the first equation, we obtain $x_1 = (2 + i)x_2$. The second equation will then be valid for all such x_1 and x_3. That is, x_2 is arbitrary. Thus the kernel of T is also spanned by the vector

$$\begin{bmatrix} x_1 \\ x_2 \\ x_3 \end{bmatrix} = \begin{bmatrix} 2+i \\ 1 \\ -1+i \end{bmatrix}$$

If we multiply this vector by $-(1 + i)/2$, then this answer agrees with the previous one.

15. **(a)** Since

$$(f+g)(1) = f(1) + g(1) = 0 + 0 = 0$$

and

$$kf(1) = k(0) = 0$$

for all functions f and g in the set and for all scalars k, this set forms a subspace.

(c) Since

$$\begin{aligned}(f+g)(-x) &= f(-x)+g(-x) = \overline{f(x)}+\overline{g(x)} \\ &= \overline{f(x)+g(x)} = \overline{(f+g)(x)}\end{aligned}$$

the set is closed under vector addition. It is closed under scalar multiplication by a real scalar, but not by a complex scalar. For instance, if $f(x) = xi$, then $f(x)$ is in the set but $if(x)$ is not.

17. **(a)** Consider the equation $k_1\mathbf{u} + k_2\mathbf{v} + k_3\mathbf{w} = (1, 1, 1)$. Equating components yields

$$\begin{aligned} k_1 + (1+i)k_2 \qquad &= 1 \\ k_2 + ik_3 &= 1 \\ -ik_1 + (1-2i)k_2 + 2k_3 &= 1 \end{aligned}$$

Solving the system yields $k_1 = -3 - 2i$, $k_2 = 3 - i$, and $k_3 = 1 + 2i$.

(c) Let A be the matrix whose first, second and third columns are the components of $\mathbf{u}$, $\mathbf{v}$, and $\mathbf{w}$, respectively. By Part (a), we know that $\det(A) \neq 0$. Hence, $k_1 = k_2 = k_3 = 0$.

19. **(a)** Recall that $e^{ix} = \cos x + i \sin x$ and that $e^{-ix} = \cos(-x) + i \sin(-x) = \cos x - i \sin x$. Therefore,

$$\cos x = \frac{e^{ix}+e^{-ix}}{2} = \frac{1}{2}\mathbf{f} + \frac{1}{2}\mathbf{g}$$

and so $\cos x$ lies in the space spanned by $\mathbf{f}$ and $\mathbf{g}$.

(b) If $a\mathbf{f} + b\mathbf{g} = \sin x$, then (see Part (a))

$$(a + b)\cos x + (a - b)i \sin x - \sin x$$

Thus, since the sine and cosine functions are linearly independent, we have

$$a + b = 0$$

and

$$a - b = -i$$

This yields $a = -i/2$, $b = i/2$, so again the vector lies in the space spanned by **f** and **g**.

(c) If $a\mathbf{f} + b\mathbf{g} = \cos x + 3i \sin x$, then (see Part (a))

$$a + b = 1$$

and

$$a - b = 3$$

Hence $a = 2$ and $b = -1$ and thus the given vector does lie in the space spanned by **f** and **g**.

21. Let A denote the matrix whose first, second, and third columns are the components of $\mathbf{u}_1$, $\mathbf{u}_2$, and $\mathbf{u}_3$, respectively.

(a) Since the last row of A consists entirely of zeros, it follows that $\det(A) = 0$ and hence $\mathbf{u}_1$, $\mathbf{u}_2$, and $\mathbf{u}_3$, are linearly dependent.

(c) Since $\det(A) = i \neq 0$, then $\mathbf{u}_1$, $\mathbf{u}_2$, and $\mathbf{u}_3$ are linearly independent.

23. Observe that $\mathbf{f} - 3\mathbf{g} - 3\mathbf{h} = \mathbf{0}$.

25. **(a)** Since $\begin{vmatrix} 2i & 4i \\ -i & 0 \end{vmatrix} = -4 \neq 0$, the vectors are linearly independent and hence form a basis for C^2.

(d) Since $\begin{vmatrix} 2-3i & 3+2i \\ i & -1 \end{vmatrix} = 0$, the vectors are linearly dependent and hence are not a basis for C^2.

27. The row-echelon form of the matrix of the system is

$$\begin{bmatrix} 1 & 1+i \\ 0 & 0 \end{bmatrix}$$

So x_2 is arbitrary and $x_1 = -(1 + i)x_2$. Hence, the dimension of the solution space is one and $\begin{bmatrix} -(1+i) \\ 1 \end{bmatrix}$ is a basis for that space.

29. The reduced row-echelon form of the matrix of the system is

$$\begin{bmatrix} 1 & 0 & -3-6i \\ 0 & 1 & 3i \\ 0 & 0 & 0 \end{bmatrix}$$

So x_3 is arbitrary, $x_2 = (-3i)x_3$, and $x_1 = (3 + 6i)x_3$. Hence, the dimension of the solution space is one and $\begin{bmatrix} 3+6i \\ -3i \\ 1 \end{bmatrix}$ is a basis for that space.

31. Let $\mathbf{u} = (u_1, u_2, \ldots, u_n)$ and $\mathbf{v} = (v_1, v_2, \ldots, v_n)$. From the definition of the Euclidean inner product in C^n, we have

$$\begin{aligned} \mathbf{u} \cdot (k\mathbf{v}) &= u_1(\overline{kv_1}) + u_2(\overline{kv_2}) + \ldots + u_n(\overline{kv_n}) \\ &= u_1(\bar{k}\bar{v}_1) + u_2(\bar{k}\bar{v}_2) + \ldots + u_n(\bar{k}\bar{v}_n) \\ &= \bar{k}(u_1\bar{v}_1) + \bar{k}(u_2\bar{v}_2) + \ldots + \bar{k}(u_n\bar{v}_n) \\ &= \bar{k}[u_1\bar{v}_1 + u_2\bar{v}_2 + \ldots + u_n\bar{v}_n] \\ &= \bar{k}(\mathbf{u} \cdot \mathbf{v}) \end{aligned}$$

33. Hint: Show that

$$\|\mathbf{u} + k\mathbf{v}\|^2 = \|\mathbf{u}\|^2 + \bar{k}(\mathbf{v} \cdot \mathbf{u}) + k(\mathbf{u} \cdot \mathbf{v}) + k\bar{k}\|\mathbf{v}\|^2$$

and apply this result to each term on the right-hand side of the identity.

EXERCISE SET 10.5

1. Let $\mathbf{u} = (u_1, u_2)$, $\mathbf{v} = (v_1, v_2)$, and $\mathbf{w} = (w_1, w_2)$. We proceed to check the four axioms

(1)
$$\overline{\langle \mathbf{v}, \mathbf{u} \rangle} = \overline{3v_1\bar{u}_1 + 2v_2\bar{u}_2}$$
$$= 3u_1\bar{v}_1 + 2u_2\bar{v}_2 = \langle \mathbf{u}, \mathbf{v} \rangle$$

(2)
$$\langle \mathbf{u} + \mathbf{v}, \mathbf{w} \rangle = 3(u_1 + v_1)\bar{w}_1 + 2(u_2 + v_2)\bar{w}_2$$
$$= [3u_1\bar{w}_1 + 2u_2\bar{w}_2] + [3v_1\bar{w}_1 + 2v_2\bar{w}_2]$$
$$= \langle \mathbf{u}, \mathbf{w} \rangle + \langle \mathbf{v}, \mathbf{w} \rangle$$

(3)
$$\langle k\mathbf{u}, \mathbf{v} \rangle = 3(ku_1)\bar{v}_1 + 2(ku_2)\bar{v}_2$$
$$= k[3u_1\bar{v}_1 + 2u_2\bar{v}_2] = k\langle \mathbf{u}, \mathbf{v} \rangle$$

(4)
$$\langle \mathbf{u}, \mathbf{u} \rangle = 3u_1\bar{u}_1 + 2u_2\bar{u}_2$$
$$= 3|u_1|^2 + 2|u_2|^2 \qquad \text{(Theorem 10.2.1)}$$
$$\geq 0$$

Indeed, $\langle \mathbf{u}, \mathbf{u} \rangle = 0 \Leftrightarrow u_1 = u_2 = 0 \Leftrightarrow \mathbf{u} = \mathbf{0}$.

Hence, this is an inner product on C^2.

3. Let $\mathbf{u} = (u_1, u_2)$ and $\mathbf{v} = (v_1, v_2)$. We check Axioms 1 and 4, leaving 2 and 3 to you.

(1)
$$\overline{\langle \mathbf{v}, \mathbf{u} \rangle} = \overline{v_1\bar{u}_1} + \overline{(1+i)v_1\bar{u}_2} + \overline{(1-i)v_2\bar{u}_1} + \overline{3v_2\bar{u}_2}$$
$$= u_1\bar{v}_1 + (1-i)u_2\bar{v}_1 + (1+i)u_1\bar{v}_2 + 3u_2\bar{v}_2$$
$$= \langle \mathbf{u}, \mathbf{v} \rangle$$

(4) Recall that $|\text{Re}(z)| \le |z|$ by Problem 37 of Section 10.2. Now

$$\begin{aligned}\langle \mathbf{u},\mathbf{u}\rangle &= u_1\bar{u}_1 + (1+i)u_1\bar{u}_2 + (1-i)u_2\bar{u}_1 + 3u_2\bar{u}_2 \\ &= |u_1|^2 + (1+i)u_1\bar{u}_2 + \overline{(1+i)u_1\bar{u}_2} + 3|u_2|^2 \\ &= |u_1|^2 + 2\text{Re}((1+i)u_1\bar{u}_2) + 3|u_2|^2 \\ &\ge |u_1|^2 - 2|(1+i)u_1\bar{u}_2| + 3|u_2|^2 \\ &= |u_1|^2 - 2\sqrt{2}|u_1|\,|u_2| + 3|u_2|^2 \\ &= \left(|u_1| - \sqrt{2}|u_2|\right)^2 + |u_2|^2\end{aligned}$$

Moreover, $\langle \mathbf{u}, \mathbf{u}\rangle = 0$ if and only if both $|u_2|$ and $|u_1| - \sqrt{2}\ |u_2| = 0$, or $\mathbf{u} = 0$.

5. (a) This is *not* an inner product on C^2. Axioms 1–3 are easily checked. Moreover,

$$\langle \mathbf{u}, \mathbf{u}\rangle = u_1\bar{u}_1 = |u_1| \ge 0$$

However, $\langle \mathbf{u}, \mathbf{u}\rangle = 0 \Leftrightarrow u_1 = 0 \not\Leftrightarrow \mathbf{u} = \mathbf{0}$. For example, $\langle i, i\rangle = 0$ although $i \ne 0$. Hence, Axiom 4 fails.

(c) This is *not* an inner product on C^2. Axioms 1 and 4 are easily checked. However, for $\mathbf{w} = (w_1, w_2)$, we have

$$\begin{aligned}\langle \mathbf{u} + \mathbf{v}, \mathbf{w}\rangle &= |u_1 + v_1|^2|w_1|^2 + |u_2 + v_2|^2|w_2|^2 \\ &\ne \left(|u_1|^2 + |v_1|^2\right)|w_1|^2 + \left(|u_2|^2 + |v_2|^2\right)|w_2|^2 \\ &= \langle \mathbf{u}, \mathbf{w}\rangle + \langle \mathbf{v}, \mathbf{w}\rangle\end{aligned}$$

For instance, $\langle 1 + 1, 1\rangle = 4$, but $\langle 1, 1\rangle + \langle 1, 1\rangle = 2$. Moreover, $\langle k\mathbf{u}, \mathbf{v}\rangle = |k|^2\langle \mathbf{u}, \mathbf{v}\rangle$, so that $\langle k\mathbf{u}, \mathbf{v}\rangle \ne k\langle \mathbf{u}, \mathbf{v}\rangle$ for most values of k, $\mathbf{u}$, and $\mathbf{v}$. Thus both Axioms 2 and 3 fail.

(e) Axiom 1 holds since

$$\begin{aligned}\overline{\langle \mathbf{v},\mathbf{u}\rangle} &= \overline{2v_1\bar{u}_1 + iv_1\bar{u}_2 - iv_2\bar{u}_1 + 2v_2\bar{u}_2} \\ &= 2u_1\bar{v}_1 - iu_2\bar{v}_1 + iu_1\bar{v}_2 + 2u_2\bar{v}_2 \\ &= \langle \mathbf{u},\mathbf{v}\rangle\end{aligned}$$

A similar argument serves to verify Axiom 2 and Axiom 3 holds by inspection. Finally, using the result of Problem 37 of Section 10.2, we have

$$\begin{aligned}\langle \mathbf{u}, \mathbf{u}\rangle &= 2u_1\bar{u}_1 + iu_1\bar{u}_2 - iu_2\bar{u}_1 + 2u_2\bar{u}_2\\ &= 2|u_1|^2 + 2\text{Re}(iu_1\bar{u}_2) + 2|u_2|^2\\ &\geq 2|u_1|^2 - 2|iu_1\bar{u}_2| + 2|u_2|^2\\ &= (|u_1| - |u_2|)^2 + |u_1|^2 + |u_2|^2\\ &\geq 0\end{aligned}$$

Moreover, $\langle \mathbf{u}, \mathbf{u}\rangle = 0 \Leftrightarrow u_1 = u_2 = 0$, or $\mathbf{u} = \mathbf{0}$. Thus all four axioms hold.

9. **(a)** $\|\mathbf{w}\| = [3(-i)(\text{i}) + 2(3i)(-3i)]^{1/2} = \sqrt{21}$

(c) $\|\mathbf{w}\| = [3(0)(0) + 2(2 - i)(2 + i)]^{1/2} = \sqrt{10}$

11. **(a)** $\|\mathbf{w}\| = [(1)(1) + (1 + i)(1)(\text{i}) + (1 - i)(-i)(1) + 3(-i)(\text{i})]^{1/2} = \sqrt{2}$

(c) $\|\mathbf{w}\| = [(3 - 4i)(3 + 4i)]^{1/2} = 5$

13. **(a)** Since $\mathbf{u} - \mathbf{v} = (1 - i, 1 + i)$, then

$$d(\mathbf{u}, \mathbf{v}) = [3(1 - i)(1 + i) + 2(1 + i)(1 - i)]^{1/2} = \sqrt{10}$$

15. **(a)** Since $\mathbf{u} - \mathbf{v} = (1 - i, 1 + i)$,

$$\begin{aligned}d(\mathbf{u}, \mathbf{v}) &= [(1 - i)(1 + i) + (1 + i)(1 - i)(1 - i)\\ &\quad + (1 - i)(1 + i)(1 + i) + 3(1 + i)(1 - i)]^{1/2}\\ &= 2\sqrt{3}\end{aligned}$$

17. **(a)** Since $\mathbf{u} \cdot \mathbf{v} = (2i)(-i) + (i)(-6i) + (3i)(\bar{k})$, then $\mathbf{u} \cdot \mathbf{v} = 0 \Leftrightarrow 8 + 3i\bar{k} = 0$ or $k = -8i/3$.

19. Since $x = \frac{1}{\sqrt{3}} e^{i\theta}(i,1,1)$, we have

$$\|\mathbf{x}\| = \frac{1}{\sqrt{3}}\left|e^{i\theta}\right|\|(i,1,1)\| = \frac{1}{\sqrt{3}}(1)(1+1+1)^{1/2} = 1$$

Also

$$\begin{aligned}\langle \mathbf{x}, (1,i,0)\rangle &= \frac{1}{\sqrt{3}} e^{i\theta}\left[(i,1,1)\cdot(1,i,0)\right] \\ &= \frac{1}{\sqrt{3}} e^{i\theta}(i-i+0) \\ &= 0\end{aligned}$$

and

$$\begin{aligned}\langle \mathbf{x}, (0,i,-i)\rangle &= \frac{1}{\sqrt{3}} e^{i\theta}\left[(i,1,1)\cdot(0,i,-i)\right] \\ &= \frac{1}{\sqrt{3}} e^{i\theta}(0-i+i) \\ &= 0\end{aligned}$$

21. **(a)** Call the vectors $\mathbf{u}_1$, $\mathbf{u}_2$, and $\mathbf{u}_3$, respectively. Then $\|\mathbf{u}_1\| = \|\mathbf{u}_2\| = \|\mathbf{u}_3\| = 1$ and $\mathbf{u}_1 \cdot \mathbf{u}_2 = \mathbf{u}_1 \cdot \mathbf{u}_3 = 0$. However, $\mathbf{u}_2 \cdot \mathbf{u}_3 = \frac{i^2}{\sqrt{6}} + \frac{(-i)^2}{\sqrt{6}} = -\frac{2}{\sqrt{6}} \neq 0$. Hence the set is not orthonormal.

25. **(a)** We have

$$\mathbf{v_1} = \left(\frac{i}{\sqrt{3}}, \frac{i}{\sqrt{3}}, \frac{i}{\sqrt{3}}\right)$$

and since $\mathbf{u}_2 \cdot \mathbf{v}_1 = 0$, then $\mathbf{u}_2 - (\mathbf{u}_2 \cdot \mathbf{v}_1)\mathbf{v}_1 = \mathbf{u}_2$. Thus,

$$\mathbf{v_2} = \left(-\frac{i}{\sqrt{2}}, \frac{i}{\sqrt{2}}, 0\right)$$

Also, $\mathbf{u}_3 \cdot \mathbf{v}_1 = 4/\sqrt{3}$ and $\mathbf{u}_3 \cdot \mathbf{v}_2 = 1/\sqrt{2}$ and hence

$$\mathbf{u}_3 - (\mathbf{u}_3 \cdot \mathbf{v}_1)\mathbf{v}_1 - (\mathbf{u}_3 \cdot \mathbf{v}_2)\mathbf{v}_2 = \left(\frac{i}{6}, \frac{i}{6}, \frac{-i}{3}\right)$$

Since the norm of the above vector is $1/\sqrt{6}$, we have

$$\mathbf{v}_3 = \left(\frac{i}{\sqrt{6}}, \frac{i}{\sqrt{6}}, \frac{-2i}{\sqrt{6}}\right)$$

27. Let $\mathbf{u}_1 = (0, i, 1-i)$ and $\mathbf{u}_2 = (-i, 0, 1+i)$. We shall apply the Gram-Schmidt process to $\{\mathbf{u}_1, \mathbf{u}_2\}$. Since $\|\mathbf{u}_1\| = \sqrt{3}$, it follows that

$$\mathbf{v}_1 = \left(0, \frac{i}{\sqrt{3}}, \frac{1-i}{\sqrt{3}}\right)$$

Since $\mathbf{u}_2 \cdot \mathbf{v}_1 = 2i/\sqrt{3}$, then

$$\mathbf{u}_2 - (\mathbf{u}_2 \cdot \mathbf{v}_1)\mathbf{v}_1 = (-i, 0, 1+i) - \left(0, -\frac{2}{3}, \frac{2}{3} + \frac{2}{3}i\right)$$

$$= \left(-i, \frac{2}{3}, \frac{1}{3} + \frac{1}{3}i\right)$$

and because the norm of the above vector is $\sqrt{15}/3$, we have

$$\mathbf{v}_2 = \left(\frac{-3i}{\sqrt{15}}, \frac{2}{\sqrt{15}}, \frac{1+i}{\sqrt{15}}\right)$$

$$\mathbf{v}_2 = \left(\frac{i}{2\sqrt{3}}, \frac{3i}{2\sqrt{3}}, -\frac{i}{2\sqrt{3}}, \frac{i}{2\sqrt{3}}\right)$$

Therefore,

$$\mathbf{w}_1 = (\mathbf{w} \cdot \mathbf{v}_1)\mathbf{v}_1 + (\mathbf{w} \cdot \mathbf{v}_2)\mathbf{v}_2$$

$$= \frac{7}{\sqrt{6}}\mathbf{v}_1 + \frac{-1}{2\sqrt{3}}\mathbf{v}_2$$

$$= \left(-\frac{5}{4}i, -\frac{1}{4}i, \frac{5}{4}i, \frac{9}{4}i\right)$$

and

$$\mathbf{w}_2 = \mathbf{w} - \mathbf{w}_1$$

$$= \left(\frac{1}{4}i, \frac{9}{4}i, \frac{19}{4}i, -\frac{9}{4}i\right)$$

29. **(a)** By Axioms (2) and (3) for inner products,

$$\langle \mathbf{u} - k\mathbf{v}, \mathbf{u} - k\mathbf{v} \rangle = \langle \mathbf{u}, \mathbf{u} - k\mathbf{v} \rangle + \langle -k\mathbf{v}, \mathbf{u} - k\mathbf{v} \rangle$$

$$= \langle \mathbf{u}, \mathbf{u} - k\mathbf{v} \rangle - k\langle \mathbf{v}, \mathbf{u} - k\mathbf{v} \rangle$$

If we use Properties (ii) and (iii) of inner products, then we obtain

$$\langle \mathbf{u} - k\mathbf{v}, \mathbf{u} - k\mathbf{v} \rangle = \langle \mathbf{u}, \mathbf{u} \rangle - \bar{k}\langle \mathbf{u}, \mathbf{v} \rangle - k\langle \mathbf{v}, \mathbf{u} \rangle + k\bar{k}\langle \mathbf{v}, \mathbf{v} \rangle$$

Finally, Axiom (1) yields

$$\langle \mathbf{u} - k\mathbf{v}, \mathbf{u} - k\mathbf{v} \rangle = \langle \mathbf{u}, \mathbf{u} \rangle - \bar{k}\langle \mathbf{u}, \mathbf{v} \rangle - k\overline{\langle \mathbf{u}, \mathbf{v} \rangle} + k\bar{k}\langle \mathbf{v}, \mathbf{v} \rangle$$

and the result is proved.

(b) This follows from Part (a) and Axiom (4) for inner products.

33. Hint: Let $\mathbf{v}$ be any nonzero vector, and consider the quantity $\langle \mathbf{v}, \mathbf{v} \rangle + \langle \mathbf{0}, \mathbf{v} \rangle$.

35. **(d)** Observe that $\|\mathbf{u} + \mathbf{v}\|^2 = \langle \mathbf{u} + \mathbf{v}, \mathbf{u} + \mathbf{v} \rangle$. As in Exercise 37,

$$\langle \mathbf{u} + \mathbf{v}, \mathbf{u} + \mathbf{v} \rangle = \langle \mathbf{u}, \mathbf{u} \rangle + 2\,\mathrm{Re}(\langle \mathbf{u}, \mathbf{v} \rangle) + \langle \mathbf{v}, \mathbf{v} \rangle$$

Since (see Exercise 37 of Section 10.2)

$$|\mathrm{Re}(\langle \mathbf{u}, \mathbf{v} \rangle)| \le |\langle \mathbf{u}, \mathbf{v} \rangle|$$

this yields

$$\langle \mathbf{u} + \mathbf{v}, \mathbf{u} + \mathbf{v} \rangle \le \langle \mathbf{u}, \mathbf{u} \rangle + 2|\langle \mathbf{u}, \mathbf{v} \rangle| + \langle \mathbf{v}, \mathbf{v} \rangle$$

By the Cauchy-Schwarz inequality and the definition of norm, this becomes

$$\|\mathbf{u} + \mathbf{v}\|^2 \le \|\mathbf{u}\|^2 + 2\|\mathbf{u}\|\,\|\mathbf{v}\| + \|\mathbf{v}\|^2 = (\|\mathbf{u}\| + \|\mathbf{v}\|)^2$$

which yields the desired result.

(h) Replace $\mathbf{u}$ by $\mathbf{u} - \mathbf{w}$ and $\mathbf{v}$ by $\mathbf{w} - \mathbf{v}$ in Theorem 6.2.2, Part (d).

37. Observe that for any complex number k,

$$\|\mathbf{u} + k\mathbf{v}\|^2 = \langle \mathbf{u} + k\mathbf{v}, \mathbf{u} + k\mathbf{v} \rangle$$

$$= \langle \mathbf{u}, \mathbf{u} \rangle + k\langle \mathbf{v}, \mathbf{u} \rangle + \bar{k}\langle \mathbf{u}, \mathbf{v} \rangle + k\bar{k}\langle \mathbf{v}, \mathbf{v} \rangle$$

$$= \langle \mathbf{u}, \mathbf{u} \rangle + 2\,\mathrm{Re}(k\langle \mathbf{v}, \mathbf{u} \rangle) + |k|^2\langle \mathbf{v}, \mathbf{v} \rangle$$

Therefore,

$$\begin{aligned}
&\|\mathbf{u}+\mathbf{v}\|^2 - \|\mathbf{u}-\mathbf{v}\|^2 + i\|\mathbf{u}+i\mathbf{v}\|^2 - i\|\mathbf{u}-i\mathbf{v}\|^2 \\
&\quad = (1-1+i-i)\langle \mathbf{u}, \mathbf{u}\rangle + 2\,\mathrm{Re}(\langle \mathbf{v}, \mathbf{u}\rangle) - 2\,\mathrm{Re}(-\langle \mathbf{v}, \mathbf{u}\rangle) + 2i\,\mathrm{Re}(i\langle \mathbf{v}, \mathbf{u}\rangle) \\
&\quad\quad -2i\,\mathrm{Re}(-i\langle \mathbf{v}, \mathbf{u}\rangle) + (1-1+i-i)\langle \mathbf{v}, \mathbf{v}\rangle \\
&\quad = 4\,\mathrm{Re}(\langle \mathbf{v}, \mathbf{u}\rangle) - 4i\,\mathrm{Im}(\langle \mathbf{v}, \mathbf{u}\rangle) \\
&\quad = 4\overline{\langle \mathbf{v}, \mathbf{u}\rangle} \\
&\quad = 4\,\langle \mathbf{u}, \mathbf{v}\rangle
\end{aligned}$$

39. We check Axioms 2 and 4. For Axiom 2, we have

$$\begin{aligned}
\langle \mathbf{f}+\mathbf{g}, \mathbf{h}\rangle &= \int_a^b (\mathbf{f}+\mathbf{g})\overline{\mathbf{h}}\,dx \\
&= \int_a^b \mathbf{f}\,\overline{\mathbf{h}}\,dx + \int_a^b \mathbf{g}\,\overline{\mathbf{h}}\,dx \\
&= \langle \mathbf{f}, \mathbf{h}\rangle + \langle \mathbf{g}, \mathbf{h}\rangle
\end{aligned}$$

For Axiom 4, we have

$$\begin{aligned}
\langle \mathbf{f}, \mathbf{f}\rangle &= \int_a^b \mathbf{f}\,\overline{\mathbf{f}}\,dx = \int_a^b |\mathbf{f}|^2\,dx \\
&= \int_a^b \left[(f_1(x))^2 + (f_2(x))^2 \right] dx
\end{aligned}$$

Since $|\mathbf{f}|^2 \geq 0$ and $a < b$, then $\langle \mathbf{f}, \mathbf{f}\rangle \geq 0$. Also, since $\mathbf{f}$ is continuous, $\int_a^b |\mathbf{f}|^2\,dx > 0$ unless $\mathbf{f} = 0$ on $[a, b]$. [That is, the integral of a nonnegative, real-valued, continuous function (which represents the area under that curve and above the x-axis from a to b) is positive unless the function is identically zero.]

41. Let $\mathbf{v}_m = e^{2\pi imx} = \cos(2\pi mx) + i\sin(2\pi mx)$. Then if $m \neq n$, we have

$$\begin{aligned}
\langle \mathbf{v}_m, \mathbf{v}_n \rangle &= \int_0^1 \left[\cos(2\pi mx) + i\sin(2\pi mx)\right]\left[\cos(2\pi nx) - i\sin(2\pi nx)\right]dx \\
&= \int_0^1 \left[\cos(2\pi mx)\cos(2\pi nx) + \sin(2\pi mx)\sin(2\pi nx)\right]dx \\
&\qquad + i\int_0^1 \left[\sin(2\pi mx)\cos(2\pi nx) - \cos(2\pi mx)\sin(2\pi nx)\right]dx \\
&= \int_0^1 \cos\left[2\pi(m-n)x\right]dx + i\int_0^1 \sin\left[2\pi(m-n)x\right]dx \\
&= \frac{1}{2\pi(m-n)}\sin[2\pi(m-n)x]\Big]_0^1 - \frac{i}{2\pi(m-n)}\cos[2\pi(m-n)x]\Big]_0^1 \\
&= -\frac{i}{2\pi(m-n)} + \frac{i}{2\pi(m-n)} \\
&= 0
\end{aligned}$$

Thus the vectors are orthogonal.

EXERCISE SET 10.6

5. **(b)** The row vectors of the matrix are

$$\mathbf{r}_1 = \left(\frac{1}{\sqrt{2}}, \frac{1}{\sqrt{2}} \right) \quad \text{and} \quad \mathbf{r}_2 = \left(-\frac{1+i}{2}, \frac{1+i}{2} \right)$$

Since $\|\mathbf{r}_1\| = \|\mathbf{r}_2\| = 1$ and

$$\mathbf{r}_1 \cdot \mathbf{r}_2 = \frac{1}{\sqrt{2}}\left(\frac{-1+i}{2} \right) + \frac{1}{\sqrt{2}}\left(\frac{1-i}{2} \right) = 0$$

the matrix is unitary by Theorem 10.6.2. Hence,

$$A^{-1} = A^* = \bar{A}^T = \begin{bmatrix} \frac{1}{\sqrt{2}} & \frac{-1+i}{2} \\ \frac{1}{\sqrt{2}} & \frac{1-i}{2} \end{bmatrix}$$

(d) The row vectors of the matrix are

$$\mathbf{r}_1 = \left(\frac{1+i}{2}, -\frac{1}{2}, \frac{1}{2} \right)$$

$$\mathbf{r}_2 = \left(\frac{i}{\sqrt{3}}, \frac{1}{\sqrt{3}}, \frac{-i}{\sqrt{3}} \right)$$

and

$$\mathbf{r}_3 = \left(\frac{3+i}{2\sqrt{15}}, \frac{4+3i}{2\sqrt{15}}, \frac{5i}{2\sqrt{15}} \right)$$

We have $\|\mathbf{r}_1\| = \|\mathbf{r}_2\| = \|\mathbf{r}_3\| = 1$,

$$\mathbf{r}_1 \bullet \mathbf{r}_2 = \left(\frac{1+i}{2}\right)\left(\frac{-i}{\sqrt{3}}\right) - \frac{1}{2} \bullet \frac{1}{\sqrt{3}} + \frac{1}{2} \bullet \frac{i}{\sqrt{3}} = 0$$

$$\mathbf{r}_2 \bullet \mathbf{r}_3 = \left(\frac{1}{\sqrt{3}}\right)\left(\frac{3-i}{2\sqrt{15}}\right) + \frac{1}{\sqrt{3}} \bullet \frac{4-3i}{2\sqrt{15}} - \frac{i}{\sqrt{3}} \bullet \frac{-5i}{2\sqrt{15}} = 0$$

and

$$\mathbf{r}_1 \bullet \mathbf{r}_3 = \left(\frac{1+i}{2}\right)\left(\frac{3-i}{2\sqrt{15}}\right) - \frac{1}{\sqrt{2}} \bullet \frac{4-3i}{2\sqrt{15}} + \frac{1}{2} \bullet \frac{-5i}{2\sqrt{15}} = 0$$

Hence, by Theorem 10.6.2, the matrix is unitary and thus

$$A^{-1} = A^* = \bar{A}^T = \begin{bmatrix} \frac{1-i}{2} & \frac{-i}{\sqrt{3}} & \frac{3-i}{2\sqrt{15}} \\ -\frac{1}{2} & \frac{1}{\sqrt{3}} & \frac{4-3i}{2\sqrt{15}} \\ \frac{1}{2} & \frac{i}{\sqrt{3}} & \frac{-5i}{2\sqrt{15}} \end{bmatrix}$$

7. The characteristic polynomial of A is

$$\det\begin{bmatrix} \lambda-4 & -1+i \\ -1-i & \lambda-5 \end{bmatrix} = (\lambda-4)(\lambda-5)-2 = (\lambda-6)(\lambda-3)$$

Therefore, the eigenvalues are $\lambda = 3$ and $\lambda = 6$. To find the eigenvectors of A corresponding to $\lambda = 3$, we let

$$\begin{bmatrix} -1 \\ -1-i \end{bmatrix}\begin{bmatrix} x_1 \\ x_2 \end{bmatrix} = \begin{bmatrix} 0 \\ 0 \end{bmatrix}$$

This yields $x_1 = -(1 - i)s$ and $x_2 = s$ where s is arbitrary. If we put $s = 1$, we see that $\begin{bmatrix} -1+i \\ 1 \end{bmatrix}$ is a basis vector for the eigenspace corresponding to $\lambda = 3$. We normalize this vector to obtain

$$\mathbf{P}_1 = \begin{bmatrix} \dfrac{-1+i}{\sqrt{3}} \\ \dfrac{1}{\sqrt{3}} \end{bmatrix}$$

To find the eigenvectors corresponding to $\lambda = 6$, we let

$$\begin{bmatrix} 2 & -1+i \\ -1-i & 1 \end{bmatrix} \begin{bmatrix} x_1 \\ x_2 \end{bmatrix} = \begin{bmatrix} 0 \\ 0 \end{bmatrix}$$

This yields $x_1 = \dfrac{1-i}{2}s$ and $x_2 = s$ where s is arbitrary. If we put $s = 1$, we have that $\begin{bmatrix} (1-i)/2 \\ 1 \end{bmatrix}$ is a basis vector for the eigenspace corresponding to $\lambda = 6$. We normalize this vector to obtain

$$\mathbf{P}_2 = \begin{bmatrix} \dfrac{1-i}{\sqrt{6}} \\ \dfrac{2}{\sqrt{6}} \end{bmatrix}$$

Thus

$$P = \begin{bmatrix} \dfrac{-1+i}{\sqrt{3}} & \dfrac{1-i}{\sqrt{6}} \\ \dfrac{1}{\sqrt{3}} & \dfrac{2}{\sqrt{6}} \end{bmatrix}$$

diagonalizes A and

$$P^{-1}AP = \begin{bmatrix} \frac{-1-i}{\sqrt{3}} & \frac{1}{\sqrt{3}} \\ \frac{1+i}{\sqrt{6}} & \frac{2}{\sqrt{6}} \end{bmatrix} \begin{bmatrix} 4 & 1-i \\ 1+i & 5 \end{bmatrix} \begin{bmatrix} \frac{-1+i}{\sqrt{3}} & \frac{1-i}{\sqrt{6}} \\ \frac{1}{\sqrt{3}} & \frac{2}{\sqrt{6}} \end{bmatrix}$$

$$= \begin{bmatrix} 3 & 0 \\ 0 & 6 \end{bmatrix}$$

9. The characteristic polynomial of A is

$$\det \begin{bmatrix} \lambda-6 & -2-2i \\ -2+2i & \lambda-4 \end{bmatrix} = (\lambda - 6)(\lambda - 4) - 8 = (\lambda - 8)(\lambda - 2)$$

Therefore the eigenvalues are $\lambda = 2$ and $\lambda = 8$. To find the eigenvectors of A corresponding to $\lambda = 2$, we let

$$\begin{bmatrix} -4 & -2-2i \\ -2+2i & -2 \end{bmatrix} \begin{bmatrix} x_1 \\ x_2 \end{bmatrix} = \begin{bmatrix} 0 \\ 0 \end{bmatrix}$$

This yields $x_1 = -\frac{1+i}{2}s$ and $x_2 = s$ where s is arbitrary. If we put $s = 1$, we have that $\begin{bmatrix} -(1+i)/2 \\ 1 \end{bmatrix}$ is a basis vector for the eigenspace corresponding to $\lambda = 2$. We normalize this vector to obtain

$$\mathbf{P}_1 = \begin{bmatrix} -\frac{1+i}{\sqrt{6}} \\ \frac{2}{\sqrt{6}} \end{bmatrix}$$

To find the eigenvectors corresponding to $\lambda = 8$, we let

$$\begin{bmatrix} 2 & -2-2i \\ -2+2i & 4 \end{bmatrix} \begin{bmatrix} x_1 \\ x_2 \end{bmatrix} = \begin{bmatrix} 0 \\ 0 \end{bmatrix}$$

This yields $x_1 = (1 + i)s$ and $x_2 = s$ where s is arbitrary. If we set $s = 1$, we have that $\begin{bmatrix} 1+i \\ 1 \end{bmatrix}$ is a basis vector for the eigenspace corresponding to $\lambda = 8$. We normalize this vector to obtain

$$\mathbf{P}_2 = \begin{bmatrix} \frac{1+i}{\sqrt{3}} \\ \frac{1}{\sqrt{3}} \end{bmatrix}$$

Thus

$$P = \begin{bmatrix} -\frac{1+i}{\sqrt{6}} & \frac{1+i}{\sqrt{6}} \\ \frac{2}{\sqrt{6}} & \frac{1}{\sqrt{3}} \end{bmatrix}$$

diagonalizes A and

$$P^{-1}AP = \begin{bmatrix} \frac{-1+i}{\sqrt{6}} & \frac{2}{\sqrt{6}} \\ \frac{1-i}{\sqrt{3}} & \frac{1}{\sqrt{3}} \end{bmatrix} \begin{bmatrix} 6 & 2+2i \\ 2-2i & 4 \end{bmatrix} \begin{bmatrix} -\frac{1+i}{\sqrt{6}} & \frac{1+i}{\sqrt{3}} \\ \frac{2}{\sqrt{6}} & \frac{1}{\sqrt{3}} \end{bmatrix}$$

$$= \begin{bmatrix} 2 & 0 \\ 0 & 8 \end{bmatrix}$$

11. The characteristic polynomial of A is

$$\det \begin{bmatrix} \lambda-5 & 0 & 0 \\ 0 & \lambda+1 & 1-i \\ 0 & 1+i & \lambda \end{bmatrix} = (\lambda-1)(\lambda-5)(\lambda+2)$$

Therefore, the eigenvalues are $\lambda = 1$, $\lambda = 5$, and $\lambda = -2$. To find the eigenvectors of A corresponding to $\lambda = 1$, we let

$$\begin{bmatrix} -4 & 0 & 0 \\ 0 & 2 & 1-i \\ 0 & 1+i & 1 \end{bmatrix} \begin{bmatrix} x_1 \\ x_2 \\ x_3 \end{bmatrix} = \begin{bmatrix} 0 \\ 0 \\ 0 \end{bmatrix}$$

This yields $x_1 = 0$, $x_2 = -\frac{1-i}{2}s$, and $x_3 = s$ where s is arbitrary. If we set $s = 1$, we have that $\begin{bmatrix} 0 \\ -(1-i)/2 \\ 1 \end{bmatrix}$ is a basis vector for the eigenspace corresponding to $\lambda = 1$. We normalize this vector to obtain

$$\mathbf{P}_1 = \begin{bmatrix} 0 \\ -\frac{1-i}{\sqrt{6}} \\ \frac{2}{\sqrt{6}} \end{bmatrix}$$

To find the eigenvectors corresponding to $\lambda = 5$, we let

$$\begin{bmatrix} 0 & 0 & 0 \\ 0 & 6 & 1-i \\ 0 & 1+i & 5 \end{bmatrix} \begin{bmatrix} x_1 \\ x_2 \\ x_3 \end{bmatrix} = \begin{bmatrix} 0 \\ 0 \\ 0 \end{bmatrix}$$

This yields $x_1 = s$ and $x_2 = x_3 = 0$ where s is arbitrary. If we let $s = 1$, we have that $\begin{bmatrix} 1 \\ 0 \\ 0 \end{bmatrix}$ is a basis vector for the eigenspace corresponding to $\lambda = 5$. Since this vector is already normal, we let

$$\mathbf{P}_2 = \begin{bmatrix} 1 \\ 0 \\ 0 \end{bmatrix}$$

To find the eigenvectors corresponding to $\lambda = -2$, we let

$$\begin{bmatrix} -7 & 0 & 0 \\ 0 & -1 & 1-i \\ 0 & 1+i & -2 \end{bmatrix} \begin{bmatrix} x_1 \\ x_2 \\ x_3 \end{bmatrix} = \begin{bmatrix} 0 \\ 0 \\ 0 \end{bmatrix}$$

This yields $x_1 = 0$, $x_2 = (1 - i)s$, and $x_3 = s$ where s is arbitrary. If we let $s = 1$, we have that $\begin{bmatrix} 0 \\ 1-i \\ 1 \end{bmatrix}$ is a basis vector for the eigenspace corresponding to $\lambda = -2$. We normalize this vector to obtain

$$\mathbf{P}_3 = \begin{bmatrix} 0 \\ \dfrac{1-i}{\sqrt{3}} \\ \dfrac{1}{\sqrt{3}} \end{bmatrix}$$

Thus

$$P = \begin{bmatrix} 0 & 1 & 0 \\ -\dfrac{1-i}{\sqrt{6}} & 0 & \dfrac{1-i}{\sqrt{3}} \\ \dfrac{2}{\sqrt{6}} & 0 & \dfrac{1}{\sqrt{3}} \end{bmatrix}$$

diagonalizes A and

$$P^{-1}AP = \begin{bmatrix} 0 & -\dfrac{1+i}{\sqrt{6}} & \dfrac{2}{\sqrt{6}} \\ 1 & 0 & 0 \\ 0 & \dfrac{1+i}{\sqrt{3}} & \dfrac{1}{\sqrt{3}} \end{bmatrix} \begin{bmatrix} 5 & 0 & 0 \\ 0 & -1 & -1+i \\ 0 & -1-i & 0 \end{bmatrix} \begin{bmatrix} 0 & 1 & 0 \\ -\dfrac{1-i}{\sqrt{6}} & 0 & \dfrac{1-i}{\sqrt{3}} \\ \dfrac{2}{\sqrt{6}} & 0 & \dfrac{1}{\sqrt{3}} \end{bmatrix}$$

$$= \begin{bmatrix} 1 & 0 & 0 \\ 0 & 5 & 0 \\ 0 & 0 & -2 \end{bmatrix}$$

13. The eigenvalues of A are the roots of the equation

$$\det\begin{bmatrix} \lambda-1 & -4i \\ -4i & \lambda-3 \end{bmatrix} = \lambda^2 - 4\lambda + 19 = 0$$

The roots of this equation, which are $\lambda = \dfrac{4 \pm \sqrt{16-4(19)}}{2}$, are not real. This shows that the eigenvalues of a symmetric matrix with nonreal entries need not be real. Theorem 10.6.6 applies only to matrices with real entries.

15. We know that det(A) is the sum of all the signed elementary products $\pm a_{1j_1}a_{2j_2}\cdots a_{nj_n}$, where a_{ij} is the entry from the ith row and j^{th} column of A. Since the ij^{th} element of $\bar{A}$ is $\bar{a}_{ij}$, then $\det(\bar{A})$ is the sum of the signed elementary products $\pm\bar{a}_{1j_1}\bar{a}_{2j_2}\cdots\bar{a}_{nj_n}$ or $\overline{\pm a_{1j_1}a_{2j_2}\cdots a_{nj_n}}$. That is, $\det(\bar{A})$ is the sum of the conjugates of the terms in det(A). But since the sum of the conjugates is the conjugate of the sum, we have $\det\left(\bar{A}\right) = \overline{\det(A)}$.

19. If A is invertible, then

$$\begin{aligned} A^*(A^{-1})^* &= (A^{-1}A)^* && \text{(by Exercise 18(d))} \\ &= I^* = \bar{I}^T = I^T = I \end{aligned}$$

Thus we have $(A^{-1})^* = (A^*)^{-1}$.

21. Let $\mathbf{r}_i$ denote the ith row of A and let $\mathbf{c}_j$ denote the jth column of A^*. Then, since $A^* = \bar{A}^T = \overline{(A^T)}$, we have $\mathbf{c}_j = \bar{\mathbf{r}}_j$ for $j = 1,\ldots,n$. Finally, let

$$\delta_{ij} = \begin{cases} 0 & \text{if } i \neq j \\ 1 & \text{if } i = j \end{cases}$$

Then A is unitary $\Leftrightarrow A^{-1} = A^*$. But

$$\begin{aligned} A^{-1} = A^* &\Leftrightarrow AA^* = I \\ &\Leftrightarrow \mathbf{r}_i \cdot \mathbf{c}_j = \delta_{ij} \text{ for all } i, j \\ &\Leftrightarrow \mathbf{r}_i \cdot \mathbf{r}_j = \delta_{ij} \text{ for all } i, j \\ &\Leftrightarrow \{\mathbf{r}_1, \ldots, \mathbf{r}_n\} \text{ is an orthonormal set} \end{aligned}$$

23. **(a)** We know that $A = A^*$, that $A\mathbf{x} = \lambda I\mathbf{x}$, and that $A\mathbf{y} = \mu I\mathbf{y}$. Therefore

$$\mathbf{x}^* A\mathbf{y} = \mathbf{x}^*(\mu I\mathbf{y}) = \mu(\mathbf{x}^* I\mathbf{y}) = \mu\mathbf{x}^*\mathbf{y}$$

and

$$\begin{aligned}\mathbf{x}^* A\mathbf{y} &= [(\mathbf{x}^* A\mathbf{y})^*]^* = [\mathbf{y}^* A^*\mathbf{x}]^* \\ &= [\mathbf{y}^* A\mathbf{x}]^* = [\mathbf{y}^* (\lambda I\mathbf{x})]^* \\ &= [\lambda\mathbf{y}^*\mathbf{x}]^* = \lambda\mathbf{x}^*\mathbf{y}\end{aligned}$$

The last step follows because λ, being the eigenvalue of an Hermitian matrix, is real.

(b) Subtracting the equations in Part (a) yields

$$(\lambda - \mu)(\mathbf{x}^* \mathbf{y}) = 0$$

Since $\lambda \neq \mu$, the above equation implies that $\mathbf{x}^*\mathbf{y}$ is the 1×1 zero matrix. Let $\mathbf{x} = (x_1,\ldots,x_n)$ and $\mathbf{y} = (y_1,\ldots,y_n)$. Then we have just shown that

$$\bar{x}_1 y_1 + \cdots + \bar{x}_n y_n = 0$$

so that

$$x_1 \bar{y}_1 + \cdots + x_n \bar{y}_n = \bar{0} = 0$$

and hence $\mathbf{x}$ and $\mathbf{y}$ are orthogonal.

SUPPLEMENTARY EXERCISES 10

3. The system of equations has solution $x_1 = -is + t$, $x_2 = s$, $x_3 = t$. Thus

$$\begin{bmatrix} x_1 \\ x_2 \\ x_3 \end{bmatrix} = \begin{bmatrix} -i \\ 1 \\ 0 \end{bmatrix} s + \begin{bmatrix} 1 \\ 0 \\ 1 \end{bmatrix} t$$

where s and t are arbitrary. Hence

$$\begin{bmatrix} -i \\ 1 \\ 0 \end{bmatrix} \quad \text{and} \quad \begin{bmatrix} 1 \\ 0 \\ 1 \end{bmatrix}$$

form a basis for the solution space.

5. The eigenvalues are the solutions, λ, of the equation

$$\det \begin{bmatrix} \lambda & 0 & -1 \\ -1 & \lambda & -\omega - 1 - \dfrac{1}{\omega} \\ 0 & -1 & \lambda + \omega + 1 + \dfrac{1}{\omega} \end{bmatrix} = 0$$

or

$$\lambda^3 + \left(\omega + 1 + \frac{1}{\omega}\right)\lambda^2 - \left(\omega + 1 + \frac{1}{\omega}\right)\lambda - 1 = 0$$

But $\dfrac{1}{\omega} = \bar{\omega}$, so that $\omega + 1 + \dfrac{1}{\omega} = 2\,\mathrm{Re}(\omega) + 1 = 0$. Thus we have

$$\lambda^3 - 1 = 0$$

or

$$(\lambda - 1)(\lambda^2 + \lambda + 1) = 0$$

Hence $\lambda = 1$, ω, or $\bar{\omega}$. Note that $\bar{\omega} = \omega^2$.

7. (c) Following the hint, we let $z = \cos\theta + i\sin\theta = e^{i\theta}$ in Part (a). This yields

$$1 + e^{i\theta} + e^{2i\theta} + \cdots + e^{ni\theta} = \frac{1 - e^{(n+1)i\theta}}{1 - e^{i\theta}}$$

If we expand and equate real parts, we obtain

$$1 + \cos\theta + \cos 2\theta + \cdots + \cos n\theta = \mathrm{Re}\left(\frac{1 - e^{(n+1)i\theta}}{1 - e^{i\theta}}\right)$$

But

$$
\begin{aligned}
\operatorname{Re}\left(\frac{1-e^{(n+1)i\theta}}{1-e^{i\theta}}\right) &= \operatorname{Re}\left(\frac{\left(1-e^{(n+1)i\theta}\right)\left(1-e^{-i\theta}\right)}{\left(1-e^{i\theta}\right)\left(1-e^{-i\theta}\right)}\right) \\
&= \operatorname{Re}\left(\frac{1-e^{(n+1)i\theta}-e^{-i\theta}+e^{ni\theta}}{2-2\operatorname{Re}(e^{i\theta})}\right) \\
&= \frac{1}{2}\left(\frac{1-\cos\left[(n+1)\theta\right]-\cos(-\theta)+\cos n\theta}{1-\cos\theta}\right) \\
&= \frac{1}{2}\left(\frac{1-\cos\theta}{1-\cos\theta}+\frac{\cos n\theta-\cos((n+1)\theta)}{1-\cos\theta}\right) \\
&= \frac{1}{2}\left(1+\frac{\cos n\theta-\left[\cos n\theta\cos\theta-\sin n\theta\sin\theta\right]}{2\sin^2\frac{\theta}{2}}\right) \\
&= \frac{1}{2}\left(1+\frac{\cos n\theta(1-\cos\theta)+2\sin n\theta\sin\frac{\theta}{2}\cos\frac{\theta}{2}}{2\sin^2\frac{\theta}{2}}\right) \\
&= \frac{1}{2}\left(1+\frac{2\cos n\theta\sin^2\frac{\theta}{2}+2\sin n\theta\sin\frac{\theta}{2}\cos\frac{\theta}{2}}{2\sin^2\frac{\theta}{2}}\right)
\end{aligned}
$$

$$
\begin{aligned}
&= \frac{1}{2}\left(1+\frac{\cos n\theta\sin\frac{\theta}{2}+\sin n\theta\cos\frac{\theta}{2}}{\sin\frac{\theta}{2}}\right) \\
&= \frac{1}{2}\left(1+\frac{\sin\left[\left(n+\frac{1}{2}\right)\theta\right]}{\sin\frac{\theta}{2}}\right)
\end{aligned}
$$

Observe that because $0 < \theta < 2\pi$, we have not divided by zero.

9. Call the diagonal matrix D. Then

$$(UD)^* = (\overline{UD})^T = \bar{D}^T \bar{U}^T = \bar{D}U^* = \bar{D}U^{-1}$$

We need only check that $(UD)^* = (UD)^{-1}$. But

$$(UD)^*(UD) = (\bar{D}U^{-1})(UD) = \bar{D}D$$

and

$$\bar{D}D = \begin{bmatrix} |z_1|^2 & 0 & \dots & 0 \\ 0 & |z_2|^2 & \dots & 0 \\ \vdots & \vdots & & \vdots \\ 0 & 0 & \dots & |z_n|^2 \end{bmatrix} = I$$

Hence $(UD)^* = (UD)^{-1}$ and so UD is unitary.

11. Show the eigenvalues of a unitary matrix have modulus one.

Proof: Let A be unitary. Then $A^{-1} = A^*$.
Let λ be an eigenvalue of A, with corresponding eigenvector x.

Then $\|Ax\|^2 = (Ax)^*(Ax) = (x^*A^*)(Ax) = x^*(A^{-1}A)x = x^*x = \|x\|^2$,

but also

$$\| Ax \|^2 = \| \lambda x \|^2 = (\lambda x)^*(\lambda x) = (\bar{\lambda}\lambda)(x^*x)$$

Since λ and $\bar{\lambda}$ are scalars $(\bar{\lambda}\lambda)(x^*x) = |\lambda|^2 \| x \|^2$. So, $|\lambda|^2 = 1$, and hence the eigenvalues of A have modulus one.

EXERCISE SET 11.1

1. **(a)** Substituting the coordinates of the points into Eq. (4) yields

$$\begin{vmatrix} x & y & 1 \\ 1 & -1 & 1 \\ 2 & 2 & 1 \end{vmatrix} = 0$$

which, upon cofactor expansion along the first row, yields $-3x + y + 4 = 0$; that is, $y = 3x - 4$.

(b) As in (a),

$$\begin{vmatrix} x & y & 1 \\ 0 & 1 & 1 \\ 1 & -1 & 1 \end{vmatrix} = 0$$

yields $2x + y - 1 = 0$ or $y = -2x + 1$.

3. Using Eq. (10) we obtain

$$\begin{vmatrix} x^2 & xy & y^2 & x & y & 1 \\ 0 & 0 & 0 & 0 & 0 & 1 \\ 0 & 0 & 1 & 0 & -1 & 1 \\ 4 & 0 & 0 & 2 & 0 & 1 \\ 4 & -10 & 25 & 2 & -5 & 1 \\ 16 & -4 & 1 & 4 & -1 & 1 \end{vmatrix} = 0$$

which is the same as

$$\begin{vmatrix} x^2 & xy & y^2 & x & y \\ 0 & 0 & 1 & 0 & -1 \\ 4 & 0 & 0 & 2 & 0 \\ 4 & -10 & 25 & 2 & -5 \\ 16 & -4 & 1 & 4 & -1 \end{vmatrix} = 0$$

by expansion along the second row (taking advantage of the zeros there). Add column five to column three and take advantage of another row of all but one zero to get

$$\begin{vmatrix} x^2 & xy & y^2+y & x \\ 4 & 0 & 0 & 2 \\ 4 & -10 & 20 & 2 \\ 16 & -4 & 0 & 4 \end{vmatrix} = 0.$$

Now expand along the first row and get $160x^2 + 320xy + 160(y^2 + y) - 320x = 0$; that is, $x^2 + 2xy + y^2 - 2x + y = 0$, which is the equation of a parabola.

7. Substituting each of the points (x_1, y_1), (x_2, y_2), (x_3, y_3), (x_4, y_4), and (x_5, y_5) into the equation

$$c_1x^2 + c_2xy + c_3y^2 + c_4x + c_5y + c_6 = 0$$

yields

$$\begin{array}{c} c_1x_1^2 + c_2x_1y_1 + c_3y_1^2 + c_4x_1 + c_5y_1 + c_6 = 0. \\ \vdots \qquad\qquad \vdots \qquad\qquad \vdots \\ c_1x_5^2 + c_2x_5y_5 + c_3y_5^2 + c_4x_5 + c_5y_5 + c_6 = 0. \end{array}$$

These together with the original equation form a homogeneous linear system with a nontrivial solution for $c_1, c_2, \cdots, c_6$. Thus the determinant of the coefficient matrix is zero, which is exactly Eq. (10).

9. Substituting the coordinates (x_i, y_i, z_i) of the four points into the equation $c_1(x^2 + y^2 + z^2) + c_2x + c_3y + c_4z + c_5 = 0$ of the sphere yields four equations, which together with the above sphere equation form a homogeneous linear system for $c_1, \cdots, c_5$ with a nontrivial solution. Thus the determinant of this system is zero, which is Eq. (12).

10. Upon substitution of the coordinates of the three points (x_1, y_1), (x_2, y_2) and (x_3, y_3), we obtain the equations:

$$\begin{aligned} c_1y + c_2x^2 + c_3x + c_4 &= 0 \\ c_1y_1 + c_2x_1^2 + c_3x_1 + c_4 &= 0 \\ c_1y_2 + c_2x_2^2 + c_3x_2 + c_4 &= 0 \\ c_1y_3 + c_2x_3^2 + c_3x_3 + c_4 &= 0. \end{aligned}$$

This is a homogeneous system with a nontrivial solution for c_1, c_2, c_3, c_4, so the determinant of the coefficient matrix is zero; that is,

$$\begin{vmatrix} y & x^2 & x & 1 \\ y_1 & x_1^2 & x_1 & 1 \\ y_2 & x_2^2 & x_2 & 1 \\ y_3 & x_3^2 & x_3 & 1 \end{vmatrix} = 0.$$

EXERCISE SET 11.2

1.

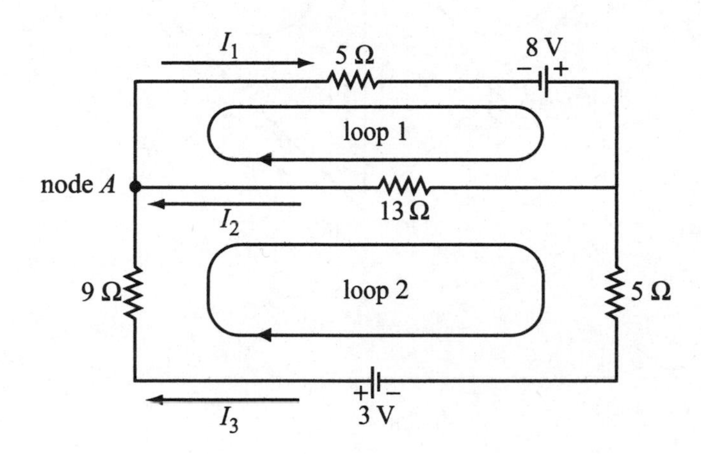

Applying Kirchhoff's current law to node A in the figure yields

$$I_1 = I_2 + I_3.$$

Applying Kirchhoff's voltage law and Ohm's law to Loops 1 and 2 yields

$$5I_1 + 13I_2 = 8$$

and

$$9I_3 - 13I_2 + 5I_3 = 3.$$

In matrix form these three equations are

$$\begin{bmatrix} 1 & -1 & -1 \\ 5 & 13 & 0 \\ 0 & -13 & 14 \end{bmatrix} \begin{bmatrix} I_1 \\ I_2 \\ I_3 \end{bmatrix} = \begin{bmatrix} 0 \\ 8 \\ 3 \end{bmatrix}$$

with solution $I_1 = \dfrac{255}{317}$, $I_2 = \dfrac{97}{317}$, $I_3 = \dfrac{158}{317}$.

3.

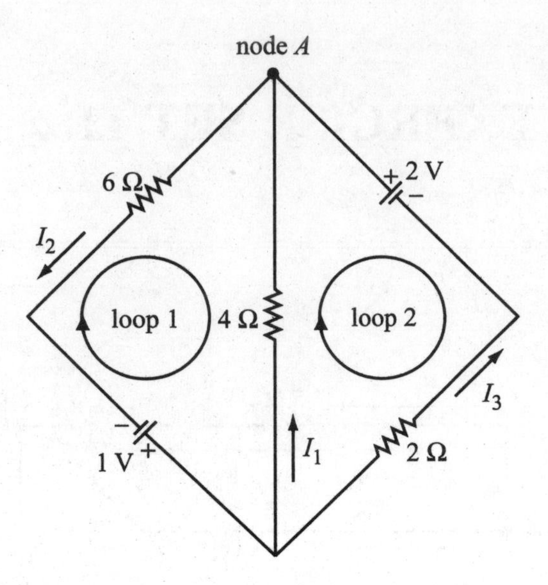

Node A gives $I_1 + I_3 = I_2$.

Loop 1 gives $-4I_1 - 6I_2 = -1$.

Loop 2 gives $4I_1 - 2I_3 = -2$.

In system form we have

$$\begin{bmatrix} 1 & -1 & 1 \\ -4 & -6 & 0 \\ 4 & 0 & -2 \end{bmatrix} \begin{bmatrix} I_1 \\ I_2 \\ I_3 \end{bmatrix} = \begin{bmatrix} 0 \\ -1 \\ -2 \end{bmatrix}$$

with solution $I_1 = -\frac{5}{22}$, $I_2 = \frac{7}{22}$, $I_3 = \frac{6}{11}$.

5.

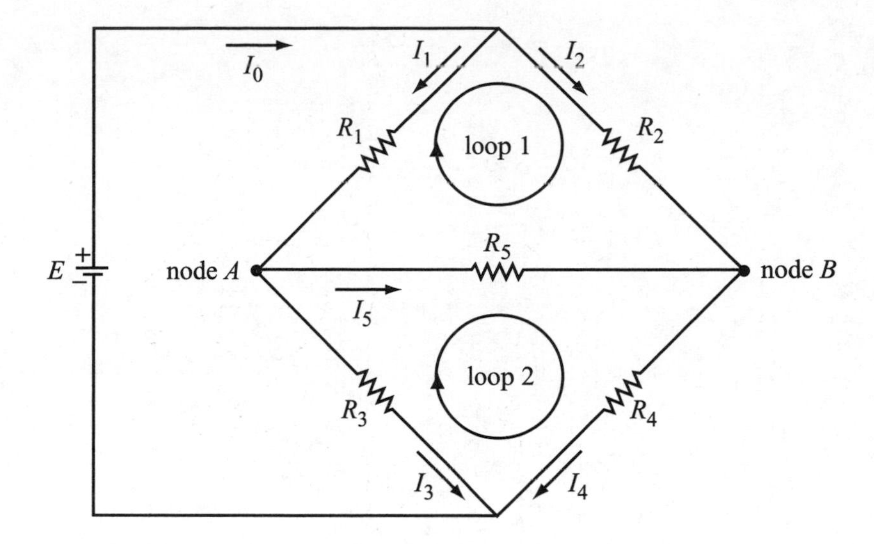

After setting $I_5 = 0$ we have that:

node A gives $I_1 = I_3$

node B gives $I_2 = I_4$

loop 1 gives $I_1R_1 = I_2R_2$

loop 2 gives $I_3R_3 = I_4R_4$.

From these four equations it easily follows that $R_4 = R_3R_2/R_1$.

EXERCISE SET 11.3

1.

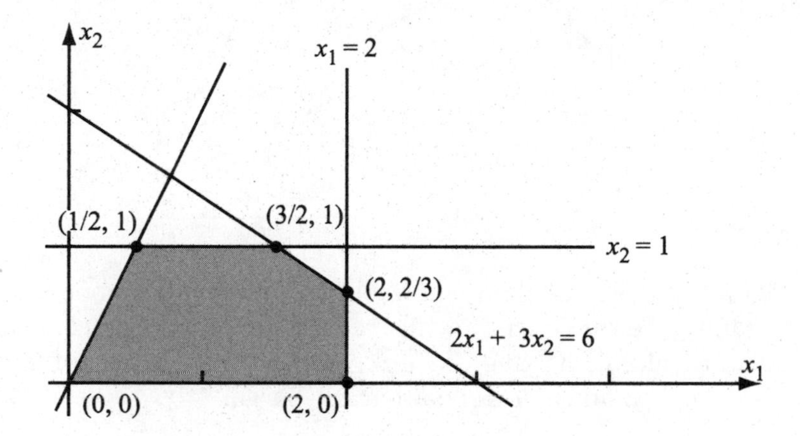

In the figure, the feasible region is shown and the extreme points are labelled. The values of the objective function are shown in the following table:

Extreme point (x_1, x_2)	Value of $z = 3x_1 + 2x_2$
(0, 0)	0
(1/2, 1)	7/2
(3/2, 1)	13/2
(2, 2/3)	22/3
(2, 0)	6

Thus the maximum, 22/3, is attained when $x_1 = 2$ and $x_2 = 2/3$.

3.

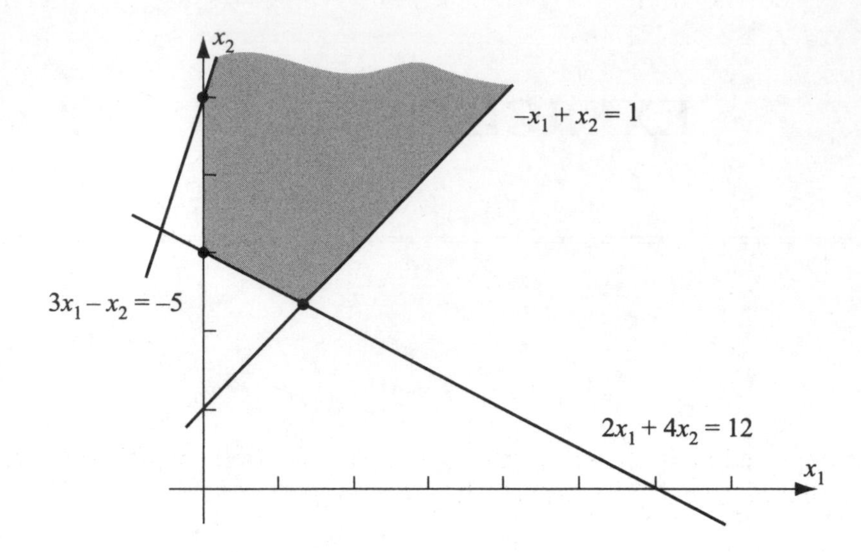

The feasible region for this problem, shown in the figure, is unbounded. The value of $z = -3x_1 + 2x_2$ cannot be minimized in this region since it becomes arbitrarily negative as we travel outward along the line $-x_1 + x_2 = 1$; i.e., the value of z is $-3x_1 + 2x_2 = -3x_1 + 2(x_1 + 1) = -x_1 + 2$ and x_1 can be arbitrarily large.

5.

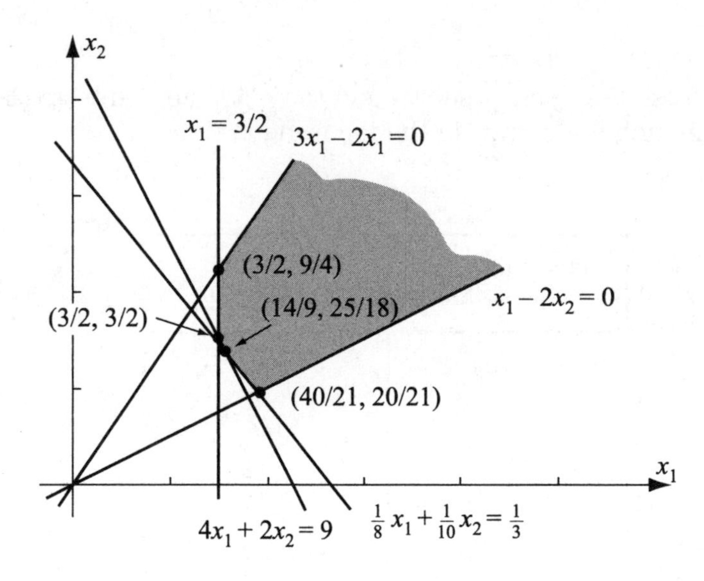

The feasible region and its extreme points are shown in the figure. Though the region is unbounded, x_1 and x_2 are always positive, so the objective function $z = 7.5x_1 + 5.0x_2$ is also. Thus, it has a minimum, which is attained at the point where $x_1 = 14/9$ and $x_2 = 25/18$. The value of z there is 335/18. In the problem's terms, if we use 7/9 cups of milk and 25/18 ounces of corn flakes, a minimum cost of 18.6¢ is realized.

7. Letting x_1 be the number of Company A's containers shipped and x_2 the number of Company B's, the problem is

$$\text{Maximize } z = 2.20x_1 + 3.00x_2$$

subject to

$$40x_1 + 50x_2 \leq 37{,}000$$

$$2x_1 + 3x_2 \leq 2{,}000$$

$$x_1 \geq 0$$

$$x_2 \geq 0.$$

The feasible region is shown in the figure. The vertex at which the maximum is attained is $x_1 = 550$ and $x_2 = 300$, where $z = 2110$.

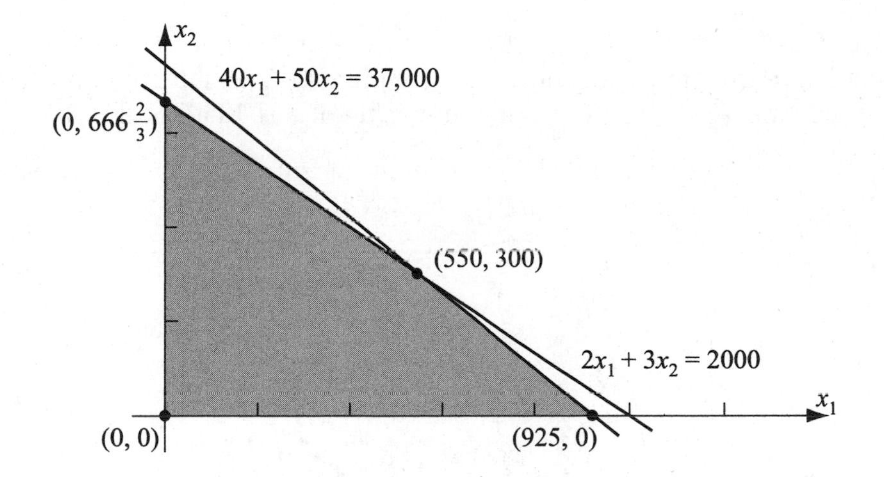

9. Let x_1 be the number of pounds of ingredient A used and x_2 the number of pounds of ingredient B. Then the problem is

$$\text{Minimize } z = 8x_1 + 9x_2$$

subject to

$$2x_1 + 5x_2 \geq 10$$

$$2x_1 + 3x_2 \geq 8$$

$$6x_1 + 4x_2 \geq 12$$

$$x_1 \geq 0$$

$$x_2 \geq 0$$

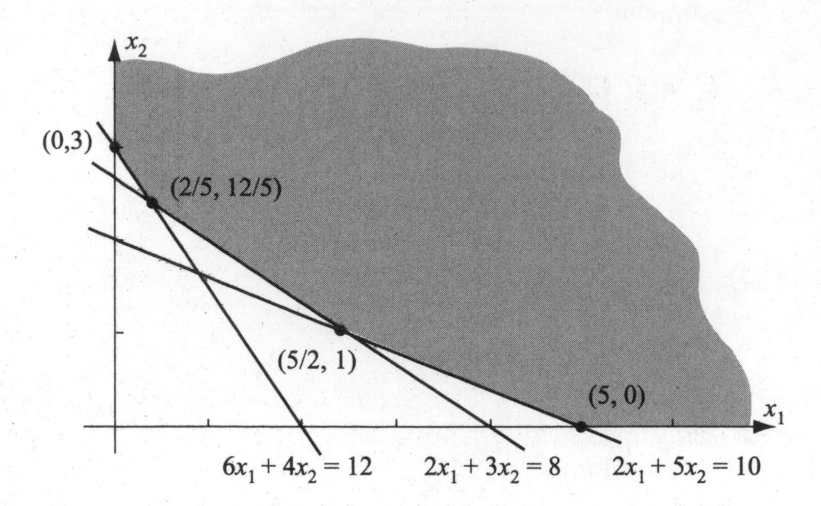

Though the feasible region shown in the figure is unbounded, the objective function is always positive there and hence must have a minimum. This minimum occurs at the vertex where $x_1 = 2/5$ and $x_2 = 12/5$. The minimum value of z is 124/5 or 24.8¢.

11.

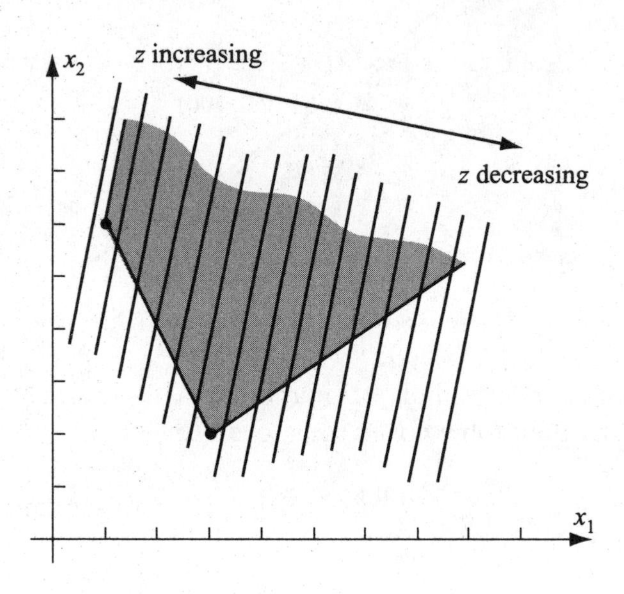

The level curves of the objective function $-5x_1 + x_2$ are shown in the figure, and it is readily seen that the objective function remains bounded in the region.

EXERCISE SET 11.4

1. The number of oxen is 50 per herd, and there are 7 herds, so there are 350 oxen. Hence the total number of oxen and sheep is 350 + 350 = 700.

3. Note that this is, effectively, Gaussian elimination applied to the augmented matrix

$$\begin{pmatrix} 1 & 1 & 10 \\ 1 & 1/4 & 7 \end{pmatrix}$$

5. **(a)** From equations 2 through n, $x_j = a_j - n$ ($j = 2, \ldots, n$). Using these equations in equation 1 gives

$$x_1 + (a_2 - x_1) + (a_3 - x_1) + \ldots + (a_n - x_1) = a_1$$

$$x_1 = (a_2 + a_3 + \ldots + a_n - a_1)/(n - 2)$$

Given x in terms of the known quantities n and the a_i. Then we can use $x_j = a_j - n$ ($j = 2, \ldots, n$) to find the other x_i.

(b) Exercise 7.(b) may be solved using this technique.

7. **(a)** The system is $x + y = 1000$, $(1/5)x - (1/4)y = 10$, with solution $x = 577$ and 7/9 staters, and $y = 422$ and 2/9 staters.

(b) The system is $G + B = (2/3)60$, $G + T = (3/4)60$, $G + I = (3/5)60$, $G + B + T + I = 60$, with solution (in minae) $G = 30.5$, $B = 9.5$, $T = 14.5$, and $I = 5.5$.

(c) The system is $A = B + (1/3)C$, $B = C + (1/3)A$, $C = (1/3)B + 10$, with solution $A = 45$, $B = 37.5$, and $C = 22.5$.

EXERCISE SET 11.5

1. We have $2b_1 = M_1, 2b_2 = M_2, \cdots, 2b_{n-1} = M_{n-1}$ from (14).

Inserting in (13) yields

$$6a_1h + M_1 = M_2$$
$$6a_2h + M_2 = M_3$$
$$\vdots$$
$$6a_{n-2}h + M_{n-2} = M_{n-1},$$

from which we obtain

$$a_1 = \frac{M_2 - M_1}{6h}$$
$$a_2 = \frac{M_3 - M_2}{6h}$$
$$\vdots$$
$$a_{n-2} = \frac{M_{n-1} - M_{n-2}}{6h}.$$

Now $S''(x_n) = M_n$, or from (14), $6a_{n-1}h + 2b_{n-1} = M_n$.

Also, $2b_{n-1} = M_{n-1}$ from (14) and so

$$6a_{n-1}h + M_{n-1} = M_n$$

or

$$a_{n-1} = \frac{M_n - M_{n-1}}{6h}.$$

Thus we have

$$a_i = \frac{M_{i+1} - M}{6h}$$

for $i = 1, 2, \ldots, n-1$. From (9) and (11) we have

$$a_i h^3 + b_i h^2 + c_i h + d_i = y_{i+1}, \quad i = 1, 2, \ldots, n-1.$$

Substituting the expressions for a_i, b_i, and d_i from (14) yields $\left(\dfrac{M_{i+1} - M_i}{6h}\right)h^3 + \dfrac{M_i}{2}h^2 + c_i h + y_i = y_{i+1}$, $i = 1, 2, \ldots, n-1$. Solving for c_i gives

$$c_i = \frac{y_{i+1} - y_i}{h} - \left(\frac{M_{i+1} - 2M_i}{6}\right)h$$

for $i = 1, 2, \ldots, n-1$.

3. **(a)** Given that the points lie on a single cubic curve, the cubic runout spline will agree exactly with the single cubic curve.

(b) Set $h = 1$ and

$$x_1 = 0, \quad y_1 = 1$$

$$x_2 = 1, \quad y_2 = 7$$

$$x_3 = 2, \quad y_3 = 27$$

$$x_4 = 3, \quad y_4 = 79$$

$$x_5 = 4, \quad y_5 = 181.$$

Then

$$6\left(y_1 - 2y_2 + y_3\right)/h^2 = 84$$

$$6\left(y_2 - 2y_3 + y_4\right)/h^2 = 192$$

$$6\left(y_3 - 2y_4 + y_5\right)/h^2 = 300$$

and the linear system (24) for the cubic runout spline becomes

$$\begin{bmatrix} 6 & 0 & 0 \\ 1 & 4 & 1 \\ 0 & 0 & 6 \end{bmatrix} \begin{bmatrix} M_2 \\ M_3 \\ M_4 \end{bmatrix} = \begin{bmatrix} 84 \\ 192 \\ 300 \end{bmatrix}.$$

Solving this system yields

$$M_2 = 14$$
$$M_3 = 32$$
$$M_4 = 50.$$

From (22) and (23) we have

$$M_1 = 2M_2 - M_3 = -4$$
$$M_5 = 2M_4 - M_3 = 68.$$

Using (14) to solve for the a_i's, b_i's, c_i's, and d_i's we have

$$a_1 = \left(M_2 - M_1\right)/6h = 3$$
$$a_2 = \left(M_3 - M_2\right)/6h = 3$$
$$a_3 = \left(M_4 - M_3\right)/6h = 3$$
$$a_4 = \left(M_5 - M_4\right)/6h = 3$$

$$b_1 = M_1/2 = -2$$
$$b_2 = M_2/2 = 7$$
$$b_3 = M_3/2 = 16$$
$$b_4 = M_4/2 = 25$$

$$c_1 = (y_2 - y_1)/h - (M_2 + 2M_1)h/6 = 5$$

$$c_2 = (y_3 - y_2)/h - (M_3 + 2M_2)h/6 = 10$$

$$c_3 = (y_4 - y_3)/h - (M_4 + 2M_3)h/6 = 33$$

$$c_4 = (y_5 - y_4)/h - (M_5 + 2M_4)h/6 = 74$$

$$d_1 = y_1 = 1$$

$$d_2 = y_2 = 7$$

$$d_3 = y_3 = 27$$

$$d_4 = y_4 = 79.$$

For $0 \le x \le 1$ we have

$$S(x) = S_1(x) = 3x^3 - 2x^2 + 5x + 1.$$

For $1 \le x \le 2$ we have

$$S(x) = S_2(x) = 3(x-1)^3 + 7(x-1)^2 + 10(x-1) + 7$$
$$= 3x^3 - 2x^2 + 5x + 1.$$

For $2 \le x \le 3$ we have

$$S(x) = S_3(x) = 3(x-2)^3 + 16(x-2)^2 + 33(x-2) + 27$$
$$= 3x^3 - 2x^2 + 5x + 1.$$

For $3 \le x \le 4$ we have

$$S(x) = S_4(x) = 3(x-3)^3 + 25(x-3)^2 + 74(x-3) + 79$$
$$= 3x^3 - 2x^2 + 5x + 1.$$

Thus $S_1(x) = S_2(x) = S_3(x) = S_4(x)$,

or $S(x) = 3x^3 - 2x^2 + 5x + 1 \quad$ for $0 \le x \le 4$.

5. The linear system (24) for the cubic runout spline becomes

$$\begin{bmatrix} 6 & 0 & 0 \\ 1 & 4 & 1 \\ 0 & 0 & 6 \end{bmatrix} \begin{bmatrix} M_2 \\ M_3 \\ M_4 \end{bmatrix} = \begin{bmatrix} -.0001116 \\ -.0000816 \\ -.0000636 \end{bmatrix}.$$

Solving this system yields

$$M_2 = -.0000186$$

$$M_3 = -.0000131$$

$$M_4 = -.0000106.$$

From (22) and (23) we have

$$M_1 = 2M_2 - M_3 = -.0000241$$

$$M_5 = 2M_4 - M_3 = -.0000081.$$

Solving for the a_i's, b_i's, c_i's and d_i's from Eqs. (14) we have

$$a_1 = (M_2 - M_1)/6h = .00000009$$

$$a_2 = (M_3 - M_2)/6h = .00000009$$

$$a_3 = (M_4 - M_3)/6h = .00000004$$

$$a_4 = (M_5 - M_4)/6h = .00000004$$

$$b_1 = M_1/2 = -.0000121$$

$$b_2 = M_2/2 = -.0000093$$

$$b_3 = M_3/2 = -.0000066$$

$$b_4 = M_4/2 = -.0000053$$

$$c_1 = \left(y_2 - y_1\right)/h - \left(M_2 + 2M_1\right)h/6 = .000282$$

$$c_2 = \left(y_3 - y_2\right)/h - \left(M_3 + 2M_2\right)h/6 = .000070$$

$$c_3 = \left(y_4 - y_3\right)/h - \left(M_4 + 2M_3\right)h/6 = .000087$$

$$c_4 = \left(y_5 - y_4\right)/h - \left(M_5 + 2M_4\right)h/6 = .000207$$

$$d_1 = y_1 = .99815$$

$$d_2 = y_2 = .99987$$

$$d_3 = y_3 = .99973$$

$$d_4 = y_4 = .99823.$$

The resulting cubic runout spline is

$$S(x) = \begin{cases} .00000009(x+10)^3 - .0000121(x+10)^2 + .000282(x+10) + .99815, & -10 \le x \le 0 \\ .00000009(x)^3 - .0000093(x)^2 + .000070(x) + .99987, & 0 \le x \le 10 \\ .00000004(x-10)^3 - .0000066(x-10)^2 + .000087(x-10) + .99973, & 10 \le x \le 20 \\ .00000004(x-20)^3 - .0000053(x-20)^2 + .000207(x-20) + .99823, & 20 \le x \le 30. \end{cases}$$

Assuming the maximum is attained in the interval [0, 10], we set $S'(x)$ equal to zero in this interval:

$$S'(x) = .00000027x^2 - .0000186x + .000070.$$

To three significant digits the root of this quadratic in the interval [0, 10] is 4.00 and

$$S(4.00) = 1.00001.$$

7. **(a)** Since $S\left(x_1\right) = y_1$ and $S\left(x_n\right) = y_n$, then from $S\left(x_1\right) = S\left(x_n\right)$ we have $y_1 = y_n$. By definition $S''\left(x_1\right) = M_1$ and $S''\left(x_n\right) = M_n$, and so from $S''\left(x_1\right) = S''\left(x_n\right)$ we have $M_1 = M_n$.

From (5) we have

$$S'(x_1) = c_1$$
$$S'(x_n) = 3a_{n-1}h^2 + 2b_{n-1}h + c_{n-1}.$$

Substituting for c_1, a_{n-1}, b_{n-1}, c_{n-1} from Eqs. (14) yields

$$S'(x_1) = (y_2 - y_1)/h - (M_2 + 2M_1)h/6$$
$$S'(x_n) = (M_n - M_{n-1})h/2 + M_{n-1}h$$
$$+ (y_n - y_{n-1})/h - (M_n + 2M_{n-1})h/6.$$

Using $M_n = M_1$ and $y_n = y_1$, the last equation becomes

$$S'(x_n) = M_1h/3 + M_{n-1}h/6 + (y_1 - y_{n-1})/h.$$

From $S'(x_1) = S'(x_n)$ we obtain

$$(y_2 - y_1)/h - (y_1 - y_{n-1})/h = M_1h/3 + M_{n-1}h/6 + (M_2 + 2M_1)h/6$$

or

$$4M_1 + M_2 + M_{n-1} = 6(y_{n-1} - 2y_1 + y_2)/h^2.$$

(b) Eqs. (15) together with the three equations in part (a) of the exercise statement give

$$\begin{aligned}
4M_1 + M_2 + M_{n-1} &= 6(y_{n-1} - 2y_1 + y_2)/h^2 \\
M_1 + 4M_2 + M_3 &= 6(y_1 - 2y_2 + y_3)/h^2 \\
M_2 + 4M_3 + M_4 &= 6(y_2 - 2y_3 + y_4)/h^2 \\
&\vdots \\
M_{n-3} + 4M_{n-2} + M_{n-1} &= 6(y_{n-3} - 2y_{n-2} + y_{n-1})/h^2 \\
M_1 + M_{n-2} + 4M_{n-1} &= 6(y_{n-2} - 2y_{n-1} + y_1)/h^2.
\end{aligned}$$

This linear system for $M_1, M_2, \ldots, M_{n-1}$ in matrix form is

$$\begin{bmatrix} 4 & 1 & 0 & 0 & \cdot & \cdot & \cdot & 0 & 0 & 0 & 1 \\ 1 & 4 & 1 & 0 & \cdot & \cdot & \cdot & 0 & 0 & 0 & 0 \\ 0 & 1 & 4 & 1 & \cdot & \cdot & \cdot & 0 & 0 & 0 & 0 \\ \vdots & \vdots & \vdots & \vdots & & & & \vdots & \vdots & \vdots & \vdots \\ 0 & 0 & 0 & 0 & \cdot & \cdot & \cdot & 0 & 1 & 4 & 1 \\ 1 & 0 & 0 & 0 & \cdot & \cdot & \cdot & 0 & 0 & 1 & 4 \end{bmatrix} \begin{bmatrix} M_1 \\ M_2 \\ M_3 \\ \vdots \\ M_{n-2} \\ M_{n-1} \end{bmatrix} = \frac{6}{h^2} \begin{bmatrix} y_{n-1} & -2y_1 & +y_2 \\ y_1 & -2y_2 & +y_3 \\ y_2 & -2y_3 & +y_4 \\ & \vdots & \\ y_{n-3} & -2y_{n-2} & +y_{n-1} \\ y_{n-2} & -2y_{n-1} & +y_1 \end{bmatrix}$$

EXERCISE SET 11.6

1. **(a)** $\mathbf{x}^{(1)} = P\mathbf{x}^{(0)} = \begin{bmatrix} .4 \\ .6 \end{bmatrix}$, $\mathbf{x}^{(2)} = P\mathbf{x}^{(1)} = \begin{bmatrix} .46 \\ .54 \end{bmatrix}$.

Continuing in this manner yields $\mathbf{x}^{(3)} = \begin{bmatrix} .454 \\ .546 \end{bmatrix}$,

$\mathbf{x}^{(4)} = \begin{bmatrix} .4546 \\ .5454 \end{bmatrix}$ and $\mathbf{x}^{(5)} = \begin{bmatrix} .45454 \\ .54546 \end{bmatrix}$.

(b) P is regular because all of the entries of P are positive. Its steady-state vector $\mathbf{q}$ solves $(P - I)\mathbf{q} = \mathbf{0}$; that is,

$$\begin{bmatrix} -.6 & .5 \\ .6 & -.5 \end{bmatrix} \begin{bmatrix} q_1 \\ q_2 \end{bmatrix} = \begin{bmatrix} 0 \\ 0 \end{bmatrix}.$$

This yields one independent equation, $.6q_1 - .5q_2 = 0$, or $q_1 = \frac{5}{6}q_2$. Solutions are thus of the form $\mathbf{q} = s\begin{bmatrix} 5/6 \\ 1 \end{bmatrix}$. Set $s = \dfrac{1}{\frac{5}{6}+1} = \dfrac{6}{11}$ to obtain $\mathbf{q} = \begin{bmatrix} 5/11 \\ 6/11 \end{bmatrix}$.

3. **(a)** Solve $(P - I)\mathbf{q} = \mathbf{0}$, i.e.,

$$\begin{bmatrix} -2/3 & 3/4 \\ 2/3 & -3/4 \end{bmatrix} \begin{bmatrix} q_1 \\ q_2 \end{bmatrix} = \begin{bmatrix} 0 \\ 0 \end{bmatrix}.$$

The only independent equation is $\frac{2}{3}q_1 = \frac{3}{4}q_2$, yielding $\mathbf{q} = \begin{bmatrix} 9/8 \\ 1 \end{bmatrix} s$. Setting $s = \frac{8}{17}$ yields $\mathbf{q} = \begin{bmatrix} 9/17 \\ 8/17 \end{bmatrix}$

(b) As in (a), solve

$$\begin{bmatrix} -.19 & .26 \\ .19 & -.26 \end{bmatrix}\begin{bmatrix} q_1 \\ q_2 \end{bmatrix} = \begin{bmatrix} 0 \\ 0 \end{bmatrix}$$

i.e., $.19q_1 = .26q_2$. Solutions have the form $\mathbf{q} = \begin{bmatrix} 26/19 \\ 1 \end{bmatrix} s$.

Set $s = \dfrac{19}{45}$ to get $\mathbf{q} = \begin{bmatrix} 26/45 \\ 19/45 \end{bmatrix}$.

(c) Again, solve

$$\begin{bmatrix} -2/3 & 1/2 & 0 \\ 1/3 & -1 & 0 \\ 1/3 & 1/2 & -1/4 \end{bmatrix}\begin{bmatrix} q_1 \\ q_2 \\ q_3 \end{bmatrix} = \begin{bmatrix} 0 \\ 0 \\ 0 \end{bmatrix}$$

by reducing the coefficient matrix to row-echelon form:

$$\begin{bmatrix} 1 & 0 & -1/4 \\ 0 & 1 & -1/3 \\ 0 & 0 & 0 \end{bmatrix}$$

yielding solutions of the form $\mathbf{q} = \begin{bmatrix} 1/4 \\ 1/3 \\ 1 \end{bmatrix} s$.

Set $s = \dfrac{12}{19}$ to get $\mathbf{q} = \begin{bmatrix} 3/19 \\ 4/19 \\ 12/19 \end{bmatrix}$.

5. Let $\mathbf{q} = \left[\frac{1}{k}\,\frac{1}{k} \cdots \frac{1}{k}\right]^T$. Then $(P\mathbf{q})_i = \sum_{j=1}^{k} p_{ij}q_j = \sum_{j=1}^{k} \frac{1}{k} p_{ij} = \frac{1}{k}\sum_{j=1}^{k} p_{ij} = \frac{1}{k}$, since the row sums of P are 1. Thus $(P\mathbf{q})_i = q_i$ for all i.

7. Let $\mathbf{x} = [x_1\ x_2]^T$ be the state vector, with x_1 = probability that John is happy and x_2 = probability that John is sad. The transition matrix P will be

$$P = \begin{bmatrix} 4/5 & 2/3 \\ 1/5 & 1/3 \end{bmatrix}$$

since the columns must sum to one. We find the steady state vector for P by solving

$$\begin{bmatrix} -1/5 & 2/3 \\ 1/5 & -2/3 \end{bmatrix} \begin{bmatrix} q_1 \\ q_2 \end{bmatrix} = \begin{bmatrix} 0 \\ 0 \end{bmatrix},$$

i.e., $\frac{1}{5}q_1 = \frac{2}{3}q_2$, so $\mathbf{q} = \begin{bmatrix} 10/3 \\ 1 \end{bmatrix} s$. Let $s = \frac{3}{13}$ and get $\mathbf{q} = \begin{bmatrix} 10/13 \\ 3/13 \end{bmatrix}$, so 10/13 is the probability that John will be happy on a given day.

EXERCISE SET 11.7

1. Note that the matrix has the same number of rows and columns as the graph has vertices, and that ones in the matrix correspond to arrows in the graph. We obtain

(a) $\begin{bmatrix} 0 & 0 & 0 & 1 \\ 1 & 0 & 1 & 1 \\ 1 & 1 & 0 & 1 \\ 0 & 0 & 0 & 0 \end{bmatrix}$

(b) $\begin{bmatrix} 0 & 1 & 1 & 0 & 0 \\ 0 & 0 & 0 & 0 & 1 \\ 1 & 0 & 0 & 1 & 0 \\ 0 & 0 & 1 & 0 & 0 \\ 0 & 0 & 1 & 0 & 0 \end{bmatrix}$

(c) $\begin{bmatrix} 0 & 1 & 0 & 1 & 0 & 0 \\ 1 & 0 & 0 & 0 & 0 & 0 \\ 0 & 1 & 0 & 1 & 1 & 1 \\ 0 & 0 & 0 & 0 & 0 & 1 \\ 0 & 0 & 0 & 0 & 0 & 1 \\ 0 & 0 & 1 & 0 & 1 & 0 \end{bmatrix}$

3. **(a)** As in problem 2, we obtain

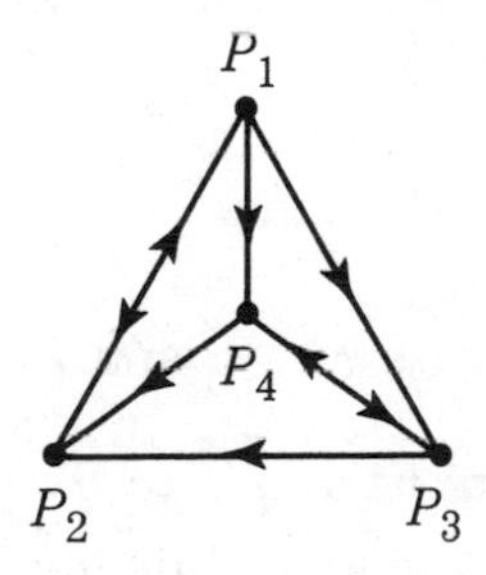

(b) $m_{12} = 1$, so there is one 1-step connection from P_1 to P_2.

$$M^2 = \begin{bmatrix} 1 & 2 & 1 & 1 \\ 0 & 1 & 1 & 1 \\ 1 & 1 & 1 & 0 \\ 1 & 1 & 0 & 1 \end{bmatrix} \quad \text{and} \quad M^3 = \begin{bmatrix} 2 & 3 & 2 & 2 \\ 1 & 2 & 1 & 1 \\ 1 & 2 & 1 & 2 \\ 1 & 2 & 2 & 1 \end{bmatrix}.$$

So $m_{12}^{(2)} = 2$ and $m_{12}^{(3)} = 3$ meaning there are two 2-step and three 3-step connections from P_1 to P_2 by Theorem 1. These are:

1-step: $P_1 \to P_2$

2-step: $P_1 \to P_4 \to P_2$ and $P_1 \to P_3 \to P_2$

3-step: $P_1 \to P_2 \to P_1 \to P_2$, $P_1 \to P_3 \to P_4 \to P_2$,

and $P_1 \to P_4 \to P_3 \to P_2$.

(c) Since $m_{14} = 1$, $m_{14}^{(2)} = 1$ and $m_{14}^{(3)} = 2$, there are one 1-step, one 2-step and two 3-step connections from P_1 to P_4. These are:

1-step: $P_1 \to P_4$

2-step: $P_1 \to P_3 \to P_4$

3-step: $P_1 \to P_2 \to P_1 \to P_4$ and $P_1 \to P_4 \to P_3 \to P_4$.

5. **(a)** Note that to be contained in a clique, a vertex must have "two-way" connections with at least two other vertices. Thus, P_4 could not be in a clique, so $\{P_1, P_2, P_3\}$ is the only possible clique. Inspection shows that this is indeed a clique.

(b) Not only must a clique vertex have two-way connections to at least two other vertices, but the vertices to which it is connected must share a two-way connection. This consideration eliminates P_1 and P_2, leaving $\{P_3, P_4, P_5\}$ as the only possible clique. Inspection shows that it is indeed a clique.

(c) The above considerations eliminate P_1, P_3 and P_7 from being in a clique. Inspection shows that each of the sets $\{P_2, P_4, P_6\}$, $\{P_4, P_6, P_8\}$, $\{P_2, P_6, P_8\}$, $\{P_2, P_4, P_8\}$ and $\{P_4, P_5, P_6\}$ satisfy conditions (i) and (ii) in the definition of a clique. But note that P_8 can be added to the first set and we still satisfy the conditions. P_5 may not be added, so $\{P_2, P_4, P_6, P_8\}$ is a clique, containing all the other possibilities except $\{P_4, P_5, P_6\}$, which is also a clique.

7.

$$M = \begin{bmatrix} 0 & 0 & 1 & 1 \\ 1 & 0 & 0 & 0 \\ 0 & 1 & 0 & 1 \\ 0 & 1 & 0 & 0 \end{bmatrix}.$$

Then

$$M^2 = \begin{bmatrix} 0 & 2 & 0 & 1 \\ 0 & 0 & 1 & 1 \\ 1 & 1 & 0 & 0 \\ 1 & 0 & 0 & 0 \end{bmatrix} \quad \text{and} \quad M + M^2 = \begin{bmatrix} 0 & 2 & 1 & 2 \\ 1 & 0 & 1 & 1 \\ 1 & 2 & 0 & 1 \\ 1 & 1 & 0 & 0 \end{bmatrix}.$$

By summing the rows of $M + M^2$, we get that the power of P_1 is $2 + 1 + 2 = 5$, the power of P_2 is 3, of P_3 is 4, and of P_4 is 2.

EXERCISE SET 11.8

1. **(a)** From Eq. (2), the expected payoff of the game is

$$\mathbf{pAq} = \begin{bmatrix} \frac{1}{2} & 0 & \frac{1}{2} \end{bmatrix} \begin{bmatrix} -4 & 6 & -4 & 1 \\ 5 & -7 & 3 & 8 \\ -8 & 0 & 6 & -2 \end{bmatrix} \begin{bmatrix} \frac{1}{4} \\ \frac{1}{4} \\ \frac{1}{4} \\ \frac{1}{4} \end{bmatrix} = -\frac{5}{8}.$$

(b) If player R uses strategy $[p_1 \quad p_2 \quad p_3]$ against player C's strategy $\begin{bmatrix} \frac{1}{4} & \frac{1}{4} & \frac{1}{4} & \frac{1}{4} \end{bmatrix}^T$

his payoff will be $\mathbf{pAq} = (-1/4)\mathbf{p}_1 + (9/4)\mathbf{p}_2 - \mathbf{p}_3$. Since p_1, p_2 and p_3 are nonnegative and add up to 1, this is a weighted average of the numbers $-1/4$, $9/4$ and -1. Clearly this is the largest if $p_1 = p_3 = 0$ and $p_2 = 1$; that is, $\mathbf{p} = [0 \quad 1 \quad 0]$.

(c) As in (*b*), if player C uses $[q_1 \quad q_2 \quad q_3 \quad q_4]^T$ against $\begin{bmatrix} \frac{1}{2} & 0 & \frac{1}{2} \end{bmatrix}$, we get $\mathbf{pAq} = -6q_1 +$

$3q_2 + q_3 - \frac{1}{2}q_4$. Clearly this is minimized over all strategies by setting $q_1 = 1$ and $q_2 = q_3 = q_4 = 0$. That is $\mathbf{q} = [1 \quad 0 \quad 0 \quad 0]^T$.

3. **(a)** Calling the matrix A, we see a_{22} is a saddle point, so the optimal strategies are pure, namely: $\mathbf{p} = [0 \quad 1]$, $\mathbf{q} = [0 \quad 1]^T$; the value of the game is $a_{22} = 3$.

(b) As in (a), a_{21} is a saddle point, so optimal strategies are $\mathbf{p} = [0 \quad 1 \quad 0]$, $\mathbf{q} = [1 \quad 0]^T$; the value of the game is $a_{21} = 2$.

(c) Here, a_{32} is a saddle point, so optimal strategies are $\mathbf{p} = [0 \quad 0 \quad 1]$, $\mathbf{q} = [0 \quad 1 \quad 0]^T$ and $v = a_{32} = 2$.

(d) Here, a_{21} is a saddle point, so $\mathbf{p} = [0 \quad 1 \quad 0 \quad 0]$, $\mathbf{q} = [1 \quad 0 \quad 0]^T$ and $v = a_{21} = -2$.

5. Let a_{11} = payoff to R if the black ace and black two are played = 3.

a_{12} = payoff to R if the black ace and red three are played = –4.

a_{21} = payoff to R if the red four and black two are played = –6.

a_{22} = payoff to R if the red four and red three are played = 7.

So, the payoff matrix for the game is $A = \begin{bmatrix} 3 & -4 \\ -6 & 7 \end{bmatrix}$.

A has no saddle points, so from Theorem 2, $\mathbf{p} = \begin{bmatrix} \frac{13}{20} & \frac{7}{20} \end{bmatrix}$,

$\mathbf{q} = \begin{bmatrix} \frac{11}{20} & \frac{9}{20} \end{bmatrix}^T$; that is, player R should play the black ace 65 percent of the time, and player C should play the black two 55 percent of the time. The value of the game is $-\frac{3}{20}$, that is, player C can expect to collect on the average 15 cents per game.

EXERCISE SET 11.9

1. **(a)** Calling the given matrix E, we need to solve

$$(I-E)\mathbf{p} = \begin{bmatrix} 1/2 & -1/3 \\ -1/2 & 1/3 \end{bmatrix} \begin{bmatrix} p_1 \\ p_2 \end{bmatrix} = \begin{bmatrix} 0 \\ 0 \end{bmatrix}.$$

This yields $\frac{1}{2}p_1 = \frac{1}{3}p_2$, that is, $\mathbf{p} = s[1 \quad 3/2]^T$. Set $s = 2$ and get $\mathbf{p} = [2 \quad 3]^T$.

(b) As in (a), solve

$$(I-E)\mathbf{p} = \begin{bmatrix} 1/2 & 0 & -1/2 \\ -1/3 & 1 & -1/2 \\ -1/6 & -1 & 1 \end{bmatrix} \begin{bmatrix} p_1 \\ p_2 \\ p_3 \end{bmatrix} = \begin{bmatrix} 0 \\ 0 \\ 0 \end{bmatrix}.$$

In row-echelon form, this reduces to

$$\begin{bmatrix} 1 & 0 & -1 \\ 0 & 1 & -5/6 \\ 0 & 0 & 0 \end{bmatrix} \begin{bmatrix} p_1 \\ p_2 \\ p_3 \end{bmatrix} = \begin{bmatrix} 0 \\ 0 \\ 0 \end{bmatrix}.$$

Solutions of this system have the form $\mathbf{p} = s[1 \quad 5/6 \quad 1]^T$. Set $s = 6$ and get $\mathbf{p} = [6 \quad 5 \quad 6]^T$.

(c) As in (a), solve

$$(I-E)\mathbf{p} = \begin{bmatrix} .65 & -.50 & -.30 \\ -.25 & .80 & -.30 \\ -.40 & -.30 & .60 \end{bmatrix} \begin{bmatrix} p_1 \\ p_2 \\ p_3 \end{bmatrix} = \begin{bmatrix} 0 \\ 0 \\ 0 \end{bmatrix},$$

which reduces to

$$\begin{bmatrix} 1 & 3/4 & -3/2 \\ 0 & 1 & -54/79 \\ 0 & 0 & 0 \end{bmatrix}\begin{bmatrix} p_1 \\ p_2 \\ p_3 \end{bmatrix} = \begin{bmatrix} 0 \\ 0 \\ 0 \end{bmatrix}.$$

Solutions are of the form $\mathbf{p} = s[78/79 \quad 54/79 \quad 1]^T$. Let $s = 79$ to obtain $\mathbf{p} = [78 \quad 54 \quad 79]^T$.

3. Theorem 2 says there will be one linearly independent price vector for the matrix E if some positive power of E is positive. Since E is not positive, try E^2:

$$E^2 = \begin{bmatrix} .2 & .34 & .1 \\ .2 & .54 & .6 \\ .6 & .12 & .3 \end{bmatrix} > 0.$$

5. Taking the CE, EE, and ME in that order, we form the consumption matrix C, where c_{ij} = the amount (per consulting dollar) of the i-th engineer's services purchased by the j-th engineer. Thus,

$$C = \begin{bmatrix} 0 & .2 & .3 \\ .1 & 0 & .4 \\ .3 & .4 & 0 \end{bmatrix}.$$

We want to solve $(I - C)\mathbf{x} = \mathbf{d}$, where $\mathbf{d}$ is the demand vector, i.e.

$$\begin{bmatrix} 1 & -.2 & -.3 \\ -.1 & 1 & -.4 \\ -.3 & -.4 & 1 \end{bmatrix}\begin{bmatrix} x_1 \\ x_2 \\ x_3 \end{bmatrix} = \begin{bmatrix} 500 \\ 700 \\ 600 \end{bmatrix}.$$

In row-echelon form this reduces to

$$\begin{bmatrix} 1 & -.2 & -.3 \\ 0 & 1 & -.43877 \\ 0 & 0 & 1 \end{bmatrix}\begin{bmatrix} x_1 \\ x_2 \\ x_3 \end{bmatrix} = \begin{bmatrix} 500 \\ 785.31 \\ 1556.19 \end{bmatrix}.$$

Back-substitution yields the solution $x = [1256.48 \quad 1448.12 \quad 1556.19]^T$.

7. The i-th column sum of E is $\sum_{j=1}^{n} e_{ji}$, and the elements of the i-th column of $I - E$ are the negatives of the elements of E, except for the ii-th, which is $1 - e_{ii}$. So, the i-th column sum of $I - E$ is $1 - \sum_{j=1}^{n} e_{ji} = 1 - 1 = 0$. Now, $(I - E)^T$ has zero row sums, so the vector $x = [1 \quad 1 \quad \cdots \quad 1]^T$ solves $(I - E)^T\mathbf{x} = 0$. This implies $\det(I - E)^T = 0$. But $\det(I - E)^T = \det(I - E)$, so $(I - E)\mathbf{p} = 0$ must have nontrivial (i.e., nonzero) solutions.

9. **(I)** Let $\mathbf{y}$ be a strictly positive vector, and $\mathbf{x} = (I - C)^{-1}\mathbf{y}$. Since C is productive $(I - C)^{-1} \geq 0$, so $\mathbf{x} = (I - C)^{-1}\mathbf{y} \geq 0$. But then $(I - C)\mathbf{x} = \mathbf{y} > 0$, i.e., $\mathbf{x} - C\mathbf{x} > 0$, i.e., $\mathbf{x} > C\mathbf{x}$.

(II) Step 1: Since both $\mathbf{x}^*$ and C are ≥ 0, so is $C\mathbf{x}^*$. Thus $\mathbf{x}^* > C\mathbf{x}^* \geq 0$.

Step 2: Since $\mathbf{x}^* > C\mathbf{x}^*$, $\mathbf{x}^* - C\mathbf{x}^* > 0$. Let ε be the smallest element in $\mathbf{x}^* - C\mathbf{x}^*$, and M the largest element in $\mathbf{x}^*$. Then $\mathbf{x}^* - C\mathbf{x}^* > \frac{\varepsilon}{2M}\mathbf{x}^* > 0$, i.e., $\mathbf{x}^* - \frac{\varepsilon}{2M}\mathbf{x}^* > C\mathbf{x}^*$. Setting $\lambda = 1 - \frac{\varepsilon}{2M} < 1$, we get $C\mathbf{x}^* < \lambda\mathbf{x}^*$.

Step 3: First, we show that if $\mathbf{x} > \mathbf{y}$, then $C\mathbf{x} > C\mathbf{y}$. But this is clear since $(Cx)_i = \sum_{j=1}^{n} c_{ij}x_j > \sum_{j=1}^{n} c_{ij}y_j = (Cy)_i$. Now we prove Step 3 by induction on n, the case $n = 1$ having been done in Step 2. Assuming the result for $n - 1$, then $C^{n-1}\mathbf{x}^* < \lambda^{n-1}\mathbf{x}^*$. But then $C^n\mathbf{x}^* = C(C^{n-1}\mathbf{x}^*) < C(\lambda^{n-1}\mathbf{x}^*) = \lambda^{n-1}(C\mathbf{x}^*) < \lambda^{n-1}(\lambda\mathbf{x}^*) = \lambda^n\mathbf{x}^*$, proving Step 3.

Step 4: Clearly, $C^n\mathbf{x}^* \geq 0$ for all n. So we have

$$0 \leq \lim_{n\to\infty} C^n \mathrm{x}^* \leq \lim_{n\to\infty} \lambda^n \mathrm{x}^* = 0, \text{ i.e., } \lim_{n\to\infty} C^n \mathrm{x}^* = 0.$$

Denote the elements of $\lim_{n\to\infty} C^n$ by $\bar{c}_{ij}$. Then we have $0 = \sum_{j=1}^{n} \bar{c}_{ij}x_j^*$ for all i. But $\bar{c}_{ij} \geq 0$ and $\mathbf{x}^* < 0$ imply $\bar{c}_{ij} = 0$ for all i and j, proving Step 4.

Step 5: By induction on n, the case $n = 1$ is trivial. Assume the result true for $n - 1$. Then

$$(I - C)(I + C + C^2 + \cdots + C^{n-1}) = (I - C)$$

$$(I + C + \cdots + C^{n-2}) + (I - C)C^{n-1} = (I - C^{n-1}) + (I - C)C^{n-1}$$

$$= I - C^{n-1} + C^{n-1} - C^n,$$

$$= I - C^n,$$

proving Step 5.

Step 6: First we show $(I - C)^{-1}$ exists. If not, then there would be a nonzero vector z such that $C\mathbf{z} = \mathbf{z}$. But then $C^n\mathbf{z} = \mathbf{z}$ for all n, so $\mathbf{z} = \lim_{n \to \infty} C^n\mathbf{z} = 0$, a contradiction, thus $I - C$ is invertible. Thus, $I + C + \cdots + C^{n-1} = (I - C)^{-1}(I - C^n)$, so $S = \lim_{n \to \infty} (I - C)^{-1}(I - C^n) = (I - C)^{-1} (I - \lim_{n \to \infty} C^n) = (I - C)^{-1}$, proving Step 6.

Step 7: Since S is the (infinite) sum of nonnegative matrices, S itself must be nonnegative.

Step 8: We have shown in Steps 6 and 7 that $(I - C)^{-1}$ exists and is nonnegative, thus C is productive.

EXERCISE SET 11.10

1. Using Eq. (18), we calculate

$$Yld_2 = \frac{30s}{2} = 15s$$

$$Yld_3 = \frac{50s}{2 + \frac{3}{2}} = \frac{100s}{7}.$$

So all the trees in the second class should be harvested for an optimal yield (since $s = 1000$) of \$15,000.

3. Assume $p_2 = 1$, then $Yld_2 = \dfrac{s}{(.28)^{-1}} = .28s$. Thus, for all the yields to be the same we must have

$$p_3 s/(.28^{-1} + .31^{-1}) = .28s$$

$$p_4 s/(.28^{-1} + .31^{-1} + .25^{-1}) = .28s$$

$$p_5 s/(.28^{-1} + .31^{-1} + .25^{-1} + .23^{-1}) = .28s$$

$$p_6 s/(.28^{-1} + .31^{-1} + .25^{-1} + .23^{-1} + .27^{-1}) = .28s$$

Solving these sucessively yields $p_3 = 1.90$, $p_4 = 3.02$, $p_5 = 4.24$ and $p_6 = 5.00$. Thus the ratio

$$p_2 : p_3 : p_4 : p_5 : p_6 = 1 : 1.90 : 3.02 : 4.24 : 5.00\ .$$

5. Since $\mathbf{y}$ is the harvest vector, $N = \sum_{i=1}^{n} y_i$ is the number of trees removed from the forest. Then Eq. (7) and the first of Eqs. (8) yield $N = g_1 x_1$, and from Eq. (17) we obtain

$$N = \frac{g_1 s}{1 + \dfrac{g_1}{g_2} + \cdots + \dfrac{g_1}{g_{k-1}}} = \frac{s}{\dfrac{1}{g_1} + \cdots + \dfrac{1}{g_{k-1}}}\ .$$

EXERCISE SET 11.11

1. **(a)** Using the coordinates of the points as the columns of a matrix we obtain

$$\begin{bmatrix} 0 & 1 & 1 & 0 \\ 0 & 0 & 1 & 1 \\ 0 & 0 & 0 & 0 \end{bmatrix}.$$

(b) The scaling is accomplished by multiplication of the coordinate matrix on the left by

$$\begin{bmatrix} \frac{3}{2} & 0 & 0 \\ 0 & \frac{1}{2} & 0 \\ 0 & 0 & 1 \end{bmatrix},$$

resulting in the matrix

$$\begin{bmatrix} 0 & \frac{3}{2} & \frac{3}{2} & 0 \\ 0 & 0 & \frac{1}{2} & \frac{1}{2} \\ 0 & 0 & 0 & 0 \end{bmatrix},$$

which represents the vertices (0, 0, 0), $\left(\frac{3}{2}, 0, 0\right)$, $\left(\frac{3}{2}, \frac{1}{2}, 0\right)$ and $\left(0, \frac{1}{2}, 0\right)$ as shown below.

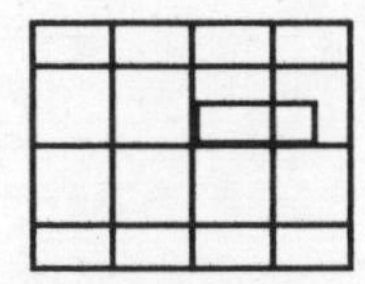

1. (c) Adding the matrix

$$\begin{bmatrix} -2 & -2 & -2 & -2 \\ -1 & -1 & -1 & -1 \\ 3 & 3 & 3 & 3 \end{bmatrix}$$

to the original matrix yields

$$\begin{bmatrix} -2 & -1 & -1 & -2 \\ -1 & -1 & 0 & 0 \\ 3 & 3 & 3 & 3 \end{bmatrix},$$

which represents the vertices (–2, –1, 3), (–1, –1, 3), (–1, 0, 3), and (–2, 0, 3) as shown below.

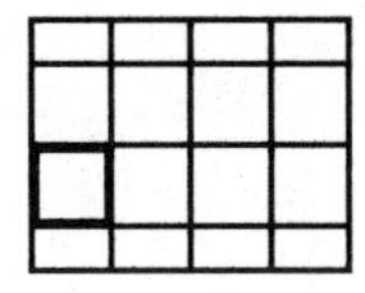

(d) Multiplying by the matrix

$$\begin{bmatrix} \cos(-30^\circ) & -\sin(-30^\circ) & 0 \\ \sin(-30^\circ) & \cos(-30^\circ) & 0 \\ 0 & 0 & 1 \end{bmatrix},$$

we obtain

$$\begin{bmatrix} 0 & \cos(-30^\circ) & \cos(-30^\circ)-\sin(-30^\circ) & -\sin(-30^\circ) \\ 0 & \sin(-30^\circ) & \cos(-30^\circ)+\sin(-30^\circ) & \cos(-30^\circ) \\ 0 & 0 & 0 & 0 \end{bmatrix} =$$

$$\begin{bmatrix} 0 & .866 & 1.366 & .500 \\ 0 & -.500 & .366 & .866 \\ 0 & 0 & 0 & 0 \end{bmatrix}.$$

The vertices are then (0, 0, 0), (.866, – .500, 0), (1.366, .366, 0), and (.500, .866, 0) as shown:

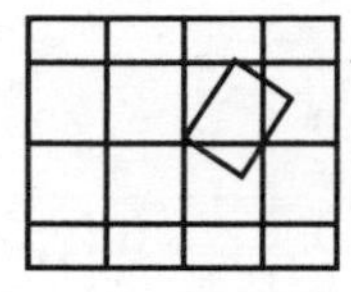

3. **(a)** This transformation looks like scaling by the factors 1, –1, 1, respectively and indeed its matrix is

$$\begin{bmatrix} 1 & 0 & 0 \\ 0 & -1 & 0 \\ 0 & 0 & 1 \end{bmatrix}.$$

(b) For this reflection we want to transform (x_i, y_i, z_i) to $(-x_i, y_i, z_i)$ with the matrix

$$\begin{bmatrix} -1 & 0 & 0 \\ 0 & 1 & 0 \\ 0 & 0 & 1 \end{bmatrix}.$$

Negating the x-coordinates of the 12 points in view 1 yields the 12 points (–1.000, –.800, .000), (–.500, –.800, –.866), etc., as shown:

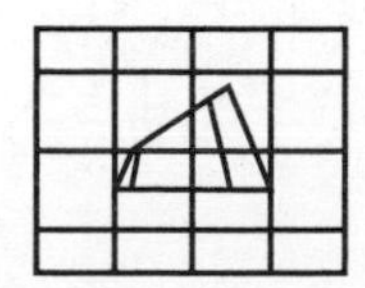

(c) Here we want to negate the z-coordinates, with the matrix

$$\begin{bmatrix} 1 & 0 & 0 \\ 0 & 1 & 0 \\ 0 & 0 & -1 \end{bmatrix}.$$

This does not change View 1.

5. **(a)** As in 4(a),

$$M_1 = \begin{bmatrix} .3 & 0 & 0 \\ 0 & .5 & 0 \\ 0 & 0 & 1 \end{bmatrix}, \qquad M_2 = \begin{bmatrix} 1 & 0 & 0 \\ 0 & \cos 45^\circ & -\sin 45^\circ \\ 0 & \sin 45^\circ & \cos 45^\circ \end{bmatrix},$$

$$M_3 = \begin{bmatrix} 1 & 1 & 1 & \cdots & 1 \\ 0 & 0 & 0 & \cdots & 0 \\ 0 & 0 & 0 & \cdots & 0 \end{bmatrix}, \quad M_4 = \begin{bmatrix} \cos 35^\circ & 0 & \sin 35^\circ \\ 0 & 1 & 0 \\ -\sin 35^\circ & 0 & \cos 45^\circ \end{bmatrix},$$

$$M_5 = \begin{bmatrix} \cos(-45^\circ) & -\sin(-45^\circ) & 0 \\ \sin(-45^\circ) & \cos(-45^\circ) & 0 \\ 0 & 0 & 1 \end{bmatrix},$$

$$M_6 = \begin{bmatrix} 0 & 0 & 0 & \cdots & 0 \\ 0 & 0 & 0 & \cdots & 0 \\ 1 & 1 & 1 & \cdots & 1 \end{bmatrix}, \quad \text{and} \quad M_7 = \begin{bmatrix} 2 & 0 & 0 \\ 0 & 1 & 0 \\ 0 & 0 & 1 \end{bmatrix}.$$

(b) As in 4(*b*), $P' = M_7(M_6 + M_5M_4(M_3 + M_2\,M_1P))$.

7. **(a)** We rewrite the formula for v_i' as

$$v_i' = \begin{bmatrix} 1 \cdot x_i + x_0 \cdot 1 \\ 1 \cdot y_i + y_0 \cdot 1 \\ 1 \cdot z_i + z_0 \cdot 1 \\ 1 \cdot 1 \end{bmatrix}.$$

So

$$v_i' = \begin{bmatrix} 1 & 0 & 0 & x_0 \\ 0 & 1 & 0 & y_0 \\ 0 & 0 & 1 & z_0 \\ 0 & 0 & 0 & 1 \end{bmatrix} \begin{bmatrix} x_i \\ y_i \\ z_i \\ 1 \end{bmatrix}.$$

(b) We want to translate x_i by –5, y_i by +9, z_i by –3, so $x_0 = -5$, $y_0 = +9$, $z_0 = -3$. The matrix is

$$\begin{bmatrix} 1 & 0 & 0 & -5 \\ 0 & 1 & 0 & 9 \\ 0 & 0 & 1 & -3 \\ 0 & 0 & 0 & 1 \end{bmatrix}.$$

EXERCISE SET 11.12

1. **(a)** The discrete mean value property yields the four equations

$$t_1 = \frac{1}{4}(t_2 + t_3)$$

$$t_2 = \frac{1}{4}(t_1 + t_4 + 1 + 1)$$

$$t_3 = \frac{1}{4}(t_1 + t_4)$$

$$t_4 = \frac{1}{4}(t_2 + t_3 + 1 + 1).$$

Translated into matrix notation, this becomes

$$\begin{bmatrix} t_1 \\ t_2 \\ t_3 \\ t_4 \end{bmatrix} = \begin{bmatrix} 0 & \frac{1}{4} & \frac{1}{4} & 0 \\ \frac{1}{4} & 0 & 0 & \frac{1}{4} \\ \frac{1}{4} & 0 & 0 & \frac{1}{4} \\ 0 & \frac{1}{4} & \frac{1}{4} & 0 \end{bmatrix} \begin{bmatrix} t_1 \\ t_2 \\ t_3 \\ t_4 \end{bmatrix} + \begin{bmatrix} 0 \\ \frac{1}{2} \\ 0 \\ \frac{1}{2} \end{bmatrix}.$$

1. **(b)** To solve the system in part (a), we solve $(I - M)\mathbf{t} = \mathbf{b}$ for $\mathbf{t}$:

$$\begin{bmatrix} 0 & -\frac{1}{4} & -\frac{1}{4} & 0 \\ -\frac{1}{4} & 1 & 0 & -\frac{1}{4} \\ -\frac{1}{4} & 0 & 1 & -\frac{1}{4} \\ 0 & -\frac{1}{4} & -\frac{1}{4} & 1 \end{bmatrix} \begin{bmatrix} t_1 \\ t_2 \\ t_3 \\ t_4 \end{bmatrix} = \begin{bmatrix} 0 \\ \frac{1}{2} \\ 0 \\ \frac{1}{2} \end{bmatrix}.$$

In row-echelon form, this is

$$\begin{bmatrix} 1 & -\frac{1}{4} & -\frac{1}{4} & 0 \\ 0 & 1 & -15 & 4 \\ 0 & 0 & 1 & -\frac{1}{2} \\ 0 & 0 & 0 & 1 \end{bmatrix} \begin{bmatrix} t_1 \\ t_2 \\ t_3 \\ t_4 \end{bmatrix} = \begin{bmatrix} 0 \\ 0 \\ -1/8 \\ 3/4 \end{bmatrix}.$$

Back substitution yields the result $\mathbf{t} = [1/4 \quad 3/4 \quad 1/4 \quad 3/4]^T$.

(c)

$$\mathbf{t}^{(1)} = M\mathbf{t}^{(0)} + \mathbf{b} = \begin{bmatrix} 0 & \frac{1}{4} & \frac{1}{4} & 0 \\ \frac{1}{4} & 0 & 0 & \frac{1}{4} \\ \frac{1}{4} & 0 & 0 & \frac{1}{4} \\ 0 & \frac{1}{4} & \frac{1}{4} & 0 \end{bmatrix} \begin{bmatrix} 0 \\ 0 \\ 0 \\ 0 \end{bmatrix} + \begin{bmatrix} 0 \\ \frac{1}{2} \\ 0 \\ \frac{1}{2} \end{bmatrix} = \begin{bmatrix} 0 \\ \frac{1}{2} \\ 0 \\ \frac{1}{2} \end{bmatrix}$$

$$\mathbf{t}^{(2)} = M\mathbf{t}^{(1)} + \mathbf{b} = [1/8 \quad 5/8 \quad 1/8 \quad 5/8]^T$$

$$\mathbf{t}^{(3)} = M\mathbf{t}^{(2)} + \mathbf{b} = [3/16 \quad 11/16 \quad 3/16 \quad 11/16]^T$$

$$\mathbf{t}^{(4)} = M\mathbf{t}^{(3)} + \mathbf{b} = [7/32 \quad 23/32 \quad 7/32 \quad 23/32]^T$$

$$\mathbf{t}^{(5)} = M\mathbf{t}^{(4)} + \mathbf{b} = [15/64 \quad 47/64 \quad 15/64 \quad 47/64]^T$$

from Eq. (10).

(d) Using percentage error $= \dfrac{\text{computed value} - \text{actual value}}{\text{actual value}} \times 100\%$ we have that the percentage error for t_1 and t_3 was $\dfrac{.0129}{.2371} \times 100\% = 5.4\%$, and for t_2 and t_4 was $\dfrac{-.0129}{.7629} \times 100\% = -1.7\%$.

3. As in 1(*c*), but using M and $\mathbf{b}$ as in the problem statement, we obtain

$$\mathbf{t}^{(1)} = M\mathbf{t}^{(0)} + \mathbf{b} = [3/4 \quad 5/4 \quad 1/2 \quad 5/4 \quad 1 \quad 1/2 \quad 5/4 \quad 1 \quad 3/4]^T$$

$$\mathbf{t}^{(2)} = M\mathbf{t}^{(1)} + \mathbf{b} = [13/16 \quad 9/8 \quad 9/16 \quad 11/8 \quad 13/16 \quad 7/16 \quad 21/16 \quad 15/8]^T.$$

EXERCISE SET 11.13

1. **(a)** Using the lines in Eq. (5) we have

$$\mathbf{a}_1 = \begin{bmatrix} 1 \\ 1 \end{bmatrix} \qquad \mathbf{a}_2 = \begin{bmatrix} 1 \\ -2 \end{bmatrix} \qquad \mathbf{a}_3 = \begin{bmatrix} 3 \\ -1 \end{bmatrix}$$

$$\mathbf{a}_1^T\mathbf{a}_1 = 2 \qquad \mathbf{a}_2^T\mathbf{a}_2 = 5 \qquad \mathbf{a}_3^T\mathbf{a}_3 = 10$$

$$b_1 = 2 \qquad b_2 = -2 \qquad b_3 = 3$$

Setting $\mathbf{x}_k^{(p)} = \begin{bmatrix} x_{k1}^{(p)} \\ x_{k2}^{(p)} \end{bmatrix}$ we have

$$\mathbf{a}_1^T\mathbf{x}_0^{(p)} = x_{01}^{(p)} + x_{02}^{(p)}$$

$$\mathbf{a}_2^T\mathbf{x}_1^{(p)} = x_{11}^{(p)} - 2x_{12}^{(p)}$$

$$\mathbf{a}_3^T\mathbf{x}_2^{(p)} = 3x_{21}^{(p)} - x_{22}^{(p)}$$

$$\frac{b_1 - \mathbf{a}_1^T\mathbf{x}_0^{(p)}}{\mathbf{a}_1^T\mathbf{a}_1} = \frac{2 - x_{01}^{(p)} - x_{02}^{(p)}}{2}$$

$$\frac{b_2 - \mathbf{a}_2^T\mathbf{x}_1^{(p)}}{\mathbf{a}_2^T\mathbf{a}_2} = \frac{-2 - x_{11}^{(p)} - x_{12}^{(p)}}{5}$$

$$\frac{b_3 - \mathbf{a}_3^T\mathbf{x}_2^{(p)}}{\mathbf{a}_3^T\mathbf{a}_3} = \frac{3 - 3x_{21}^{(p)} - x_{22}^{(p)}}{10}.$$

Writing $\mathbf{x}_k^{(p)} = \mathbf{x}_{k-1}^{(p)} + \left(\dfrac{b_k - \mathbf{a}_k^T\mathbf{x}_{k-1}^{(p)}}{\mathbf{a}_k^T\mathbf{a}_k} \right) \mathbf{a}_k$

in the form:

$$\begin{bmatrix} x_{k1}^{(p)} \\ x_{k2}^{(p)} \end{bmatrix} = \begin{bmatrix} x_{k-1,1}^{(p)} \\ x_{k-1,2}^{(p)} \end{bmatrix} + \left(\frac{b_k - \mathbf{a}_k^T \mathbf{x}_{k-1}^{(p)}}{\mathbf{a}_k^T \mathbf{a}_k} \right) \begin{bmatrix} a_{k1} \\ a_{k2} \end{bmatrix}$$

we have

$$x_{k1}^{(p)} = x_{k-1,1}^{(p)} + \left(\frac{b_k - \mathbf{a}_k^T \mathbf{x}_{k-1}^{(p)}}{\mathbf{a}_k^T \mathbf{a}_k} \right) a_{k1}$$

$$x_{k2}^{(p)} = x_{k-1,2}^{(p)} + \left(\frac{b_k - \mathbf{a}_k^T \mathbf{x}_{k-1}^{(p)}}{\mathbf{a}_k^T \mathbf{a}_k} \right) a_{k2}.$$

Substituting for the expressions $\left(\frac{b_k - \mathbf{a}_k^T \mathbf{x}_{k-1}^{(p)}}{\mathbf{a}_k^T \mathbf{a}_k} \right)$ gives for $k = 1$:

$$x_{11}^{(p)} = x_{01}^{(p)} + \frac{2 - x_{01}^{(p)} - x_{02}^{(p)}}{2} \bullet 1 = \frac{1}{2}\left[2 + x_{01}^{(p)} - x_{02}^{(p)} \right]$$

for $k = 2$:

$$x_{12}^{(p)} = x_{02}^{(p)} + \frac{2 - x_{01}^{(p)} - x_{02}^{(p)}}{2} \bullet 1 = \frac{1}{2}\left[2 + x_{01}^{(p)} - x_{02}^{(p)} \right]$$

$$x_{21}^{(p)} = x_{11}^{(p)} + \frac{-2 - x_{11}^{(p)} + 2x_{12}^{(p)}}{5} \bullet 1 = \frac{1}{5}\left[-2 + 4x_{11}^{(p)} + 2x_{12}^{(p)} \right]$$

$$x_{22}^{(p)} = x_{12}^{(p)} + \frac{-2 - x_{11}^{(p)} + 2x_{12}^{(p)}}{5} \bullet (-2) = \frac{1}{5}\left[4 + 2x_{11}^{(p)} + 2x_{12}^{(p)} \right]$$

for $k = 3$:

$$x_{31}^{(p)} = x_{21}^{(p)} + \frac{3 - 3x_{21}^{(p)} + x_{22}^{(p)}}{10} \bullet 3 = \frac{1}{10}\left[9 + x_{21}^{(p)} + 3x_{22}^{(p)} \right]$$

$$x_{32}^{(p)} = x_{22}^{(p)} + \frac{3 - 3x_{21}^{(p)} + x_{22}^{(p)}}{10} \bullet (-1) = \frac{1}{10}\left[-3 + 3x_{21}^{(p)} + 9x_{22}^{(p)} \right]$$

where $x_0^{(p+1)} = x_3^{(p)}$, which means

$$\left(x_{01}^{(p+1)}, x_{02}^{(p+1)} \right) = \left(x_{31}^{(p)}, x_{32}^{(p)} \right).$$

1. (b) From part (a) we have

$$x_{31}^{(p)} = \frac{1}{10}\left[\, 9 + x_{21}^{(p)} + 3x_{22}^{(p)} \,\right].$$

Subsituting the expressions for $x_{21}^{(p)}$ and $x_{22}^{(p)}$ from part (a) gives

$$\begin{aligned} x_{31}^{(p)} &= \frac{1}{10}\left[\, 9 + \frac{1}{5}\left(-2 + 4x_{11}^{(p)} + 2x_{12}^{(p)}\right) + \frac{3}{5}\left(4 + 2x_{11}^{(p)} + x_{12}^{(p)}\right)\right] \\ &= \frac{1}{50}\left[\, 55 + 10x_{11}^{(p)} + 5x_{12}^{(p)} \,\right]. \end{aligned}$$

Subsituting the expressions for $x_{11}^{(p)}$ and $x_{12}^{(p)}$ from part (a) gives

$$\begin{aligned} x_{31}^{(p)} &= \frac{1}{50}\left[\, 55 + \frac{10}{2}\left(2 + x_{01}^{(p)} - x_{02}^{(p)}\right) + \frac{5}{2}\left(2 - x_{01}^{(p)} + x_{02}^{(p)}\right)\right] \\ &= \frac{1}{20}\left[\, 28 + x_{01}^{(p)} - x_{02}^{(p)} \,\right]. \end{aligned}$$

From part (a) we have

$$x_{32}^{(p)} = \frac{1}{10}\left[\, -3 + 3x_{21}^{(p)} + 9x_{22}^{(p)} \,\right].$$

Subsituting for $x_{21}^{(p)}$ and $x_{22}^{(p)}$ from part (a) gives

$$\begin{aligned} x_{32}^{(p)} &= \frac{1}{10}\left[\, -3 + \frac{3}{5}\left(-2 + 4x_{11}^{(p)} + 2x_{12}^{(p)}\right) + \frac{9}{5}\left(4 + 2x_{11}^{(p)} + x_{12}^{(p)}\right)\right] \\ &= \frac{3}{10}\left[\, 1 + 2x_{11}^{(p)} + x_{12}^{(p)} \,\right]. \end{aligned}$$

Subsituting for $x_{11}^{(p)}$ and $x_{12}^{(p)}$ from part (a) gives

$$\begin{aligned} x_{32}^{(p)} &= \frac{3}{10}\left[\, 1 + \frac{2}{2}\left(2 + x_{01}^{(p)} - x_{02}^{(p)}\right) + \frac{1}{2}\left(2 - x_{01}^{(p)} + x_{02}^{(p)}\right)\right] \\ &= \frac{1}{20}\left[\, 24 + 3x_{01}^{(p)} - 3x_{02}^{(p)} \,\right]. \end{aligned}$$

Now $x_0^{(p)} = x_3^{(p-1)}$ or $\left(x_{01}^{(p)},\, x_{02}^{(p)}\right) = \left(x_{31}^{(p-1)},\, x_{32}^{(p-1)}\right)$,

and so

$$x_{31}^{(p)} = \frac{1}{20}\left[28 + x_{31}^{(p-1)} - x_{32}^{(p-1)}\right]$$

$$x_{32}^{(p)} = \frac{1}{20}\left[24 + 3x_{31}^{(p-1)} - 3x_{02}^{(p-1)}\right].$$

1. **(c)** The linear system

$$x_{31}^{*} = \frac{1}{20}\left[28 + x_{31}^{*} - x_{32}^{*}\right]$$

$$x_{32}^{*} = \frac{1}{20}\left[24 + 3x_{31}^{*} - 3x_{32}^{*}\right]$$

can be rewritten as

$$19x_{31}^{*} + x_{32}^{*} = 28$$

$$-3x_{31}^{*} + 23x_{32}^{*} = 24,$$

which has the solution

$$x_{31}^{*} = 31/22$$

$$x_{32}^{*} = 27/22.$$

3. **(a)** The three lines are

$$L_1 : x_1 + x_2 = 2$$

$$L_2 : x_1 - 2x_2 = -2$$

$$L_3 : 3x_1 - x_2 = 3$$

x_1^{*} lies on L_1 since $\dfrac{12}{11} + \dfrac{10}{11} = \dfrac{22}{11} = 2$

x_2^{*} lies on L_2 since $\dfrac{46}{55} - 2\left(\dfrac{78}{55}\right) = \dfrac{-110}{55} = -2$

x_3^{*} lies on L_3 since $3\left(\dfrac{31}{22}\right) - \dfrac{27}{22} = \dfrac{66}{22} = 3.$

The slope of line L_1 is –1.

The slope of the line $\overline{x_1^* x_3^*}$ is $\dfrac{27/22 - 10/11}{31/22 - 12/11} = \dfrac{7}{7} = 1$.

Since (–1) • 1 = –1, it follows that $\overline{x_1^* x_2^*}$ is perpendicular to L_1.

The slope of line L_2 is $\frac{1}{2}$.

The slope of the line $\overline{x_1^* x_2^*}$ is $\dfrac{78/55 - 10/11}{46/55 - 12/11} = \dfrac{28}{-14} = -2$.

Since $\frac{1}{2}$ • (–2) = –1, it follows that $\overline{x_1^* x_2^*}$ is perpendicular to L_2.

The slope of line L_3 is 3.

The slope of the line $\overline{x_2^* x_3^*}$ is $\dfrac{27/22 - 78/55}{31/22 - 46/55} = \dfrac{-21}{63} = \dfrac{-1}{3}$.

Since $3 \bullet \left(-\dfrac{1}{3}\right) = -1$, it follows that $\overline{x_2^* x_3^*}$ is perpendicular to L_3.

(b) Substituting $\mathbf{x}_0^{(1)} = \left(\dfrac{31}{22}, \dfrac{27}{22}\right)$ in the equations for $x_1^{(1)}$, part (a) of Exercise 1 gives

$$
\begin{aligned}
x_{11}^{(1)} &= \frac{1}{2}\left[2 + \frac{31}{22} - \frac{27}{22}\right] = \frac{48}{44} = \frac{12}{11} \\
x_{12}^{(1)} &= \frac{1}{2}\left[2 - \frac{31}{22} + \frac{27}{22}\right] = \frac{40}{44} = \frac{10}{11}
\end{aligned}
$$

Then $\mathbf{x}_1^{(1)} = \left(\dfrac{12}{11}, \dfrac{10}{11}\right)$ gives

$$
\begin{aligned}
x_{21}^{(1)} &= \frac{1}{5}\left[-2 + \left(4 \bullet \frac{12}{11}\right) + \left(2 \bullet \frac{10}{11}\right)\right] = \frac{46}{55} \\
x_{22}^{(1)} &= \frac{1}{5}\left[4 + \left(2 \bullet \frac{12}{11}\right) + \frac{10}{11}\right] = \frac{78}{55}
\end{aligned}
$$

and $\mathbf{x}_2^{(1)} = \left(\frac{46}{55}, \frac{78}{55}\right)$ gives

$$x_{31}^{(1)} = \frac{1}{10}\left[9 + \frac{46}{55} + \left(3 \cdot \frac{78}{55}\right)\right] = \frac{775}{550} = \frac{31}{22}$$

$$x_{32}^{(1)} = \frac{1}{10}\left[-3 + \left(3 \cdot \frac{46}{55}\right) + \left(9 \cdot \frac{78}{55}\right)\right] = \frac{675}{550} = \frac{27}{22}$$

or $\mathbf{x}_3^{(1)} = \left(\frac{31}{22}, \frac{27}{22}\right)$.

5. **(a)** From Theorem 1:

$$\mathbf{x}_p = \mathbf{x}^* + \left(\frac{b - \mathbf{a}^T\mathbf{x}^*}{\mathbf{a}^T\mathbf{a}}\right)\mathbf{a}$$

and so

$$\mathbf{a}^T\mathbf{x}_p = \mathbf{a}^T\mathbf{x}^* + \left(\frac{b - \mathbf{a}^T\mathbf{x}^*}{\mathbf{a}^T\mathbf{a}}\right)\mathbf{a}^T\mathbf{a}$$

$$= \mathbf{a}^T\mathbf{x}^* + b - \mathbf{a}^T\mathbf{x}^* = b.$$

Therefore $\mathbf{x}_p$ lies on the line $\mathbf{a}^t\mathbf{x} = b$.

(b) Let $\mathbf{z}$ be a point on the line L, so that $\mathbf{a}^T\mathbf{z} = b$. We will prove that $\mathbf{x}_p - \mathbf{x}^*$ is perpendicular to $\mathbf{x}_p - \mathbf{z}$ by showing that $(\mathbf{x}_p - \mathbf{z})^T(\mathbf{x}_p - \mathbf{x}^*) = 0$ (i.e., their dot product is 0). Now

$$(\mathbf{x}_p - \mathbf{z})^T(\mathbf{x}_p - \mathbf{x}^*) = (\mathbf{x}_p - \mathbf{z})^T\left(\frac{b - \mathbf{a}^T\mathbf{x}^*}{\mathbf{a}^T\mathbf{a}}\right)\mathbf{a}$$

$$= \left(\frac{b - \mathbf{a}^T\mathbf{x}^*}{\mathbf{a}^T\mathbf{a}}\right)(\mathbf{x}_p - \mathbf{z})^T\mathbf{a}$$

Since $\mathbf{x}_p$ and $\mathbf{z}$ are on L, we have

$$\mathbf{a}^T\mathbf{x}_p = b$$

$$\mathbf{a}^T\mathbf{z} = b,$$

and so $\mathbf{a}^T(\mathbf{x}_p - \mathbf{z}) = 0$ or $(\mathbf{x}_p - \mathbf{z})^T\mathbf{a} = 0$.

Thus $(\mathbf{x}_p - \mathbf{z})^T(\mathbf{x}_p - \mathbf{x}^*) = 0$, and so $\mathbf{x}_p - \mathbf{x}^*$ is perpendicular to L.

7. Let us choose units so that each pixel is one unit wide. Then

a_{ij} = *length of the center line of the i-th beam that lies in the j-th pixel*

If the i-th beam crosses the j-th pixel squarely, it follows that $a_{ij} = 1$. From Fig. 11.13.11 in the text, it is then clear that

$$a_{17} = a_{18} = a_{19} = 1$$

$$a_{24} = a_{25} = a_{26} = 1$$

$$a_{31} = a_{32} = a_{33} = 1$$

$$a_{73} = a_{76} = a_{79} = 1$$

$$a_{82} = a_{85} = a_{88} = 1$$

$$a_{91} = a_{94} = a_{97} = 1$$

since beams 1, 2, 3, 7, 8, and 9 cross the pixels squarely.

Next, the centerlines of beams 5 and 11 lie along the diagonals of pixels 3, 5, 7 and 1, 5, 9, respectively. Since these diagonals have length $\sqrt{2}$, we have

$$a_{53} = a_{55} = a_{57} = \sqrt{2} = 1.41421$$

$$a_{11,1} = a_{11,5} = a_{11,9} = \sqrt{2} = 1.41421.$$

In the following diagram, the hypotenuse of triangle A is the portion of the centerline of the 10th beam that lies in the 2nd pixel. The length of this hypotenuse is twice the height of triangle A, which in turn is $\sqrt{2} - 1$. Thus,

$$a_{10,2} = 2(\sqrt{2} - 1) = .82843.$$

By symmetry we also have

$$a_{10,2} = a_{10,6} = a_{12,4} = a_{12,8}$$

$$= a_{62} = a_{64} = a_{46} = a_{48} = .82843.$$

Also from the diagram, we see that the hypotenuse of triangle B is the portion of the centerline of the 10th beam that lies in the 3rd pixel. Thus,

$$a_{10,3} = 2 - \sqrt{2} = .58579.$$

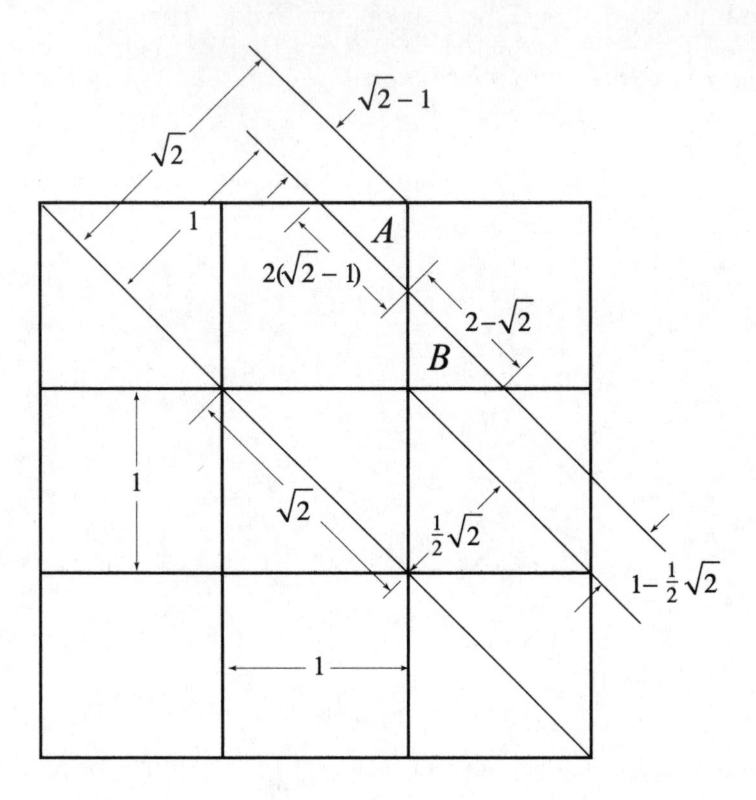

By symmetry we have

$$a_{10,3} = a_{12,7} = a_{61} = a_{49} = .58579.$$

The remaining a_{ij}'s are all zero, and so the 12 beam equations (4) are

$$
\begin{aligned}
x_7 + x_8 + x_9 &= 13.00 \\
x_4 + x_5 + x_6 &= 15.00 \\
x_1 + x_2 + x_3 &= 8.00 \\
.82843\,(x_6 + x_8) + .58579x_9 &= 14.79 \\
1.41421\,(x_3 + x_5 + x_7) &= 14.31 \\
.82843\,(x_2 + x_4) + .58579\,x_1 &= 3.81 \\
x_3 + x_6 + x_9 &= 18.00 \\
x_2 + x_5 + x_8 &= 12.00 \\
x_1 + x_4 + x_7 &= 6.00 \\
.82843\,(x_2 + x_6) + .58579x_3 &= 10.51 \\
1.41421\,(x_1 + x_5 + x_9) &= 16.13 \\
.82843\,(x_4 + x_8) + .58579x_7 &= 7.04
\end{aligned}
$$

EXERCISE SET 11.14

1. Each of the subsets S_1, S_2, S_3, S_4 in the figure is congruent to the entire set scaled by a factor of 12/25. Also, the rotation angles for the four subsets are all 0°. The displacement distances can be determined from the figure to find the four similitudes that map the entire set onto the four subsets S_1, S_2, S_3, S_4. These are, respectively,

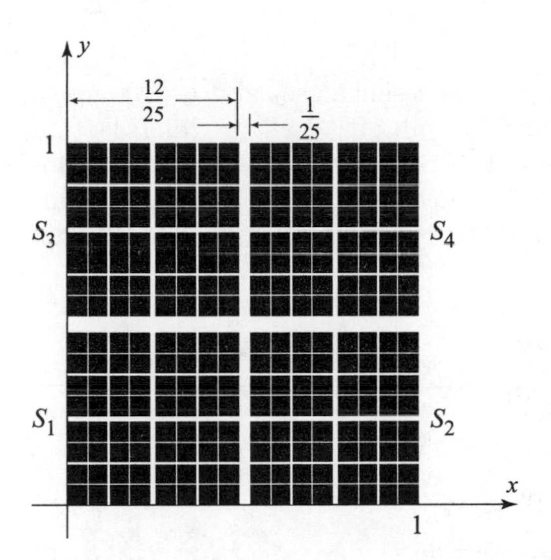

$$T_i\left(\begin{bmatrix} x \\ y \end{bmatrix}\right) = \frac{12}{25}\begin{bmatrix} 1 & 0 \\ 0 & 1 \end{bmatrix}\begin{bmatrix} x \\ y \end{bmatrix} + \begin{bmatrix} e_i \\ f_i \end{bmatrix},$$ $i = 1, 2, 3, 4$, where the four values of

$\begin{bmatrix} e_i \\ f_i \end{bmatrix}$ are $\begin{bmatrix} 0 \\ 0 \end{bmatrix}$, $\begin{bmatrix} 13/25 \\ 0 \end{bmatrix}$, $\begin{bmatrix} 0 \\ 13/25 \end{bmatrix}$, and $\begin{bmatrix} 13/25 \\ 13/25 \end{bmatrix}$.

Because $s = 12/25$ and $k = 4$ in the definition of a self-similar set, the Hausdorff dimension of the set is $d_H(S) = \ln(k)/\ln(1/s) = \ln(4)/\ln(25/12) = 1.889\ldots$. The set is a fractal because its Hausdorff dimension is not an integer.

3. **(a)** The figure shows the original self-similar set and a decomposition of the set into seven nonoverlapping congruent subsets, each of which is congruent to the original set scaled by a factor $s = 1/3$. By inspection, the rotations angles are 0° for all seven subsets. The Hausdorff dimension of the set is $d_H(S) = \ln(k)/\ln(1/s) = \ln(7)/\ln(3) = 1.771 \ldots$. Because its Hausdorff dimension is not an integer, the set is a fractal.

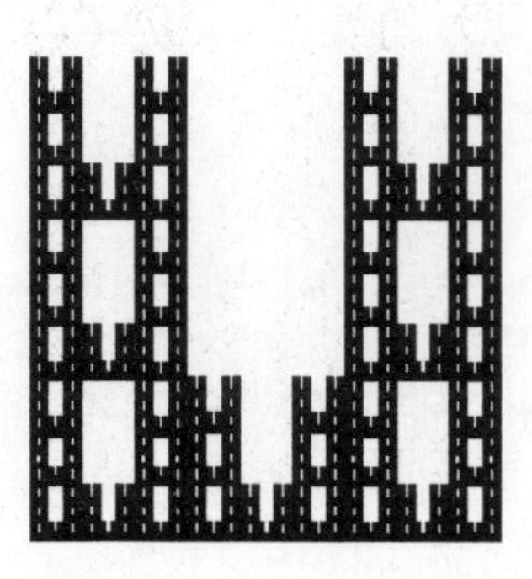

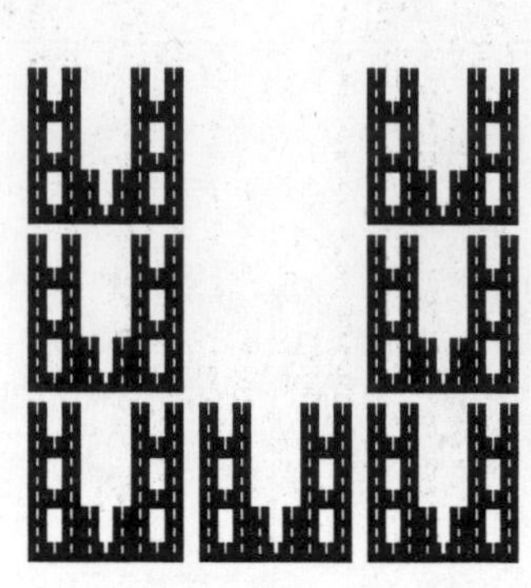

(b) The figure shows the original self-similar set and a decomposition of the set into three nonoverlapping congruent subsets, each of which is congruent to the original set scaled by a factor $s = 1/2$. By inspection, the rotation angles are 180° for all three subsets. The Hausdorff dimension of the set is $d_H(S) = \ln(k)/\ln(1/s) = \ln(3)/\ln(2) = 1.584 \ldots$. Because its Hausdorff dimension is not an integer, the set is a fractal.

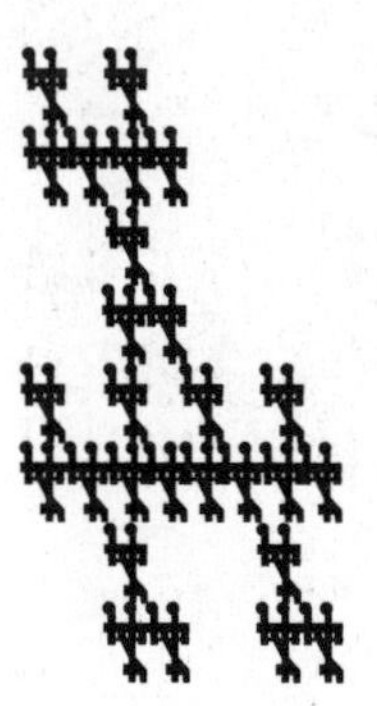

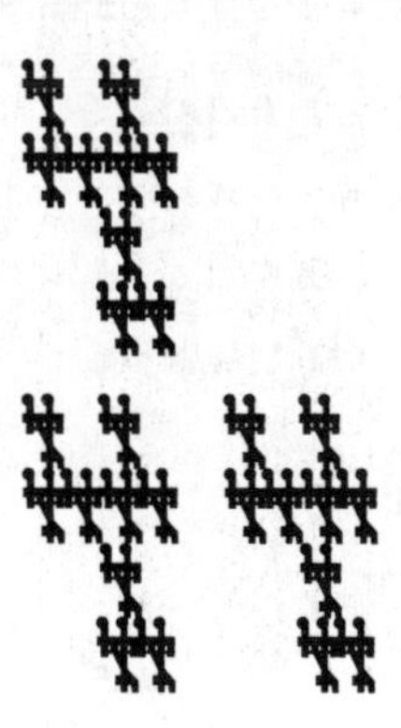

3. **(c)** The figure shows the original self-similar set and a decomposition of the set into three nonoverlapping congruent subsets, each of which is congruent to the original set scaled by a factor s = 1/2. By inspection, the rotation angles are 180°, 180°, and –90° for S_1, S_2, and S_3, respectively. The Hausdorff dimension of the set is $d_H(s) = \ln(k)/\ln(1/s) = \ln(3)/\ln(2) = 1.584 \ldots$. Because its Hausdorff dimension is not an integer, the set is a fractal.

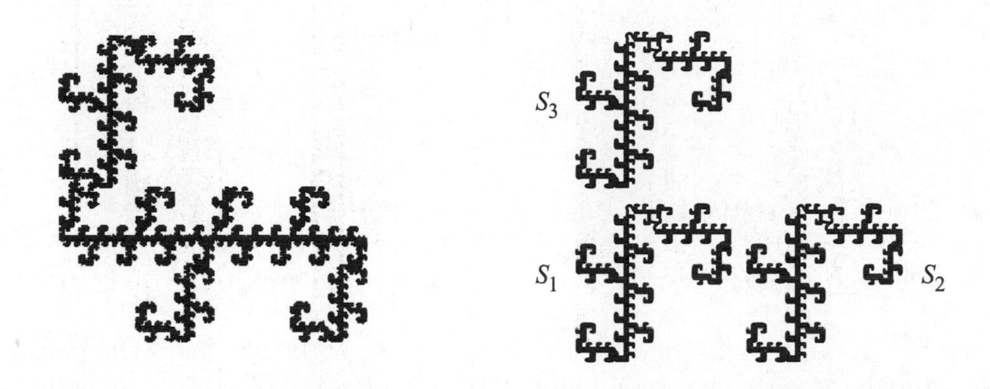

(d) The figure shows the original self-similar set and a decomposition of the set into three nonoverlapping congruent subsets, each of which is congruent to the original set scaled by a factor s = 1/2. By inspection, the rotation angles are 180°, 180°, and –90° for S_1, S_2, and S_3, respectively. The Hausdorff dimension of the set is $d_H(S) = \ln(k)/\ln(1/s) = \ln(3)/\ln(2) = 1.584 \ldots$. Because its Hausdorff dimension is not an integer, the set is a fractal.

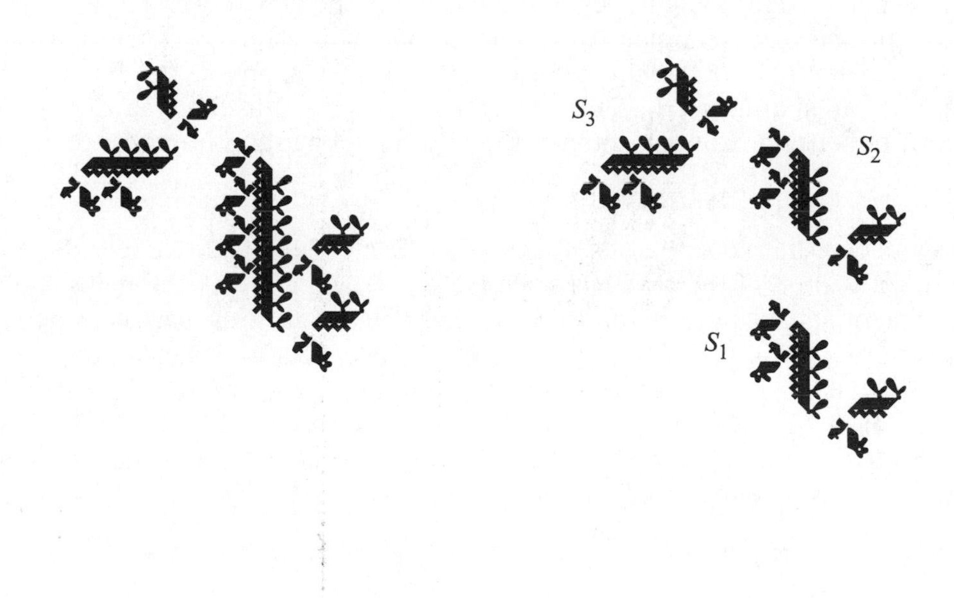

5. Letting $\begin{bmatrix} x \\ y \end{bmatrix}$ be the vector to the tip of the fern and using the hint, we have $\begin{bmatrix} x \\ y \end{bmatrix} =$

$T_2\left(\begin{bmatrix} x \\ y \end{bmatrix}\right)$ or $\begin{bmatrix} x \\ y \end{bmatrix} = \begin{bmatrix} .85 & .04 \\ -.04 & .85 \end{bmatrix}\begin{bmatrix} x \\ y \end{bmatrix} + \begin{bmatrix} .075 \\ .180 \end{bmatrix}$. Solving this matrix equation gives

$\begin{bmatrix} x \\ y \end{bmatrix} = \begin{bmatrix} .15 & -.04 \\ .04 & .15 \end{bmatrix}^{-1}\begin{bmatrix} .075 \\ .180 \end{bmatrix} = \begin{bmatrix} .766 \\ .996 \end{bmatrix}$ rounded to three decimal places.

7. The similitude T_1 maps the unit square (whose vertices are (0, 0), (1, 0), (1, 1), and (0, 1)) onto the square whose vertices are (0, 0), (3/4, 0), (3/4, 3/4), and (0, 3/4). The similitude T_2 maps the unit square onto the square whose vertices are (1/4, 0), (1, 0), (1, 3/4), and (1/4, 3/4). The similitude T_3 maps the unit square onto the square whose vertices are (0, 1/4), (3/4, 1/4), (3/4, 1), and (0, 1). Finally, the similitude T_4 maps the unit square onto the square whose vertices are (1/4, 1/4), (1, 1/4), (1, 1), and (1/4, 1). The union of these four smaller squares is the unit square, but the four smaller squares overlap. Each of the four smaller squares has side length of 3/4, so that the common scale factor of the similitudes is $s = 3/4$. The right-hand side of Equation (2) of the text gives $\ln(k)/\ln(1/s) = \ln(4)/\ln(4/3) = 4.818 \ldots$. This is not the correct Hausdorff dimension of the square (which is 2) because the four smaller squares overlap.

9. A careful examination of Figure Ex-9 on page 703 of the text shows that the Menger sponge can be expressed as the union of 20 smaller nonoverlapping congruent Menger sponges each of side length 1/3. Consequently, $k = 20$ and $s = 1/3$, and so the Hausdorff dimension of the Menger sponge is $d_H(S) = \ln(k)/\ln(1/s) = \ln(20)/\ln(3) = 2.726 \ldots$. Because its Hausdorff dimension is not an integer, the Menger sponge is a fractal.

11. The area of the unit square S_0 is, of course, 1. Each of the eight similitudes $T_1, T_2, \ldots, T_8$ given in Equation 8 of the text has scale factor $s = 1/3$, and so each maps the unit square onto a smaller square of area 1/9. Because these eight smaller squares are nonoverlapping, their total area is 8/9, which is then the area of the set S_1. By a similar argument, the area of the set S_2 is 8/9-th the area of the set S_1. Continuing the argument further, we find that the areas of $S_0, S_1, S_2, S_3, S_4, \ldots,$ form the geometric sequence 1, 8/9, $(8/9)^2$, $(8/9)^3$, $(8/9)^4, \ldots$. (Notice that this implies that the area of the Sierpinski carpet is 0, since the limit of $(8/9)^n$ as n tends to infinity is 0.)

EXERCISE SET 11.15

1. Because $250 = 2 \cdot 5^3$ it follows from (i) that $\prod(250) = 3 \cdot 250 = 750$.

Because $25 = 5^2$ it follows from (ii) that $\prod(25) = 2 \cdot 25 = 50$.

Because $125 = 5^3$ it follows from (ii) that $\prod(125) = 2 \cdot 125 = 250$.

Because $30 = 6 \cdot 5$ it follows from (ii) that $\prod(30) = 2 \cdot 30 = 60$.

Because $10 = 2 \cdot 5$ it follows from (i) that $\prod(10) = 3 \cdot 10 = 30$.

Because $50 = 2 \cdot 5^2$ it follows from (i) that $\prod(50) = 3 \cdot 50 = 150$.

Because $3750 = 6 \cdot 5^4$ it follows from (ii) that $\prod(3750) = 2 \cdot 3750 = 7500$.

Because $6 = 6 \cdot 5^0$ it follows from (ii) that $\prod(6) = 2 \cdot 6 = 12$.

Because $5 = 5^1$ it follows from (ii) that $\prod(5) = 2 \cdot 5 = 10$.

3. **(a)** We are given that $x_0 = 3$ and $x_1 = 7$. With $p = 15$ we have

$$\begin{aligned}
x_3 &= x_2 + x_1 \mod 15 = 7 + 3 \mod 15 = 10 \mod 15 = 10,\\
x_4 &= x_3 + x_2 \mod 15 = 10 + 7 \mod 15 = 17 \mod 15 = 2,\\
x_5 &= x_4 + x_3 \mod 15 = 2 + 10 \mod 15 = 12 \mod 15 = 12,\\
x_6 &= x_5 + x_4 \mod 15 = 12 + 2 \mod 15 = 14 \mod 15 = 14,\\
x_7 &= x_6 + x_5 \mod 15 = 14 + 12 \mod 15 = 26 \mod 15 = 11,\\
x_8 &= x_7 + x_6 \mod 15 = 11 + 14 \mod 15 = 25 \mod 15 = 10,\\
x_9 &= x_8 + x_7 \mod 15 = 10 + 11 \mod 15 = 21 \mod 15 = 6,\\
x_{10} &= x_9 + x_8 \mod 15 = 6 + 10 \mod 15 = 16 \mod 15 = 1,\\
x_{11} &= x_{10} + x_9 \mod 15 = 1 + 6 \mod 15 = 7 \mod 15 = 7,
\end{aligned}$$

$$x_{12} = x_{11} + x_{10} \mod 15 = 7 + 1 \mod 15 = 8 \mod 15 = 8,$$
$$x_{13} = x_{12} + x_{11} \mod 15 = 8 + 7 \mod 15 = 15 \mod 15 = 0,$$
$$x_{14} = x_{13} + x_{12} \mod 15 = 0 + 8 \mod 15 = 8 \mod 15 = 8,$$
$$x_{15} = x_{14} + x_{13} \mod 15 = 8 + 0 \mod 15 = 8 \mod 15 = 8,$$
$$x_{16} = x_{15} + x_{14} \mod 15 = 8 + 8 \mod 15 = 16 \mod 15 = 1,$$
$$x_{17} = x_{16} + x_{15} \mod 15 = 1 + 8 \mod 15 = 9 \mod 15 = 9,$$
$$x_{18} = x_{17} + x_{16} \mod 15 = 9 + 1 \mod 15 = 10 \mod 15 = 10,$$
$$x_{19} = x_{18} + x_{17} \mod 15 = 10 + 9 \mod 15 = 19 \mod 15 = 4,$$
$$x_{20} = x_{19} + x_{18} \mod 15 = 4 + 10 \mod 15 = 14 \mod 15 = 14,$$
$$x_{21} = x_{20} + x_{19} \mod 15 = 14 + 4 \mod 15 = 18 \mod 15 = 3,$$
$$x_{22} = x_{21} + x_{20} \mod 15 = 3 + 14 \mod 15 = 17 \mod 15 = 2,$$
$$x_{23} = x_{22} + x_{21} \mod 15 = 2 + 3 \mod 15 = 5 \mod 15 = 5,$$
$$x_{24} = x_{23} + x_{22} \mod 15 = 5 + 2 \mod 15 = 7 \mod 15 = 7,$$
$$x_{25} = x_{24} + x_{23} \mod 15 = 7 + 5 \mod 15 = 12 \mod 15 = 12,$$
$$x_{26} = x_{25} + x_{24} \mod 15 = 12 + 7 \mod 15 = 19 \mod 15 = 4,$$
$$x_{27} = x_{26} + x_{25} \mod 15 = 4 + 12 \mod 15 = 16 \mod 15 = 1,$$
$$x_{28} = x_{27} + x_{26} \mod 15 = 1 + 4 \mod 15 = 5 \mod 15 = 5,$$
$$x_{29} = x_{28} + x_{27} \mod 15 = 5 + 1 \mod 15 = 6 \mod 15 = 6,$$
$$x_{30} = x_{29} + x_{28} \mod 15 = 6 + 5 \mod 15 = 11 \mod 15 = 11,$$
$$x_{31} = x_{30} + x_{29} \mod 15 = 11 + 6 \mod 15 = 17 \mod 15 = 2,$$
$$x_{32} = x_{31} + x_{30} \mod 15 = 2 + 11 \mod 15 = 13 \mod 15 = 13,$$
$$x_{33} = x_{32} + x_{31} \mod 15 = 13 + 2 \mod 15 = 15 \mod 15 = 0,$$
$$x_{34} = x_{33} + x_{32} \mod 15 = 0 + 13 \mod 15 = 13 \mod 15 = 13,$$
$$x_{35} = x_{34} + x_{33} \mod 15 = 13 + 0 \mod 15 = 13 \mod 15 = 13,$$
$$x_{36} = x_{35} + x_{34} \mod 15 = 13 + 13 \mod 15 = 26 \mod 15 = 11,$$
$$x_{37} = x_{36} + x_{35} \mod 15 = 11 + 13 \mod 15 = 24 \mod 15 = 9,$$
$$x_{38} = x_{37} + x_{36} \mod 15 = 9 + 11 \mod 15 = 20 \mod 15 = 5,$$

$$x_{39} = x_{38} + x_{37} \mod 15 = 5 + 9 \mod 15 = 14 \mod 15 = 14,$$

$$x_{40} = x_{39} + x_{38} \mod 15 = 14 + 5 \mod 15 = 19 \mod 15 = 4,$$

$$x_{41} = x_{40} + x_{39} \mod 15 = 4 + 14 \mod 15 = 18 \mod 15 = 3,$$

$$x_{42} = x_{41} + x_{40} \mod 15 = 3 + 4 \mod 15 = 7 \mod 15 = 7,$$

and finally: $x_{41} = x_0$ and $x_{42} = x_1$. Thus this sequence is periodic with period 41.

(b) Step (ii) of the algorithm is

$$x_{n+1} = x_n + x_{n-1} \mod p. \qquad \text{(A)}$$

Replacing n in this formula by $n + 1$ gives

$$x_{n+2} = x_{n+1} + x_n \mod p = (x_n + x_{n-1}) + x_n \mod p = 2x_n + x_{n-1} \mod p. \qquad \text{(B)}$$

Equations (A) and (B) can be written as

$$x_{n+1} = x_{n-1} + x_n \mod p$$

$$x_{n+2} = x_{n-1} + 2x_n \mod p$$

which in matrix form are

$$\begin{bmatrix} x_{n+1} \\ x_{n+2} \end{bmatrix} = \begin{bmatrix} 1 & 1 \\ 1 & 2 \end{bmatrix} \begin{bmatrix} x_{n-1} \\ x_n \end{bmatrix} \mod p.$$

3. **(c)** Beginning with $\begin{bmatrix} x_0 \\ x_1 \end{bmatrix} = \begin{bmatrix} 5 \\ 5 \end{bmatrix}$, we obtain

$$\begin{bmatrix} x_2 \\ x_3 \end{bmatrix} = \begin{bmatrix} 1 & 1 \\ 1 & 2 \end{bmatrix} \begin{bmatrix} 5 \\ 5 \end{bmatrix} \bmod 21 = \begin{bmatrix} 10 \\ 15 \end{bmatrix} \bmod 21 = \begin{bmatrix} 10 \\ 15 \end{bmatrix}$$

$$\begin{bmatrix} x_4 \\ x_5 \end{bmatrix} = \begin{bmatrix} 1 & 1 \\ 1 & 2 \end{bmatrix} \begin{bmatrix} 10 \\ 15 \end{bmatrix} \bmod 21 = \begin{bmatrix} 25 \\ 40 \end{bmatrix} \bmod 21 = \begin{bmatrix} 4 \\ 19 \end{bmatrix}$$

$$\begin{bmatrix} x_6 \\ x_7 \end{bmatrix} = \begin{bmatrix} 1 & 1 \\ 1 & 2 \end{bmatrix} \begin{bmatrix} 4 \\ 19 \end{bmatrix} \bmod 21 = \begin{bmatrix} 23 \\ 42 \end{bmatrix} \bmod 21 = \begin{bmatrix} 2 \\ 0 \end{bmatrix}$$

$$\begin{bmatrix} x_8 \\ x_9 \end{bmatrix} = \begin{bmatrix} 1 & 1 \\ 1 & 2 \end{bmatrix} \begin{bmatrix} 2 \\ 0 \end{bmatrix} \bmod 21 = \begin{bmatrix} 2 \\ 2 \end{bmatrix} \bmod 21 = \begin{bmatrix} 2 \\ 2 \end{bmatrix}$$

$$\begin{bmatrix} x_{10} \\ x_{11} \end{bmatrix} = \begin{bmatrix} 1 & 1 \\ 1 & 2 \end{bmatrix} \begin{bmatrix} 2 \\ 2 \end{bmatrix} \bmod 21 = \begin{bmatrix} 4 \\ 6 \end{bmatrix} \bmod 21 = \begin{bmatrix} 4 \\ 6 \end{bmatrix}$$

$$\begin{bmatrix} x_{12} \\ x_{13} \end{bmatrix} = \begin{bmatrix} 1 & 1 \\ 1 & 2 \end{bmatrix} \begin{bmatrix} 4 \\ 6 \end{bmatrix} \bmod 21 = \begin{bmatrix} 10 \\ 16 \end{bmatrix} \bmod 21 = \begin{bmatrix} 10 \\ 16 \end{bmatrix}$$

$$\begin{bmatrix} x_{14} \\ x_{15} \end{bmatrix} = \begin{bmatrix} 1 & 1 \\ 1 & 2 \end{bmatrix} \begin{bmatrix} 10 \\ 16 \end{bmatrix} \bmod 21 = \begin{bmatrix} 26 \\ 42 \end{bmatrix} \bmod 21 = \begin{bmatrix} 5 \\ 0 \end{bmatrix}$$

$$\begin{bmatrix} x_{16} \\ x_{17} \end{bmatrix} = \begin{bmatrix} 1 & 1 \\ 1 & 2 \end{bmatrix} \begin{bmatrix} 5 \\ 0 \end{bmatrix} \bmod 21 = \begin{bmatrix} 5 \\ 5 \end{bmatrix} \bmod 21 = \begin{bmatrix} 5 \\ 5 \end{bmatrix}.$$

and we see that $\begin{bmatrix} x_{16} \\ x_{17} \end{bmatrix} = \begin{bmatrix} x_0 \\ x_1 \end{bmatrix}$.

5. If $0 \le x < 1$ and $0 \le y < 1$, then $T(x, y) = (x + 5/12, y) \bmod 1$, and so $T^2(x, y) = (x + 10/12, y) \bmod 1$, $T^3(x, y) = (x + 15/12, y) \bmod 1, \ldots, T^{12}(x, y) = (x + 60/12, y) \bmod 1 = (x + 5, y) \bmod 1 = (x, y)$. Thus every point in S returns to its original position after 12 iterations and so every point in S is a periodic point with period at most 12. Because every point is a periodic point, no point can have a dense set of iterates, and so the mapping cannot be chaotic.

7. That Arnold's cat map is a one-to-one mapping over the unit square S and that its range is S is fairly obvious geometrically. An analytical proof would proceed as follows: To show that it is one-to-one, suppose that (x_1, y_1) and (x_2, y_2), with $0 \le x_i < 1$ and $0 \le y_i < 1$, are two points in S such that $\Gamma(x_1, y_1) = \Gamma(x_2, y_2)$. From the definition of Γ it follows that

$$(x_1 + y_1) \bmod 1 = (x_2 + y_2) \bmod 1$$

$$(x_1 + 2\,y_1) \bmod 1 = (x_2 + 2\,y_2) \bmod 1.$$

From the definition of "mod 1", there are integers r_1, s_1, r_2, and s_2 such that

$$x_1 + y_1 - r_1 = x_2 + y_2 - r_2 \qquad \text{(A)}$$

$$x_1 + 2y_1 - s_1 = x_2 + 2y_2 - s_2. \qquad \text{(B)}$$

Subtracting the first equation from the second gives

$$y_1 - s_1 + r_1 = y_2 - s_2 + r_2$$

or

$$y_1 - y_2 = s_1 - s_2 + r_2 - r_1. \qquad \text{(C)}$$

The right-hand side of Eq. (C) is an integer and the left-hand side satisfies $-1 < y_1 - y_2 < 1$. Consequently, the integer on the right-hand side must be zero, and so $y_1 = y_2$. Putting $y_1 = y_2$ into Eq. (A) gives $x_1 - r_1 = x_2 - r_2$ and a similar argument shows that $x_1 = x_2$. Thus $(x_1, y_1) = (x_2, y_2)$, which shows that the mapping is one-to-one.

To show that the range of Γ is S, let (u, v) be any point in S, so that $0 \le u < 1$ and $0 \le v < 1$. We want to show that there exists a point (x, y) in S such that $\Gamma(x, y) = (u, v)$; that is, such that

$$(x + y) \bmod 1 = u$$

$$(x + 2y) \bmod 1 = v.$$

From the definition of "mod 1", this means that we must find integers r and s such that

$$x + y - r = u$$

$$x + 2y - s = v$$

and for which x and y lie in $[0, 1)$. Solving for x and y in this linear system gives

$$x = (2u - v) + (2r - s)$$
$$y = (-u + 2v) + (-r + s).$$

Let a and b be integers such that $(2u - v) + a$ and $(-u + 2v) + b$ lie in the interval $[0, 1)$. Solving the system

$$2r - s = a$$

$$-r + s = b$$

gives $r = a + b$ and $s = a + 2b$, both of which are integers. This choice of r and s then provides the desired values of x and y.

9. As per the hint, we wish to find the regions in S that map onto the four indicated regions in the figure below.

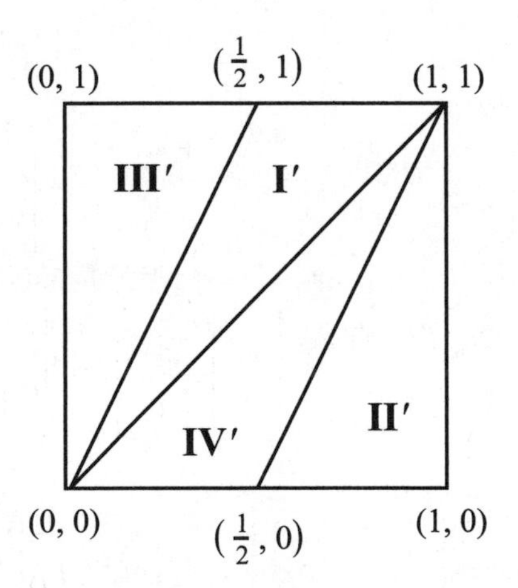

We first consider region **I′** with vertices (0, 0), (1/2, 1), and (1, 1). We seek points (x_1, y_1), (x_2, y_2), and (x_3, y_3), with entries that lie in [0, 1], that map onto these three points under the mapping $\begin{bmatrix} x \\ y \end{bmatrix} \to \begin{bmatrix} 1 & 2 \\ 1 & 2 \end{bmatrix}\begin{bmatrix} x \\ y \end{bmatrix} + \begin{bmatrix} a \\ b \end{bmatrix}$ for certain integer values of a and b to be determined. This leads to the three equations

$$\begin{bmatrix} 1 & 1 \\ 1 & 2 \end{bmatrix}\begin{bmatrix} x_1 \\ y_1 \end{bmatrix} + \begin{bmatrix} a \\ b \end{bmatrix} = \begin{bmatrix} 0 \\ 0 \end{bmatrix},$$

$$\begin{bmatrix} 1 & 1 \\ 1 & 2 \end{bmatrix}\begin{bmatrix} x_2 \\ y_2 \end{bmatrix} + \begin{bmatrix} a \\ b \end{bmatrix} = \begin{bmatrix} 1/2 \\ 1 \end{bmatrix},$$

$$\begin{bmatrix} 1 & 1 \\ 1 & 2 \end{bmatrix}\begin{bmatrix} x_3 \\ y_3 \end{bmatrix} + \begin{bmatrix} a \\ b \end{bmatrix} = \begin{bmatrix} 1 \\ 1 \end{bmatrix}.$$

The inverse of the matrix $\begin{bmatrix} 1 & 1 \\ 1 & 2 \end{bmatrix}$ is $\begin{bmatrix} 2 & -1 \\ -1 & 1 \end{bmatrix}$. We multiply the above three matrix equations by this inverse and set $\begin{bmatrix} c \\ d \end{bmatrix} = \begin{bmatrix} 2 & -1 \\ -1 & 1 \end{bmatrix}\begin{bmatrix} a \\ b \end{bmatrix}$. Notice that c and d must be integers. This leads to

$$\begin{bmatrix} x_1 \\ y_1 \end{bmatrix} = \begin{bmatrix} 2 & -1 \\ -1 & 1 \end{bmatrix}\begin{bmatrix} 0 \\ 0 \end{bmatrix} - \begin{bmatrix} c \\ d \end{bmatrix} = -\begin{bmatrix} c \\ d \end{bmatrix},$$

$$\begin{bmatrix} x_2 \\ y_2 \end{bmatrix} = \begin{bmatrix} 2 & -1 \\ -1 & 1 \end{bmatrix}\begin{bmatrix} 1/2 \\ 1 \end{bmatrix} - \begin{bmatrix} c \\ d \end{bmatrix} = \begin{bmatrix} 0 \\ 1/2 \end{bmatrix} - \begin{bmatrix} c \\ d \end{bmatrix},$$

$$\begin{bmatrix} x_3 \\ y_3 \end{bmatrix} = \begin{bmatrix} 2 & -1 \\ -1 & 1 \end{bmatrix}\begin{bmatrix} 1 \\ 1 \end{bmatrix} - \begin{bmatrix} c \\ d \end{bmatrix} = \begin{bmatrix} 1 \\ 0 \end{bmatrix} - \begin{bmatrix} c \\ d \end{bmatrix}.$$

The only integer values of c and d that will give values of x_i and y_i in the interval [0, 1] are $c = d = 0$. This then gives $a = b = 0$ and the mapping $\begin{bmatrix} x \\ y \end{bmatrix} \to \begin{bmatrix} 1 & 1 \\ 1 & 2 \end{bmatrix}\begin{bmatrix} x \\ y \end{bmatrix}$ that maps the three points (0, 0), (0, 1/2), and (1, 0) to the three points (0, 0), (0, 1/2), and

(1, 1), respectively. The three points (0, 0), (0, 1/2), and (1, 0) define the triangular region labeled **I** in the diagram below, which then maps onto the region **I′** above.

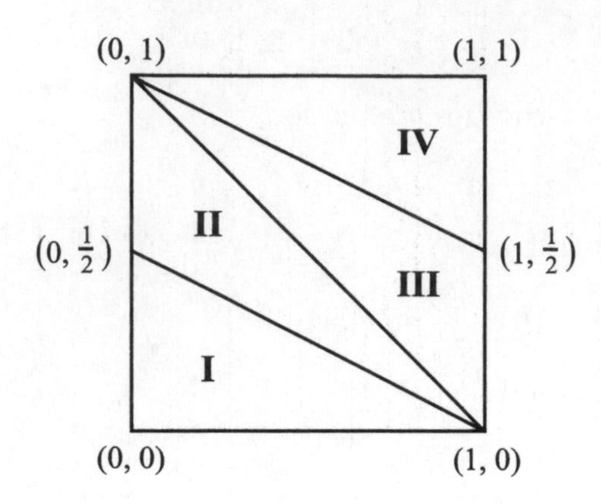

For region **II′**, the calculations are as follows:

$$\begin{bmatrix} 1 & 1 \\ 1 & 2 \end{bmatrix}\begin{bmatrix} x_1 \\ y_1 \end{bmatrix} + \begin{bmatrix} a \\ b \end{bmatrix} = \begin{bmatrix} 1/2 \\ 0 \end{bmatrix},$$

$$\begin{bmatrix} 1 & 1 \\ 1 & 2 \end{bmatrix}\begin{bmatrix} x_2 \\ y_2 \end{bmatrix} + \begin{bmatrix} a \\ b \end{bmatrix} = \begin{bmatrix} 1 \\ 1 \end{bmatrix},$$

$$\begin{bmatrix} 1 & 1 \\ 1 & 2 \end{bmatrix}\begin{bmatrix} x_3 \\ y_3 \end{bmatrix} + \begin{bmatrix} a \\ b \end{bmatrix} = \begin{bmatrix} 1 \\ 0 \end{bmatrix};$$

$$\begin{bmatrix} x_1 \\ y_1 \end{bmatrix} = \begin{bmatrix} 2 & -1 \\ -1 & 1 \end{bmatrix}\begin{bmatrix} 1/2 \\ 0 \end{bmatrix} - \begin{bmatrix} c \\ d \end{bmatrix} = \begin{bmatrix} 1 \\ -1/2 \end{bmatrix} - \begin{bmatrix} c \\ d \end{bmatrix},$$

$$\begin{bmatrix} x_2 \\ y_2 \end{bmatrix} = \begin{bmatrix} 2 & -1 \\ -1 & 1 \end{bmatrix}\begin{bmatrix} 1 \\ 1 \end{bmatrix} - \begin{bmatrix} c \\ d \end{bmatrix} = \begin{bmatrix} 0 \\ 1 \end{bmatrix} - \begin{bmatrix} c \\ d \end{bmatrix},$$

$$\begin{bmatrix} x_3 \\ y_3 \end{bmatrix} = \begin{bmatrix} 2 & -1 \\ -1 & 1 \end{bmatrix}\begin{bmatrix} 1 \\ 0 \end{bmatrix} - \begin{bmatrix} c \\ d \end{bmatrix} = \begin{bmatrix} 2 \\ -1 \end{bmatrix} - \begin{bmatrix} c \\ d \end{bmatrix}.$$

Only $c = 1$ and $d = -1$ will work. This leads to $a = 0$, $b = -1$ and the mapping $\begin{bmatrix} x \\ y \end{bmatrix} \to \begin{bmatrix} 1 & 1 \\ 1 & 2 \end{bmatrix}\begin{bmatrix} x \\ y \end{bmatrix} + \begin{bmatrix} 0 \\ -1 \end{bmatrix}$ that maps region **II** with vertices (0, 1/2), (0, 1), and (1, 0) onto region **II′**.

For region **III′**, the calculations are as follows:

$$\begin{bmatrix} 1 & 1 \\ 1 & 2 \end{bmatrix}\begin{bmatrix} x_1 \\ y_1 \end{bmatrix} + \begin{bmatrix} a \\ b \end{bmatrix} = \begin{bmatrix} 0 \\ 0 \end{bmatrix},$$

$$\begin{bmatrix} 1 & 1 \\ 1 & 2 \end{bmatrix}\begin{bmatrix} x_2 \\ y_2 \end{bmatrix} + \begin{bmatrix} a \\ b \end{bmatrix} = \begin{bmatrix} 1/2 \\ 1 \end{bmatrix},$$

$$\begin{bmatrix} 1 & 1 \\ 1 & 2 \end{bmatrix}\begin{bmatrix} x_3 \\ y_3 \end{bmatrix} + \begin{bmatrix} a \\ b \end{bmatrix} = \begin{bmatrix} 0 \\ 1 \end{bmatrix};$$

$$\begin{bmatrix} x_1 \\ y_1 \end{bmatrix} = \begin{bmatrix} 2 & -1 \\ -1 & 1 \end{bmatrix}\begin{bmatrix} 0 \\ 0 \end{bmatrix} - \begin{bmatrix} c \\ d \end{bmatrix} = -\begin{bmatrix} c \\ d \end{bmatrix}$$

$$\begin{bmatrix} x_2 \\ y_2 \end{bmatrix} = \begin{bmatrix} 2 & -1 \\ -1 & 1 \end{bmatrix}\begin{bmatrix} 1/2 \\ 1 \end{bmatrix} - \begin{bmatrix} c \\ d \end{bmatrix} = \begin{bmatrix} 0 \\ 1/2 \end{bmatrix} - \begin{bmatrix} c \\ d \end{bmatrix},$$

$$\begin{bmatrix} x_3 \\ y_3 \end{bmatrix} = \begin{bmatrix} 2 & -1 \\ -1 & 1 \end{bmatrix}\begin{bmatrix} 0 \\ 1 \end{bmatrix} - \begin{bmatrix} c \\ d \end{bmatrix} = \begin{bmatrix} -1 \\ 1 \end{bmatrix} - \begin{bmatrix} c \\ d \end{bmatrix}.$$

Only $c = -1$ and $d = 0$ will work. This leads to $a = -1$, $b = -1$ and the mapping $\begin{bmatrix} x \\ y \end{bmatrix} \to \begin{bmatrix} 1 & 1 \\ 1 & 2 \end{bmatrix}\begin{bmatrix} x \\ y \end{bmatrix} + \begin{bmatrix} -1 \\ -1 \end{bmatrix}$ that maps region **III** with vertices (1, 0), (1, 1/2), and (0, 1) onto region **III′**.

For region **IV′**, the calculations are as follows:

$$\begin{bmatrix} 1 & 1 \\ 1 & 2 \end{bmatrix}\begin{bmatrix} x_1 \\ y_1 \end{bmatrix} + \begin{bmatrix} a \\ b \end{bmatrix} = \begin{bmatrix} 0 \\ 0 \end{bmatrix},$$

$$\begin{bmatrix} 1 & 1 \\ 1 & 2 \end{bmatrix}\begin{bmatrix} x_2 \\ y_2 \end{bmatrix} + \begin{bmatrix} a \\ b \end{bmatrix} = \begin{bmatrix} 1 \\ 1 \end{bmatrix},$$

$$\begin{bmatrix} 1 & 1 \\ 1 & 2 \end{bmatrix}\begin{bmatrix} x_3 \\ y_3 \end{bmatrix} + \begin{bmatrix} a \\ b \end{bmatrix} = \begin{bmatrix} 1/2 \\ 0 \end{bmatrix};$$

$$\begin{bmatrix} x_1 \\ y_1 \end{bmatrix} = \begin{bmatrix} 2 & -1 \\ -1 & 1 \end{bmatrix}\begin{bmatrix} 0 \\ 0 \end{bmatrix} - \begin{bmatrix} c \\ d \end{bmatrix} = -\begin{bmatrix} c \\ d \end{bmatrix}$$

$$\begin{bmatrix} x_2 \\ y_2 \end{bmatrix} = \begin{bmatrix} 2 & -1 \\ -1 & 1 \end{bmatrix}\begin{bmatrix} 1 \\ 1 \end{bmatrix} - \begin{bmatrix} c \\ d \end{bmatrix} = \begin{bmatrix} 1 \\ 0 \end{bmatrix} - \begin{bmatrix} c \\ d \end{bmatrix},$$

$$\begin{bmatrix} x_3 \\ y_3 \end{bmatrix} = \begin{bmatrix} 2 & -1 \\ -1 & 1 \end{bmatrix}\begin{bmatrix} 1/2 \\ 0 \end{bmatrix} - \begin{bmatrix} c \\ d \end{bmatrix}\begin{bmatrix} 1 \\ -1/2 \end{bmatrix} - \begin{bmatrix} c \\ d \end{bmatrix}.$$

Only $c = 0$ and $d = -1$ will work. This leads to $a = -1$, $b = -2$ and the mapping $\begin{bmatrix} x \\ y \end{bmatrix} \rightarrow \begin{bmatrix} 1 & 1 \\ 1 & 2 \end{bmatrix}\begin{bmatrix} x \\ y \end{bmatrix} + \begin{bmatrix} -1 \\ -2 \end{bmatrix}$ that maps region **IV** with vertices (0, 0), (1, 1), and (1/2, 0) onto region **IV′**.

11. As per the hint, we want to show that the only solution of the matrix equation

$$\begin{bmatrix} x_0 \\ y_0 \end{bmatrix} = \begin{bmatrix} 1 & 1 \\ 1 & 2 \end{bmatrix}\begin{bmatrix} x_0 \\ y_0 \end{bmatrix} - \begin{bmatrix} r \\ s \end{bmatrix}$$

with $0 \le x_0 < 1$, $0 \le y_0 < 1$, and integers r and s is $x_0 = y_0 = 0$. This matrix equation is equivalent to the system

$$x_0 = x_0 + y_0 - r$$

$$y_0 = x_0 + 2y_0 - s.$$

The first equation is $y_0 = r$, which can only have the solution $y_0 = r = 0$ with the given constraints on y_0 and r. Putting $y_0 = 0$ into the second equation yields $x_0 = s$, which likewise has only the solution $x_0 = s = 0$.

13. In the matrix equation $\begin{bmatrix} x_0 \\ y_0 \end{bmatrix} = \begin{bmatrix} 1 & 1 \\ 1 & 2 \end{bmatrix}^n \begin{bmatrix} x_0 \\ y_0 \end{bmatrix}$ mod 1, let the matrix $\begin{bmatrix} 1 & 1 \\ 1 & 2 \end{bmatrix}^n$ be denoted by $\begin{bmatrix} a_n & b_n \\ c_n & d_n \end{bmatrix}$, all of whose entries are positive integers. The matrix equation can then be written as

$$\begin{bmatrix} x_0 \\ y_0 \end{bmatrix} = \begin{bmatrix} a_n & b_n \\ c_n & d_n \end{bmatrix} \begin{bmatrix} x_0 \\ y_0 \end{bmatrix} - \begin{bmatrix} r \\ s \end{bmatrix} \quad \text{or} \quad \begin{bmatrix} a_n-1 & b_n \\ c_n & d_n-1 \end{bmatrix} \begin{bmatrix} x_0 \\ y_0 \end{bmatrix} = \begin{bmatrix} r \\ s \end{bmatrix}$$

where r and s are integers. By Cramer's rule, this last equation has the solution

$$x_0 = \frac{(d_n - 1)r - b_n s}{(a_n - 1)(d_n - 1) - b_n c_n} \quad \text{and} \quad y_0 = \frac{(a_n - 1)r - c_n r}{(a_n - 1)(d_n - 1) - b_n c_n}.$$

These two expressions identify x_0 and y_0 as the quotients of integers, and hence as rational numbers. One point remains: to verify that the determinant of the matrix $\begin{bmatrix} a_n - 1 & b_n \\ c_n & d_n - 1 \end{bmatrix}$ is not equal to zero, so that Cramer's rule is valid. This is equivalent to showing that the matrix $\begin{bmatrix} a_n & b_n \\ c_n & d_n \end{bmatrix}$ does not have the eigenvalue 1. Now, the eigenvalues of this matrix are λ_1^n and λ_2^n, where λ_1 and λ_2 are the eigenvalues of $\begin{bmatrix} 1 & 1 \\ 1 & 2 \end{bmatrix}$. Because these eigenvalues are $\lambda_1 = (3 + \sqrt{5}/2) = 2.6180\ldots$ and $\lambda_2 = (3 - \sqrt{5}/2) = 0.3819\ldots$, it is not possible that their nth powers can be 1 for $n = 1, 2, \ldots$

EXERCISE SET 11.16

1. First we group the plaintext into pairs and add the dummy letter T:

$$DA \quad RK \quad NI \quad GH \quad TT$$

or equivalently from Table 1:

$$4\ 1 \quad 18\ 11 \quad 14\ 9 \quad 7\ 8 \quad 20\ 20$$

(a) For the enciphering matrix $A = \begin{bmatrix} 1 & 3 \\ 2 & 1 \end{bmatrix}$ we have

$$\begin{aligned}
\begin{bmatrix} 1 & 3 \\ 2 & 1 \end{bmatrix}\begin{bmatrix} 4 \\ 1 \end{bmatrix} &= \begin{bmatrix} 7 \\ 9 \end{bmatrix} & & G\ I \\
\begin{bmatrix} 1 & 3 \\ 2 & 1 \end{bmatrix}\begin{bmatrix} 18 \\ 11 \end{bmatrix} &= \begin{bmatrix} 51 \\ 47 \end{bmatrix} = \begin{bmatrix} 25 \\ 21 \end{bmatrix} & & Y\ U \\
\begin{bmatrix} 1 & 3 \\ 2 & 1 \end{bmatrix}\begin{bmatrix} 14 \\ 9 \end{bmatrix} &= \begin{bmatrix} 41 \\ 37 \end{bmatrix} = \begin{bmatrix} 15 \\ 11 \end{bmatrix} & & O\ K \qquad \text{(mod 26)} \\
\begin{bmatrix} 1 & 3 \\ 2 & 1 \end{bmatrix}\begin{bmatrix} 7 \\ 8 \end{bmatrix} &= \begin{bmatrix} 31 \\ 22 \end{bmatrix} = \begin{bmatrix} 5 \\ 22 \end{bmatrix} & & E\ V \\
\begin{bmatrix} 1 & 3 \\ 2 & 1 \end{bmatrix}\begin{bmatrix} 20 \\ 20 \end{bmatrix} &= \begin{bmatrix} 80 \\ 60 \end{bmatrix} = \begin{bmatrix} 2 \\ 8 \end{bmatrix} & & B\ H
\end{aligned}$$

The Hill cipher is

$$GIYUOKEV\ BH$$

1. **(b)** For the enciphering matrix $A = \begin{bmatrix} 4 & 3 \\ 1 & 2 \end{bmatrix}$ we have

$$\begin{bmatrix} 4 & 3 \\ 1 & 2 \end{bmatrix}\begin{bmatrix} 4 \\ 1 \end{bmatrix} = \begin{bmatrix} 19 \\ 6 \end{bmatrix} \qquad \begin{matrix} S \\ F \end{matrix}$$

$$\begin{bmatrix} 4 & 3 \\ 1 & 2 \end{bmatrix}\begin{bmatrix} 18 \\ 11 \end{bmatrix} = \begin{bmatrix} 105 \\ 40 \end{bmatrix} = \begin{bmatrix} 1 \\ 14 \end{bmatrix} \qquad \begin{matrix} A \\ N \end{matrix}$$

$$\begin{bmatrix} 4 & 3 \\ 1 & 2 \end{bmatrix}\begin{bmatrix} 14 \\ 9 \end{bmatrix} = \begin{bmatrix} 83 \\ 32 \end{bmatrix} = \begin{bmatrix} 5 \\ 6 \end{bmatrix} \qquad \begin{matrix} E \\ F \end{matrix} \qquad \text{(mod 26)}$$

$$\begin{bmatrix} 4 & 3 \\ 1 & 2 \end{bmatrix}\begin{bmatrix} 7 \\ 8 \end{bmatrix} = \begin{bmatrix} 52 \\ 23 \end{bmatrix} = \begin{bmatrix} 0 \\ 23 \end{bmatrix} \qquad \begin{matrix} Z \\ W \end{matrix}$$

$$\begin{bmatrix} 4 & 3 \\ 1 & 2 \end{bmatrix}\begin{bmatrix} 20 \\ 20 \end{bmatrix} = \begin{bmatrix} 140 \\ 60 \end{bmatrix} = \begin{bmatrix} 10 \\ 8 \end{bmatrix} \qquad \begin{matrix} J \\ H \end{matrix}$$

The Hill cipher is

$$SFANEFZW\ JH$$

3. From Table 1 the numerical equivalent of this ciphertext is

19 1   11 14   15 24   1 15   10 24

Now we have to find the inverse of $A = \begin{bmatrix} 4 & 1 \\ 3 & 2 \end{bmatrix}$.

Since $\det(A) = 8 - 3 = 5$, we have from Corollary 2:

$$A^{-1} = (5)^{-1}\begin{bmatrix} 2 & -1 \\ -3 & 4 \end{bmatrix} = 21\begin{bmatrix} 2 & -1 \\ -3 & 4 \end{bmatrix} = \begin{bmatrix} 42 & -21 \\ -63 & 84 \end{bmatrix} = \begin{bmatrix} 16 & 5 \\ 15 & 6 \end{bmatrix} \qquad \text{(mod 26)}.$$

To obtain the plaintext, we multiply each ciphertext vector by A^{-1}:

$$\begin{bmatrix} 16 & 5 \\ 15 & 6 \end{bmatrix}\begin{bmatrix} 19 \\ 1 \end{bmatrix} = \begin{bmatrix} 309 \\ 291 \end{bmatrix} = \begin{bmatrix} 23 \\ 5 \end{bmatrix} \quad \begin{matrix} W \\ E \end{matrix}$$

$$\begin{bmatrix} 16 & 5 \\ 15 & 6 \end{bmatrix}\begin{bmatrix} 11 \\ 14 \end{bmatrix} = \begin{bmatrix} 246 \\ 249 \end{bmatrix} = \begin{bmatrix} 12 \\ 15 \end{bmatrix} \quad \begin{matrix} L \\ O \end{matrix}$$

$$\begin{bmatrix} 16 & 5 \\ 15 & 6 \end{bmatrix}\begin{bmatrix} 15 \\ 24 \end{bmatrix} = \begin{bmatrix} 360 \\ 369 \end{bmatrix} = \begin{bmatrix} 22 \\ 6 \end{bmatrix} \quad \begin{matrix} V \\ E \end{matrix} \qquad \pmod{26}$$

$$\begin{bmatrix} 16 & 5 \\ 15 & 6 \end{bmatrix}\begin{bmatrix} 1 \\ 15 \end{bmatrix} = \begin{bmatrix} 91 \\ 105 \end{bmatrix} = \begin{bmatrix} 13 \\ 1 \end{bmatrix} \quad \begin{matrix} M \\ A \end{matrix}$$

$$\begin{bmatrix} 16 & 5 \\ 15 & 6 \end{bmatrix}\begin{bmatrix} 10 \\ 24 \end{bmatrix} = \begin{bmatrix} 280 \\ 294 \end{bmatrix} = \begin{bmatrix} 20 \\ 8 \end{bmatrix} \quad \begin{matrix} T \\ H \end{matrix}$$

The plaintext is thus

$$WE \ LOVE \ MATH$$

5. From Table 1 the numerical equivalent of the known plaintext is

$$\begin{matrix} & AT & & & OM & \\ 1 & & 20 & 15 & & 13 \end{matrix}$$

and the numerical equivalent of the corresponding ciphertext is

$$\begin{matrix} & JY & & & QO & \\ 10 & & 25 & 17 & & 15 \end{matrix}$$

The corresponding plaintext and ciphertext vectors are:

$$\mathbf{p}_1 = \begin{bmatrix} 1 \\ 20 \end{bmatrix} \longleftrightarrow \mathbf{c}_1 = \begin{bmatrix} 10 \\ 25 \end{bmatrix}$$

$$\mathbf{p}_2 = \begin{bmatrix} 15 \\ 13 \end{bmatrix} \longleftrightarrow \mathbf{c}_2 = \begin{bmatrix} 17 \\ 15 \end{bmatrix}$$

We want to reduce

$$C = \begin{bmatrix} 10 & 25 \\ 17 & 15 \end{bmatrix}$$

to I by elementary row operations and simultaneously apply these operations to

$$P = \begin{bmatrix} 1 & 20 \\ 15 & 13 \end{bmatrix}.$$

The calculations are as follows:

$$\left[\begin{array}{cc|cc} 10 & 25 & 1 & 20 \\ 17 & 15 & 15 & 13 \end{array}\right]$$

We formed the matrix $\left[C \mid P\right]$.

$$\left[\begin{array}{cc|cc} 27 & 40 & 16 & 33 \\ 17 & 15 & 15 & 13 \end{array}\right]$$

We added the second row to the first (since 10^{-1} does not exist mod 26).

$$\left[\begin{array}{cc|cc} 1 & 14 & 16 & 7 \\ 17 & 15 & 15 & 13 \end{array}\right]$$ We replaced the entries in the first row by their residues modulo 26.

$$\left[\begin{array}{cc|cc} 1 & 14 & 16 & 7 \\ 0 & -223 & -257 & -106 \end{array}\right]$$ We added –17 times the first row to the second.

$$\left[\begin{array}{cc|cc} 1 & 14 & 16 & 7 \\ 0 & 11 & 3 & 24 \end{array}\right]$$ We replaced the entries in the second row by their residues modulo 26.

$$\left[\begin{array}{cc|cc} 1 & 14 & 16 & 7 \\ 0 & 1 & 57 & 456 \end{array}\right]$$ We multiplied the second row by $11^{-1} = 19 \pmod{26}$.

$$\left[\begin{array}{cc|cc} 1 & 14 & 16 & 7 \\ 0 & 1 & 5 & 14 \end{array}\right]$$ We replaced the entries in the second row by their residues modulo 26.

$$\left[\begin{array}{cc|cc} 1 & 0 & -54 & -189 \\ 0 & 1 & 5 & 14 \end{array}\right]$$ We added –14 times the second row to the first.

$$\left[\begin{array}{cc|cc} 1 & 0 & 24 & 19 \\ 0 & 1 & 5 & 14 \end{array}\right]$$ We replaced –54 and –189 by their residues modulo 26.

Thus $\left(A^{-1}\right)^T = \begin{bmatrix} 24 & 19 \\ 5 & 14 \end{bmatrix}$, and so the deciphering matrix is

$$A^{-1} = \begin{bmatrix} 24 & 5 \\ 19 & 14 \end{bmatrix}.$$

From Table 1 the numerical equivalent of the given ciphertext is

LN	*GI*	*HG*	*YB*	*VR*	*EN*	*JY*	*QO*
12 14	7 9	8 7	25 2	22 18	5 14	10 25	17 15

To obtain the plaintext pairs, we multiply each ciphertext vector by A^{-1}:

$$\begin{bmatrix} 24 & 5 \\ 19 & 14 \end{bmatrix}\begin{bmatrix} 12 \\ 14 \end{bmatrix} = \begin{bmatrix} 358 \\ 424 \end{bmatrix} = \begin{bmatrix} 20 \\ 8 \end{bmatrix} \quad \begin{matrix} T \\ H \end{matrix}$$

$$\begin{bmatrix} 24 & 5 \\ 19 & 14 \end{bmatrix}\begin{bmatrix} 7 \\ 9 \end{bmatrix} = \begin{bmatrix} 213 \\ 259 \end{bmatrix} = \begin{bmatrix} 5 \\ 25 \end{bmatrix} \quad \begin{matrix} E \\ Y \end{matrix}$$

$$\begin{bmatrix} 24 & 5 \\ 19 & 14 \end{bmatrix}\begin{bmatrix} 8 \\ 7 \end{bmatrix} = \begin{bmatrix} 227 \\ 250 \end{bmatrix} = \begin{bmatrix} 19 \\ 16 \end{bmatrix} \quad \begin{matrix} S \\ P \end{matrix}$$

$$\begin{bmatrix} 24 & 5 \\ 19 & 14 \end{bmatrix}\begin{bmatrix} 25 \\ 2 \end{bmatrix} = \begin{bmatrix} 610 \\ 503 \end{bmatrix} = \begin{bmatrix} 12 \\ 9 \end{bmatrix} \quad \begin{matrix} L \\ I \end{matrix} \qquad \text{(mod 26)}$$

$$\begin{bmatrix} 24 & 5 \\ 19 & 14 \end{bmatrix}\begin{bmatrix} 22 \\ 18 \end{bmatrix} = \begin{bmatrix} 618 \\ 670 \end{bmatrix} = \begin{bmatrix} 20 \\ 20 \end{bmatrix} \quad \begin{matrix} T \\ T \end{matrix}$$

$$\begin{bmatrix} 24 & 5 \\ 19 & 14 \end{bmatrix}\begin{bmatrix} 5 \\ 14 \end{bmatrix} = \begin{bmatrix} 190 \\ 291 \end{bmatrix} = \begin{bmatrix} 8 \\ 5 \end{bmatrix} \quad \begin{matrix} H \\ E \end{matrix}$$

$$\begin{bmatrix} 24 & 5 \\ 19 & 14 \end{bmatrix}\begin{bmatrix} 10 \\ 25 \end{bmatrix} = \begin{bmatrix} 365 \\ 540 \end{bmatrix} = \begin{bmatrix} 1 \\ 20 \end{bmatrix} \quad \begin{matrix} A \\ T \end{matrix}$$

$$\begin{bmatrix} 24 & 5 \\ 19 & 14 \end{bmatrix}\begin{bmatrix} 17 \\ 15 \end{bmatrix} = \begin{bmatrix} 483 \\ 533 \end{bmatrix} = \begin{bmatrix} 15 \\ 13 \end{bmatrix} \quad \begin{matrix} O \\ M \end{matrix}$$

which yields the message

THEY SPLIT THE ATOM

9. Testing $x = 0, 1, 2, \ldots, 25$ in $4x = 1 \pmod{26}$ gives

$$
\begin{aligned}
4(0) &= 0 && \neq 1 \\
4(1) &= 4 && \neq 1 \\
4(2) &= 8 && \neq 1 \\
4(3) &= 12 && \neq 1 \\
4(4) &= 16 && \neq 1 \\
4(5) &= 20 && \neq 1 \\
4(6) &= 24 && \neq 1 \\
4(7) &= 28 = 2 && \neq 1 \\
4(8) &= 32 = 6 && \neq 1 \\
4(9) &= 36 = 10 && \neq 1 \\
4(10) &= 40 = 14 && \neq 1 \qquad \pmod{26} \\
4(11) &= 44 = 18 && \neq 1 \\
4(12) &= 48 = 22 && \neq 1 \\
4(13) &= 52 = 0 && \neq 1 \\
4(14) &= 56 = 4 && \neq 1 \\
4(15) &= 60 = 8 && \neq 1 \\
4(16) &= 64 = 12 && \neq 1 \\
4(17) &= 68 = 16 && \neq 1 \\
4(18) &= 72 = 20 && \neq 1 \\
4(19) &= 76 = 24 && \neq 1 \\
4(20) &= 80 = 2 && \neq 1 \\
4(21) &= 84 = 6 && \neq 1 \\
4(22) &= 88 = 10 && \neq 1 \\
4(23) &= 92 = 14 && \neq 1 \\
4(24) &= 96 = 18 && \neq 1 \\
4(25) &= 100 = 22 && \neq 1
\end{aligned}
$$

EXERCISE SET 11.17

1. Use induction on n, the case $n = 1$ being already given. If the result is true for $n - 1$, then $M^n = M^{n-1}M = (PD^{n-1}P^{-1})(PDP^{-1}) = PD^{n-1}(P^{-1}P)DP^{-1} = PD^{n-1}DP^{-1} = PD^nP^{-1}$, proving the result.

3. Call M_1 the matrix of Example 1, and M_2 the matrix of Exercise 2. Then $\mathbf{x}^{(2n)} = (M_2M_1)^n\,\mathbf{x}^{(0)}$ and $\mathbf{x}^{(2n+1)} = M_1(M_2M_1)^n\,\mathbf{x}^{(0)}$. We have

$$M_2M_1 = \begin{bmatrix} \frac{1}{2} & \frac{1}{4} & 0 \\ \frac{1}{2} & \frac{1}{2} & \frac{1}{2} \\ 0 & \frac{1}{4} & \frac{1}{2} \end{bmatrix}\begin{bmatrix} 1 & \frac{1}{2} & 0 \\ 0 & \frac{1}{2} & 1 \\ 0 & 0 & 0 \end{bmatrix} = \begin{bmatrix} \frac{1}{2} & \frac{3}{8} & \frac{1}{4} \\ \frac{1}{2} & \frac{1}{2} & \frac{1}{2} \\ 0 & \frac{1}{8} & \frac{1}{4} \end{bmatrix}.$$

The characteristic polynomial of this matrix is $\lambda^3 - \frac{5}{4}\lambda^2 + \frac{1}{4}\lambda$, so the eigenvalues are $\lambda_1 = 1$, $\lambda_2 = \frac{1}{4}$, $\lambda_3 = 0$. Corresponding eigenvectors are $\mathbf{e}_1 = \begin{bmatrix}5 & 6 & 1\end{bmatrix}^T$, $\mathbf{e}_2 = \begin{bmatrix}-1 & 0 & 1\end{bmatrix}^T$, and $\mathbf{e}_3 = \begin{bmatrix}1 & -2 & 1\end{bmatrix}^T$. Thus,

$$(M_2M_1)^n = PD^nP^{-1} = \begin{bmatrix} 5 & -1 & 1 \\ 6 & 0 & -2 \\ 1 & 1 & 1 \end{bmatrix}\begin{bmatrix} 1 & 0 & 0 \\ 0 & \left(\frac{1}{4}\right)^n & 0 \\ 0 & 0 & 0 \end{bmatrix}\begin{bmatrix} \frac{1}{12} & \frac{1}{12} & \frac{1}{12} \\ -\frac{1}{3} & \frac{1}{6} & \frac{2}{3} \\ \frac{1}{4} & -\frac{1}{4} & \frac{1}{4} \end{bmatrix}$$

Using the notation of Example 1 (recall that $a_0 + b_0 + c_0 = 1$), we obtain

$$a_{2n} = \frac{5}{12} + \frac{1}{6 \cdot 4^n}(2a_0 - b_0 + 4c_0)$$
$$b_{2n} = \frac{1}{2}$$
$$c_{2n} = \frac{1}{12} - \frac{1}{6 \cdot 4^n}(2a_0 - b_0 + 4c_0).$$

and

$$a_{2n+1} = \frac{2}{3} + \frac{1}{6 \cdot 4^n}(2a_0 - b_0 + 4c_0)$$
$$b_{2n+1} = \frac{1}{3} - \frac{1}{6 \cdot 4^n}(2a_0 - b_0 + 4c_0)$$
$$c_{2n+1} = 0.$$

5. From Eq. (9), if $b_0 = .25 = \frac{1}{4}$, we get $b_1 \frac{1/4}{9/8} = \frac{2}{9}$, then $b_2 = \frac{2/9}{10/9} = \frac{1}{5}$, $b_3 = \frac{1/5}{11/10} = \frac{2}{11}$, and, in general, $b_n = \frac{2}{8+n}$. We will reach $\frac{2}{20} = .10$ in 12 generations. According to Eq. (8), under the controlled program the percentage would be $\frac{1}{2^{14}}$ in 12 generations, or $\frac{1}{16384} = .00006 = .006\%$.

7. From (13) we have that the probability that the limiting sibling-pairs will be type (A, AA) is

$$a_0 + \frac{2}{3}b_0 + \frac{1}{3}c_0 + \frac{2}{3}d_0 + e_0.$$

The proportion of A genes in the population at the outset is as follows: all the type (A, AA) genes, 2/3 of the type (A, Aa) genes, 1/3 the type (A, aa) genes, etc.... yielding

$$a_0 + \frac{2}{3}b_0 + \frac{1}{3}c_0 + \frac{2}{3}d_0 + e_0.$$

9. For the first column of M we realize that parents of type (A, AA) can produce offspring only of that type, and similarly for the last column. The fifth column is like the second column, and follows the analysis in the text. For the middle two columns, say the third, note that male offspring from (A, aa) must be of type a, and females are of type Aa, because of the way the genes are inherited.

EXERCISE SET 11.18

1. **(a)** The characteristic polynomial of L is $\lambda^2 - \lambda - 1/3$, so the eigenvalues are $\frac{1}{2}+\frac{\sqrt{21}}{6}$, $\lambda_2 = \frac{\sqrt{21}}{6}$. The eigenvector corresponding to $\begin{bmatrix} \frac{2}{3\left(-\frac{1}{2}+\frac{\sqrt{21}}{6}\right)} \\ 1 \end{bmatrix}$.

(b) $\mathbf{x}^{(1)} = L\mathbf{x}^{(0)} = \begin{bmatrix} 100 \\ 50 \end{bmatrix}$, $\mathbf{x}^{(2)} = \begin{bmatrix} 175 \\ 50 \end{bmatrix}$, $\mathbf{x}^{(3)} = \begin{bmatrix} 250 \\ 88 \end{bmatrix}$,

$\mathbf{x}^{(4)} = \begin{bmatrix} 382 \\ 125 \end{bmatrix}$, $\mathbf{x}^{(5)} = \begin{bmatrix} 570 \\ 191 \end{bmatrix}$.

(c) $\mathbf{x}^{(6)} = L\mathbf{x}^{(5)} = \begin{bmatrix} 857 \\ 285 \end{bmatrix}$. $\lambda_1\mathbf{x}^{(5)} = \begin{bmatrix} 855 \\ 287 \end{bmatrix}$.

3. From the solution in Exercise 2, we see that the derivative of the characteristic polynomial evaluated at λ_1 is

$$n\lambda_1^{n-1} - (n-1)a_1\lambda_1^{n-2} - (n-2)a_2b_1\lambda_1^{n-3} - \cdots - a_{n-1}b_1b_2\cdots b_{n-2}.$$

Recalling that $a_i \geq 0$, at least one $a_i > 0$, and $b_i > 0$, we see that the above is strictly greater than

$$n\left(\lambda_1^{n-1} - a_1\lambda_1^{n-2} - a_2b_1\lambda_1^{n-3} - \cdots - a_{n-1}b_1\cdots b_{n-2}\right)$$
$$\geq \frac{n}{\lambda_-}a_nb_1\cdots b_{n-1} \geq 0.$$

So the derivative is positive and λ_1 is a simple root.

5. a_1 is the average number of offspring produced in the first age period. a_2b_1 is the number of offspring produced in the second period times the probability that the female will live into the second period, i.e., it is the expected number of offspring per female during the second period, and so on for all the periods. Thus, the sum of these, which is the net reproduction rate, is the expected number of offspring produced by a given female during her expected lifetime.

7. $R=0+4\left(\frac{1}{2}\right)+3\left(\frac{1}{2}\right)\left(\frac{1}{4}\right)=19/8.$

9. Let λ be any eigenvalue of L, and write $\lambda=re^{i\theta}=r(\cos\theta+i\sin\theta)$. We know $\lambda^n = r^n e^{in\theta}$, so $1=q(\lambda)=\frac{a_1}{r}(\cos\ \theta-i\sin\ \theta)+\cdots+\frac{a_n b_1 b_2\cdots b_{n-1}}{r^n}(\cos\ n\theta-i\sin\ n\theta)$ Since $q(\lambda)$ is real, we can ignore all the sine terms in the formula.

Thus

$$1 = q(\lambda) = \frac{a_1}{r}\cos\theta + \frac{a_2 b_1}{r^2}\cos 2\theta + \cdots + \frac{a_n b_1 b_2 \cdots b_{n-1}}{r^2}\cos n\theta$$

$$\leq \frac{a_1}{r} + \frac{a_2 b_1}{r} + \cdots + \frac{a_n b_1 b_2 \cdots b_{n-1}}{r^n} = q(r)$$

i.e., $q(r)\geq 1$. Since q is a decreasing function, we have $r\leq\lambda_1$.

EXERCISE SET 11.19

1. **(a)** The characteristic polynomial of L is $\lambda^3 - 2\lambda - 3/8 = (\lambda - 3/2)[\lambda^2 + (3/2)\lambda + 1/4]$, so $\lambda_1 = 3/2$. Thus h, the fraction harvested of each age group, is $1 - \frac{2}{3} = \frac{1}{3}$ from Eq. (6), so the yield is $33\frac{1}{3}\%$ of the population. The eigenvector corresponding to $\lambda_1 = 3/2$ is $[1 \quad 1/3 \quad 1/18]^T$; this is the age distribution vector after each harvest.

(b) From Eq. (10), the age distribution vector $\mathbf{x}_1$ is $[1 \quad 1/2 \quad 1/8]^T$. Eq. (9) tells us that $h_1 = 1 - 1/(19/8) = 11/19$, so we harvest 11/19 or 57.9% of the youngest age class. Since $L\mathbf{x}_1 = [19/8 \quad 1/2 \quad 1/8]^T$, the youngest class contains 79.2% of the population. Thus the yield is 57.9% of 79.2%, or 45.8% of the population.

3. Using the L of Eq. (3) and the $\mathbf{x}_1$ of Eq. (10) we obtain for the first coordinate of $L\mathbf{x}_1$,

$$a_1 + a_2b_1 + a_3b_1b_2 + \cdots + a_nb_1b_2b_3 \cdots b_{n-1}.$$

The other coordinates of $L\mathbf{x}_1$ are, respectively, $b_1, b_1b_2, b_1b_2b_3, \cdots, b_1b_2 \cdots b_{n-1}$. Thus,

$$L\mathbf{x}_1 - \mathbf{x}_1 = \begin{bmatrix} a_1 + a_2b_1 + a_3b_1b_2 + \cdots + a_nb_1b_2 \cdots b_{n-1} - 1 \\ 0 \\ 0 \\ \vdots \\ 0 \end{bmatrix}$$

$$= \begin{bmatrix} R-1 \\ 0 \\ 0 \\ \vdots \\ 0 \end{bmatrix}.$$

5. Here $h_J = 1$, $h_I \neq 0$, and all the other h_k's are zero. Then Eq. (4) becomes

$$a_1 + a_2b_1 + \cdots + a_{I-1}b_1b_2 \cdots b_{I-2} + (1 - h_I)[a_Ib_1b_2 \cdots b_{I-1} + \cdots + a_{J-1}b_1b_2 \cdots b_{J-2}] = 1.$$

We solve for h_I to obtain

$$h_I = \frac{a_1 + a_2b_1 + \cdots + a_{I-1}b_1b_2 \cdots b_{I-2} - 1}{a_Ib_1b_2 \cdots b_{I-1} + \cdots + a_{J-1}b_1b_2 \cdots b_{J-1}} + 1$$

$$= \frac{a_1 + a_2b_1 + \cdots + a_{J-1}b_1b_2 \cdots b_{J-2} - 1}{a_Ib_1b_2 \cdots b_{I-1} + \cdots + a_{J-1}b_1b_2 \cdots b_{J-1}}.$$

EXERCISE SET 11.20

1. From Theorem 2, we compute $a_0 = \frac{1}{\pi}\int_0^{2\pi}(t-\pi)^2 dt = \frac{2}{3}\pi^2, a_k = \frac{1}{\pi}\int_0^{2\pi}(t-\pi)^2 \cos kt\, dt =$ $\frac{4}{k^2}$, and $b_k = \frac{1}{\pi}\int_0^{2\pi}(t-\pi)^2 \sin kt\, dt = 0$. So the least-squares trigonometric polynomial of order 3 is

$$\frac{\pi^2}{3} + 4\cos t + \cos 2t + \frac{4}{9}\cos 3t.$$

3. As in Exercise 1, $a_0 = \frac{1}{\pi}\int_0^{2\pi} f(t)\, dt = \frac{1}{\pi}\int_0^{\pi} \sin t\, dt = \frac{2}{\pi}$

(Note the upper limit on the second integral),

$$\begin{aligned}
a_k &= \frac{1}{\pi}\int_0^{\pi} \sin t \cos kt\, dt \\
&= \frac{1}{\pi}\left(\frac{1}{k^2-1}[k \sin kt \sin t + \cos kt \cos t]\right)\Bigg]_0^{\pi} \\
&= \frac{1}{\pi}\left(\frac{1}{k^2-1}[0 + (-1)^{k-1} - 1]\right) \\
&= \begin{cases} 0 & \text{if } k \text{ is odd} \\ -\dfrac{2}{\pi(k^2-1)} & \text{if } k \text{ is even.} \end{cases}
\end{aligned}$$

$$b_k = \frac{1}{\pi}\int_0^{\pi} \sin kt \sin t\, dt = \begin{cases} \dfrac{1}{2} & \text{if } k = 1 \\ 0 & \text{if } k > 1 \end{cases}.$$

So the least-squares trigonometric polynomial of order 4 is

$$\frac{1}{\pi}+\frac{1}{2}\sin t-\frac{2}{3\pi}\cos 2t-\frac{2}{15\pi}\cos 4t.$$

5. As in Exercise 2,

$$a_0=\frac{2}{T}\int_0^T f(t)\,dt=\frac{2}{T}\int_0^{\frac{1}{2}T} t\,dt+\frac{2}{T}\int_{\frac{1}{2}T}^{T}(T-t)\,dt=\frac{T}{2}.$$

$$a_k=\frac{2}{T}\int_0^{\frac{1}{2}T} t\cos\frac{2k\pi t}{T}dt+\frac{2}{T}\int_{\frac{1}{2}T}^{T}(T-t)\cos\frac{2k\pi t}{T}\,dt$$

$$=\frac{4T}{4k^2\pi^2}((-1)^k-1)=\begin{cases}0 & \text{if } k \text{ is even}\\ \dfrac{8T}{(2k)^2\pi^2} & \text{if } k \text{ is odd}\end{cases}.$$

$$b_k=\frac{2}{T}\int_0^{\frac{1}{2}T} t\cos\frac{2k\pi t}{T}dt+\frac{2}{T}\int_{\frac{1}{2}T}^{T}(T-t)\cos\frac{2k\pi t}{T}dt=0.$$

So the least-squares trigonometric polynomial of order n is:

$$\frac{T}{4}-\frac{8T}{\pi^2}\left(\frac{1}{2^2}\cos\frac{2\pi t}{T}+\frac{1}{6^2}\cos\frac{6\pi t}{T}+\frac{1}{10^2}\cos\frac{10\pi t}{T}+\cdots+\frac{1}{(2n)^2}\cos\frac{2\pi n t}{T}\right)$$

if n is even; the last term involves $n-1$ if n is odd.

7. There are several cases to consider. First, the function 1 is orthogonal to all the others because

$$\int_0^{2\pi}\cos kt\,dt=\int_0^{2\pi}\sin kt\,dt=0.$$

Then consider $\int_0^{2\pi}\cos mt\cos nt\,dt=0$ if $m\neq n$. Similarly $\int_0^{2\pi}\sin mt\sin nt\,dt=0$ if $m\neq n$. Finally $\int_0^{2\pi}\cos mt\sin nt\,dt=0$ for all values of m and n.

EXERCISE SET 11.21

1. (a) Equation (2) is

$$c_1\begin{bmatrix}1\\1\end{bmatrix}+c_2\begin{bmatrix}3\\5\end{bmatrix}+c_3\begin{bmatrix}4\\2\end{bmatrix}=\begin{bmatrix}3\\3\end{bmatrix}$$

and Equation (3) is $c_1 + c_2 + c_3 = 1$. These equations can be written in combined matrix form as

$$\begin{bmatrix}1&3&4\\1&5&2\\1&1&1\end{bmatrix}\begin{bmatrix}c_1\\c_2\\c_3\end{bmatrix}=\begin{bmatrix}3\\3\\1\end{bmatrix}.$$

This system has the unique solution $c_1 = 1/5$, $c_2 = 2/5$, and $c_3 = 2/5$. Because these coefficients are all nonnegative, it follows that $\mathbf{v}$ is a convex combination of the vectors $\mathbf{v}_1$, $\mathbf{v}_2$, and $\mathbf{v}_3$.

(b) As in part (a) the system for c_1, c_2 and c_3 is

$$\begin{bmatrix}1&3&4\\1&5&2\\1&1&1\end{bmatrix}\begin{bmatrix}c_1\\c_2\\c_3\end{bmatrix}=\begin{bmatrix}2\\4\\1\end{bmatrix}$$

which has the unique solution $c_1 = 2/5$, $c_2 = 4/5$, and $c_3 = -1/5$. Because one of these coefficients is negative, it follows that $\mathbf{v}$ is not a convex combination of the vectors $\mathbf{v}_1$, $\mathbf{v}_2$, and $\mathbf{v}_3$.

1. **(c)** As in part (a) the system for c_1, c_2 and c_3 is

$$\begin{bmatrix} 3 & -2 & 3 \\ 3 & -2 & 0 \\ 1 & 1 & 1 \end{bmatrix} \begin{bmatrix} c_1 \\ c_2 \\ c_3 \end{bmatrix} = \begin{bmatrix} 0 \\ 0 \\ 1 \end{bmatrix}$$

which has the unique solution $c_1 = 2/5$, $c_2 = 3/5$, and $c_3 = 0$. Because these coefficients are all nonnegative, it follows that $\mathbf{v}$ is a convex combination of the vectors $\mathbf{v}_1$, $\mathbf{v}_2$, and $\mathbf{v}_3$.

(d) As in part (a) the system for c_1, c_2 and c_3 is

$$\begin{bmatrix} 3 & -2 & 3 \\ 3 & -2 & 0 \\ 1 & 1 & 1 \end{bmatrix} \begin{bmatrix} c_1 \\ c_2 \\ c_3 \end{bmatrix} = \begin{bmatrix} 1 \\ 0 \\ 1 \end{bmatrix}$$

which has the unique solution $c_1 = 4/15$, $c_2 = 6/15$, and $c_3 = 5/15$. Because these coefficients are all nonnegative, it follows that $\mathbf{v}$ is a convex combination of the vectors $\mathbf{v}_1$, $\mathbf{v}_2$, and $\mathbf{v}_3$.

3. Combining everything that is given in the statement of the problem, we obtain:

$$\begin{aligned} \mathbf{w} &= M\mathbf{v} + \mathbf{b} = M(c_1\mathbf{v}_1 + c_2\mathbf{v}_2 + c_3\mathbf{v}_3) + (c_1 + c_2 + c_3)\mathbf{b} \\ &= c_1(M\mathbf{v}_1 + \mathbf{b}) + c_2(M\mathbf{v}_2 + \mathbf{b}) + c_3(M\mathbf{v}_3 + \mathbf{b}) = c_1\mathbf{w}_1 + c_2\mathbf{w}_2 + c_3\mathbf{w}_3. \end{aligned}$$

5. **(a)** Let $M = \begin{bmatrix} m_{11} & m_{12} \\ m_{21} & m_{22} \end{bmatrix}$ and $\mathbf{b} = \begin{bmatrix} b_1 \\ b_2 \end{bmatrix}$. Then the three matrix equations $M\mathbf{v}_i + \mathbf{b} = \mathbf{w}_i$, $i = 1, 2, 3$, can be written as the six scalar equations

$$m_{11} + m_{12} + b_1 = 4$$
$$m_{21} + m_{22} + b_2 = 3$$

$$2m_{11} + 3m_{12} + b_1 = 9$$
$$2m_{21} + 3m_{22} + b_2 = 5$$

$$2m_{11} + m_{12} + b_1 = 5$$
$$2m_{21} + m_{22} + b_2 = 3$$

The first, third, and fifth equations can be written in matrix form as

$$\begin{bmatrix} 1 & 1 & 1 \\ 2 & 3 & 1 \\ 2 & 1 & 1 \end{bmatrix} \begin{bmatrix} m_{11} \\ m_{12} \\ b_1 \end{bmatrix} = \begin{bmatrix} 4 \\ 9 \\ 5 \end{bmatrix}.$$

and the second, fourth, and sixth equations as

$$\begin{bmatrix} 1 & 1 & 1 \\ 2 & 3 & 1 \\ 2 & 1 & 1 \end{bmatrix} \begin{bmatrix} m_{21} \\ m_{22} \\ b_2 \end{bmatrix} = \begin{bmatrix} 3 \\ 5 \\ 3 \end{bmatrix}.$$

The first system has the solution $m_{11} = 1$, $m_{12} = 2$, $b_1 = 1$ and the second system has the solution $m_{21} = 0$, $m_{22} = 1$, $b_2 = 2$. Thus we obtain $M = \begin{bmatrix} 1 & 2 \\ 0 & 1 \end{bmatrix}$ and $\mathbf{b} = \begin{bmatrix} 1 \\ 2 \end{bmatrix}$.

(b) As in part (a), we are led to the following two linear systems:

$$\begin{bmatrix} -2 & 2 & 1 \\ 0 & 0 & 1 \\ 2 & 1 & 1 \end{bmatrix} \begin{bmatrix} m_{11} \\ m_{12} \\ b_1 \end{bmatrix} = \begin{bmatrix} -8 \\ 0 \\ 5 \end{bmatrix} \quad \text{and} \quad \begin{bmatrix} -2 & 2 & 1 \\ 0 & 0 & 1 \\ 2 & 1 & 1 \end{bmatrix} \begin{bmatrix} m_{21} \\ m_{22} \\ b_2 \end{bmatrix} = \begin{bmatrix} 1 \\ 1 \\ 4 \end{bmatrix}.$$

Solving these two linear systems leads to $M = \begin{bmatrix} 3 & -1 \\ 1 & 1 \end{bmatrix}$ and $\mathbf{b} = \begin{bmatrix} 0 \\ 1 \end{bmatrix}$.

(c) As in part (a), we are led to the following two linear systems:

$$\begin{bmatrix} -2 & 1 & 1 \\ 3 & 5 & 1 \\ 1 & 0 & 1 \end{bmatrix} \begin{bmatrix} m_{11} \\ m_{12} \\ b_1 \end{bmatrix} = \begin{bmatrix} 0 \\ 5 \\ 3 \end{bmatrix}$$

and

$$\begin{bmatrix} -2 & 1 & 1 \\ 3 & 5 & 1 \\ 1 & 0 & 1 \end{bmatrix} \begin{bmatrix} m_{21} \\ m_{22} \\ b_2 \end{bmatrix} = \begin{bmatrix} -2 \\ 2 \\ -3 \end{bmatrix}.$$

Solving these two linear systems leads to $M = \begin{bmatrix} 1 & 0 \\ 0 & 1 \end{bmatrix}$ and $\mathbf{b} = \begin{bmatrix} 2 \\ -3 \end{bmatrix}$.

(d) As in part (a), we are led to the following two linear systems:

$$\begin{bmatrix} 0 & 2 & 1 \\ 2 & 2 & 1 \\ -4 & -2 & 1 \end{bmatrix} \begin{bmatrix} m_{11} \\ m_{12} \\ b_1 \end{bmatrix} = \begin{bmatrix} 5/2 \\ 7/2 \\ -7/2 \end{bmatrix}$$

and

$$\begin{bmatrix} 0 & 2 & 1 \\ 2 & 2 & 1 \\ -4 & -2 & 1 \end{bmatrix} \begin{bmatrix} m_{21} \\ m_{22} \\ b_2 \end{bmatrix} = \begin{bmatrix} -1 \\ 3 \\ -9 \end{bmatrix}.$$

Solving these two linear systems leads to $M = \begin{bmatrix} 1/2 & 1 \\ 2 & 0 \end{bmatrix}$ and $\mathbf{b} = \begin{bmatrix} 1/2 \\ -1 \end{bmatrix}$.

7. **(a)** The vertices $\mathbf{v}_1$, $\mathbf{v}_2$, and $\mathbf{v}_3$ of a triangle can be written as the convex combinations $\mathbf{v}_1 = (1)\mathbf{v}_1 + (0)\mathbf{v}_2 + (0)\mathbf{v}_3$, $\mathbf{v}_2 = (0)\mathbf{v}_1 + (1)\mathbf{v}_2 + (0)\mathbf{v}_3$, and $\mathbf{v}_3 = (0)\mathbf{v}_1 + (0)\mathbf{v}_2 + (1)\mathbf{v}_3$. In each of these cases, precisely two of the coefficients are zero and one coefficient is one.

(b) If, for example, $\mathbf{v}$ lies on the side of the triangle determined by the vectors $\mathbf{v}_1$ and $\mathbf{v}_2$ then from Exercise 6(a) we must have that $\mathbf{v} = c_1\mathbf{v}_1 + c_2\mathbf{v}_2 + (0)\mathbf{v}_3$ where $c_1 + c_2 = 1$. Thus at least one of the coefficients, in this example c_3, must equal zero.

(c) From part (b), if at least one of the coefficients in the convex combination is zero, then the vector must lie on one of the sides of the triangle. Consequently, none of the coefficients can be zero if the vector lies in the interior of the triangle.